ASTRONOMY AND
ASTROPHYSICS LIBRARY

Series Editors: M. Harwit, R. Kippenhahn, V. Trimble, J.-P. Zahn

ASTRONOMY AND
ASTROPHYSICS LIBRARY

Series Editors: M. Harwit, R. Kippenhahn, V. Trimble, J.-P. Zahn

Walter F. Huebner (Ed.)

Physics and Chemistry of Comets

With a Foreword by Fred L.Whipple

With 147 Figures

Springer-Verlag
Berlin Heidelberg New York London
Paris Tokyo Hong Kong Barcelona

Dr. Walter F. Huebner

Southwest Research Institute, Div. 15, P.O. Drawer 28510,
6220 Culebra Road, San Antonio, TX 78228-0510, USA

Series Editors

Martin Harwit

The National Air and Space
Museum, Smithsonian Institution
7th St. and Independence Ave. S.W.
Washington, DC 20560, USA

Virginia Trimble

Astronomy Program
University of Maryland
College Park, MD 20742, USA
and Department of Physics
University of California
Irvine, CA 92717, USA

Rudolf Kippenhahn

Max-Planck-Institut für
Physik und Astrophysik
Institut für Astrophysik
Karl-Schwarzschild-Straße 1
D-8046 Garching, Fed. Rep. of Germany

Jean-Paul Zahn

Université Paul Sabatier
Observatoires du Pie-du-Midi
et de Toulouse
14, Avenue Edouard-Belin
F-31400 Toulouse, France

Cover picture: The nucleus of Comet Halley as observed from the Giotto spacecraft. The pseudocolor image is composed of six images with increasing resolution toward the brightest part, and the visibility of the dark side is enhanced by a masking technique.

ISBN 3-540-51228-4 Springer-Verlag Berlin Heidelberg New York
ISBN 0-387-51228-4 Springer-Verlag New York Berlin Heidelberg

Library of Congress Cataloging-in-Publication Data. Physics and chemistry of comets / Walter F. Huebner, ed. p. cm. – (Astronomy and astrophysics library). Includes bibliographical references (p.) and index. ISBN 3-540-51228-4 (Springer-Verlag Berlin Heidelberg New York) – ISBN 0-387-51228-4 (Springer-Verlag New York Berlin Heidelberg). 1. Comets. 2. Astrophysics. I. Huebner, W. F. (Walter F.), 1928–. II. Series. QB721.P49 1990 523.6 – dc20 90-10100

© Springer-Verlag Berlin Heidelberg 1990
Printed in Germany

The contents was processed by the authors using the T_EX macro package.

2156/3150-543210 – Printed on acid-free paper

Foreword

As this excellent book demonstrates, the study of comets has now reached the fascinating stage where we understand comets in general simple terms while, at the same time, we are uncertain about practically all the details of cometary nature, structure, processes, and origin. In every aspect, even including dynamics, a choice among several or many competing theories is made impossible simply by the lack of detailed knowledge. The space missions, snapshot studies of two comets, particularly the one that immortalizes the name of Sir Edmund Halley, have produced a huge mass of valuable new information and a number of surprises. Nonetheless, we face the tantalizing realization that we have obtained only a fleeting glance at two of perhaps a hundred billion (10^{11}) or more comets with possibly differing natures, origins, and physical histories.

To my personal satisfaction, comets seem to have discrete nuclei made up of dirty snowballs, as I concluded four decades ago, but perhaps they are more like frozen rubbish piles. Almost certainly they are debris left over from the building of the solar system, but just how? And just where? Are comets frozen froth or a mixture of solids and voids? How big are the hunks, if there are hunks? Were they formed from silicate dust cores with icy mantles or are they just a mishmash of colliding atoms and ions? Have the centers of the biggest ones been melted by radioactivity? Do some or many end up as inert Earth-crossing asteroids, a perpetual hazard to life on Earth? Even the density of the Comet P/Halley nucleus still remains uncertain although it is our best determination.

This book is a thorough presentation of our current knowledge of comets, covering all ramifications, including the one area in which comets are actually useful to modern physics. Comets are, indeed, space laboratories for the study of magnetohydrodynamics, a frightful word standing for the processes occurring in hot gases or plasmas containing charged particles or ions along with magnetic fields such that the clouds of gas are partly or primarily controlled by the magnetic fields. Because most of the known universe consists of plasmas, the subject is vitally important in modern astronomy.

Nearer to home, comets are intimately related to the origin of the Earth, our Solar System, and probably to star formation generally. Perhaps they made life possible on Earth, but perhaps not. They may be prevalent in galaxies, possibly next to stars in total mass although no one seriously suspects them of providing the mysterious extra mass needed to hold our universe together.

When I say that this book is a thorough presentation, I note that the serious student who wishes to proceed farther with cometary studies will be able, if he or she chooses, to consult nearly eight hundred references in the literature, listed in an appendix where they do not clutter the clear-cut subject presentations. For the hasty reader a concise summary at the end provides an overview which may also be an alert for topics of individual special interest.

We all hope that the next big step in understanding comets will be space missions, probing not only their tails and comae but the actual nuclei themselves.

Fred L. Whipple

Director Emeritus
Smithsonian Astrophysical Observatory,
Phillips Professor of Astronomy Emeritus
Harvard University

Preface

This book provides a comprehensive review of our understanding of comets, following the spacecraft investigations of Comets Giacobini–Zinner and Halley. The relevant physical and chemical processes are discussed, starting with the quasi-permanent celestial body, the comet nucleus. This is followed by discussions, ordered according to the ephemeral development of a comet, of the neutral coma, the dust coma and tail, and the plasma components in the coma and tail, including the solar wind interaction. Furthermore, the properties of a comet nucleus are deduced. The traditional description of the Oort cloud is supplemented by much-debated theoretical discussions connecting the measured physico-chemical data of comets via their observed orbits to the more speculative origin and evolution of comet nuclei. The book concludes with an overview of the conditions of the nebula in which comets formed, implications for the origins of life, and prospects for future research.

Many of the processes described are interlinked so that parts of descriptions may be found in several places. However, cross references and an exhaustive index aid the reader in finding his or her way. Although researchers will find the book useful as a reference text, it is intended for advanced undergraduate and graduate students in space science, planetary science, astrophysics, and astrochemistry. New entrants to the field of comet observations and comet science will appreciate the treatment of conceptual material. The book will be useful for the scientific implications and the planning of new comet missions. It can serve as a guide to the type of observation and work that are still needed, with some suggestions on how to approach them and where to find additional data.

Special thanks go to all the members of the Halley Multicolour Camera team for the exciting cooperation and discussions in the preparation of the chapter on "The Nucleus"; W. Curdt, R. Kramm, and N. Thomas helped to produce most of the images of the nucleus and its dusty environment. Thanks go also to the many colleagues at the Max-Planck-Institut für Aeronomie who provided useful information for the chapter on "The Plasma" and engaged in many stimulating discussions with the authors. Indebtedness is expressed to E. van den Heuvel for a number of valuable suggestions on the chapter on the "Orbital Distribution of Comets".

On a personal note, it is a pleasure to thank my colleagues and coauthors for their patience and their collaborative spirit, without which the coordination of the material for this book would not have been possible. I wish to express my sincere appreciation to the many dedicated "behind-the-scenes" reviewers for their invaluable counsel and for correcting errors and misconceptions in the text. Comments and critical remarks on especially difficult sections were made by the authors on each other's chapters as well as by Claude Arpigny, Daniel C. Boice, Bertram D. Donn, Tamas I. Gombosi, Ichishiro Konno, and Stuart J. Weidenschilling. Sincere thanks are also due to the many colleagues who supplied figures from their own research and made special efforts to revise them for inclusion in this book. Hermann U. Schmidt and R. Wegmann deserve special mention in this regard.

Very special thanks are due to Mrs. Inge Gehne for her extraordinary efforts in preparing the manuscript on "The Plasma" and to Richard Spinks for his skill and good humor in proofreading and correcting the text and for his help in setting it into TEX.

Much of my own research has been generously supported by the Planetary Atmospheres Program of NASA's Solar System Exploration Division. I especially wish to acknowledge the enthusiastic support I received from the management and staff of the Southwest Research Institute. Considering all the help that I received, I wonder how much I really contributed to this book.

San Antonio, June 1990   *Walter F. Huebner*

Table of Contents

3. The Neutral Coma

Index of Contributors

A'Hearn, Michael F.
Astronomy Program, University of Maryland,
College Park, MD 20742, USA

Axford, W.I.
Max-Planck-Institut für Aeronomie, Postfach 20,
D-3411 Katlenburg-Lindau, Fed. Rep. of Germany

Festou, Michel C.
Observatoire de Besançon, 41 bis av. de l'Observatoire,
F-25044 Besançon Cedex, France

Grün, E.
Max-Planck-Institut für Kernphysik, Postfach 10 39 80, Saupfercheckweg 1,
D-6900 Heidelberg, Fed. Rep. of Germany

Huebner, Walter F.
Southwest Research Institute, Division 15, P.O. Drawer 28510,
6220 Culebra Road, San Antonio, TX 78228–0510, USA

Ip, W.-H.
Max-Planck-Institut für Aeronomie, Postfach 20,
D-3411 Katlenburg-Lindau, Fed. Rep. of Germany

Jessberger, E.K.
Max-Planck-Institut für Kernphysik, Postfach 10 39 80, Saupfercheckweg 1,
D-6900 Heidelberg, Fed. Rep. of Germany

Keller, H.Uwe
Max-Planck-Institut für Aeronomie, Postfach 20, Max-Planck-Straße 2,
D-3411 Katlenburg-Lindau, Fed. Rep. of Germany

MacKay, C.P.
Ames Research Center, Life Science Div., Solar System Exploration Branch,
Moffett Field, CA 94035, USA

Oort, J.H.
Sterrewacht Leiden, Postbus 9513, NL-2300 RA Leiden, Netherlands

Rickman, Hans
Astronomiska Observatoriet, Box 515, S-75120 Uppsala, Sweden

1. Introduction

Walter F. Huebner

1.1 Plan of the Book

In the past few years the enthusiasm for observing comets and for analyzing and interpreting comet data, mostly kindled through the encouragement and coordination of the International Halley Watch, has reached a new peak of activity. The brainpower brought to bear has led to a plethora of new and exciting results, and our understanding of comet phenomena has increased in an unprecedented manner. In this book we review the physics and chemistry of comets as interpreted from modern observations, but in particular the spacecraft missions to Comets Giacobini-Zinner and Halley that have so dramatically enriched our knowledge. Pre-mission theories and hypotheses that have been confirmed by *in situ* measurements from the spacecraft and coordinated observations using modern satellite, rocket-borne, and ground-based instruments are summarized briefly in the context of global understanding and to accent the recent advances in comet science. Emphasis is placed on new observations that changed older theories or were unexpected. The goal of the book is to present a coherent interpretation of the wealth of new scientific data, resolve controversies, propose new questions that might be answered through further observations and future spacecraft missions to comets, and project our knowledge to the place and time of the origin of comets. New constraints based on the recent advances are applied to the continuity of physical and chemical processes and causality scenarios that lead to a better understanding of the conditions in the nebula in which comets formed and that impact theories about the origins of life.

The book begins with the quasi-permanent celestial body, the comet nucleus. This is followed by chapters discussing, in order of ephemeral development, the neutral coma, the dust coma and tail, and the plasma components in the coma and tail, including the solar wind interaction. These chapters not only reveal properties of interplanetary space and the ephemeral parts of a comet, but also lead to deductions about properties of the nucleus. The traditional description of the Oort cloud is supplemented by much-debated theoretical discussions connecting the measured physico-chemical data of comets via their observed orbits to the more speculative origin and evolution of comet nuclei. The book concludes with an overview about the conditions of the nebula in which comets formed, implications on the origins of life, and a perspective for future research.

1.2 General Concepts

Comet nuclei are primitive bodies of the Solar System containing a mixture of frozen gases, carbonaceous particles with complex hydrocarbons, and refractory grains. This icy conglomerate nucleus, first proposed in a more rudimentary form by Whipple in 1950, is the source of a comet's diffusely appearing atmosphere (coma) and two morphologically different tails when its orbit brings it close to the Sun. The coma and subsequently plasma and dust tails develop during the approach to the Sun and subside and disappear in reverse order after perihelion passage. Since the nucleus is of low density and only about ten kilometers in size, it has insufficient mass to bind its atmosphere gravitationally, contrary to what is typically the case for planets. The escape velocity is of the order of 1 m/s, depending somewhat on the size of the nucleus. The escaping dusty atmosphere causes the ephemeral, visually observable effects that define a comet. The word comet derives from the Greek "kometes," meaning long-haired, and refers to a comet's appearance as a hairy star or a tail star.

Comets lose matter when heated. Their fragility associated with the progressive mass loss suggests that comets have not been heated significantly during formation or their early existence before they entered the inner Solar System. The volatile materials in the nucleus are primordial ices (frozen gases) that condensed before or during formation of the Solar System far from the hot, central region. Therefore it has been said that comets contain the most pristine material in the Solar System. However, we must be cautious about this interpretation: pristine implies an extreme state of unaltered purity. Just as vacuum is pristine emptiness only to the philosopher, so pristine material has only a meaning relative to the unaltered state of the ices in a comet nucleus. However, the frozen gases in the surface layer of the nucleus have been altered by ultraviolet (UV) radiation and cosmic rays during the 4.5 Gy that a nucleus is part of the distant comet cloud. The interior may have undergone similar changes from residual nuclear activity, from the conversion of kinetic energy to energy of deformation and heat during collisions in the aggregation process, and from the release of energy during the phase change from amorphous to crystalline water ice. Once a comet enters the inner Solar System, it decays quite rapidly. Whether this leads to complete evaporation or to a dead, asteroid-like body has not been determined.

A large cloud of comet nuclei, arranged in a spherical distribution with mean radius of about 50000 astronomical units (AU) around the Sun, as deduced by Oort in 1950 from aphelia positions of long-period comets, serves as a reservoir for the new comets that visit the inner Solar System. The orbits of these comets differ from those of the planets and the asteroids not only by the large aphelion distances and hence long period of revolution of several million years, but also in that they are not confined to the region of the ecliptic. When they are perturbed by the galaxy or a passing star in such a way that they come into the inner Solar System where comets become observable, their elliptic orbits are so long that they are referred to as "nearly parabolic." Aside from some clustering, caused by perturbations from passing stars or interstellar clouds, and a depletion in a narrow band along the galactic equator, apparently caused by galactic tide effects, their aphelia distribution on the sky is isotropic. Although comets have been expelled from the Solar System into interstellar space, the chance that one of these interstellar comets passes close to the Sun is extremely small. This may explain why none have been observed with certainty.

At the other extreme are the short-period comets ($P < 200$ years). They have been captured by planets, mostly Jupiter, into orbits that prevalently lie close to the ecliptic, or they had their origin in an "inner" cloud in the ecliptic trans-Neptunian region. The mass of a comet is less than 10^{-10} times that of the Earth; hence planets can perturb the orbits of comets, but the reverse effect is negligible.

The completed missions to Comets Giacobini-Zinner and Halley and all presently planned comet missions, including the Comet Rendezvous and Asteroid Flyby (CRAF) and the Comet Nucleus Sample Return (CNSR) missions, are visits to short-period comets. These comets have passed many times through the inner Solar System. A cosmic struggle between the heat from the Sun and the icy masses of the comets has progressively eroded the surfaces of their nuclei. They have aged, becoming depleted of their more volatile constituents, and are therefore less active than the comets with nearly parabolic orbits that come directly from the Oort cloud. To advance our knowledge, it would be more rewarding to explore comets with nearly parabolic orbits, because they relate more closely to the conditions under which comets were formed and thus more directly provide clues about the early history of our Solar System. However, orbits for such comets cannot be determined accurately and early enough during their approach to the inner Solar System so that a spacecraft of the type available now could be directed to intercept them. Spacecraft with high-thrust engines in orbit around the Sun would be required for a start of such a mission. The energy requirements are too large to accomplish this feat in the foreseeable future. Thus we are limited to the exploration of short-period comets. Fortunately, to first order, all comets have similar properties. Much can be learned from short-period comets through *in situ* measurements on the development of the coma (CRAF mission) and the solid phase of the nucleus (CNSR mission, also called the Rosetta mission). However, to relate the data from the comets to the nebula in which they were formed requires modeling on computers supported by laboratory measurements of the physical properties of comet materials, comet simulations, and space release experiments. The full impact of this will not be known for many more years.

1.3 Orbits of Some Comets for Spacecraft Missions

Figures 1.1 and 1.2 show the orbits of short-period comets that have been visited by spacecraft – Comets[1] P/Giacobini-Zinner and P/Halley – and the comets that are possible targets for future Giotto and CRAF missions – P/Grigg-Skjellerup, P/Tempel 2, P/Kopff, P/Wild 2, and P/Wirtanen – in relation to the planetary system. The fully drawn lines indicate the portions of the comet orbits above the ecliptic and the

---

[1] The symbol P/ designates a short-period comet.

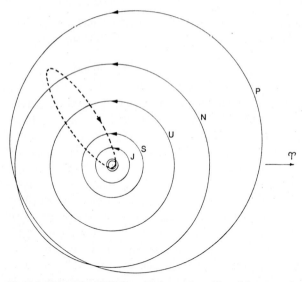

Fig. 1.1. The orbit of P/Halley relative to the orbits of the outer planets. The orbits of Jupiter (J), Saturn (S), Uranus (U), Neptune (N), and Pluto (P) are shown. The smallest circle represents the orbit of the Earth and the next one Mars. Most of the orbit of P/Halley is below the ecliptic (dashed part of the ellipse). The vernal equinox is indicated by ♈

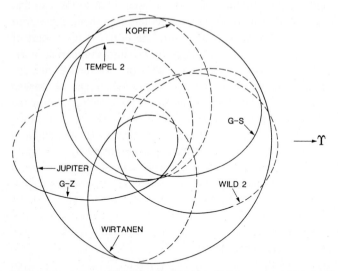

Fig. 1.2. The orbits of Comets P/Kopff, P/Grigg-Skjellerup (G-S), P/Wild 2, P/Wirtanen, P/Giacobini-Zinner (G-Z), and P/Tempel 2, relative to the orbit of Jupiter. Dashed lines indicate those parts of the orbits that are below the ecliptic, solid lines above the ecliptic

dashed lines indicate those below the ecliptic. No such distinction is made for the planetary orbits, since they are within a few degrees of the ecliptic. The orbit of P/Halley is clearly different from those of the other comets. Not only is its aphelion distance larger (35 AU), its inclination (the angle of its orbital plane with respect to the ecliptic) is also larger (162°) because its motion is retrograde with respect

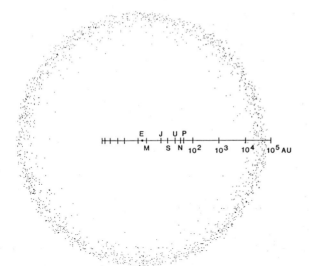

Fig. 1.3. A schematic presentation of the spherical Oort cloud relative to the planetary system, projected on a plane perpendicular to the ecliptic. The radial dimensions of the projection are logarithmic in astronomical units. On such a scale the Earth orbit (E) and everything inside it is presented by the dot in the center. The symmetrical positions of the orbits of Mars (M), Jupiter (J), Saturn (S), Uranus (U), Neptune (N), and Pluto (P) are shown

to the motion of the planets. Comet P/Giacobini-Zinner is also different, although this is not apparent from the presentation of its orbit in Fig. 1.2. Its inclination is 31°, foreshortening its minor axis. The orbits of the other comets are more closely related to those of the planets, making a spacecraft rendezvous much easier.

If the instruments, in particular the camera, are still operational on the Giotto spacecraft when it comes close to the Earth in 1990, it will be sent on another comet mission. The target for this mission is most likely P/Grigg-Skjellerup. Since the instrumentation for the exploration of P/Halley and P/Grigg-Skjellerup would be the same, this mission will give direct evidence for the similarities and differences of two comets.

The CRAF mission, on the other hand, will be a long-term mission to a comet. The spacecraft will stay with the comet in its orbit, making various excursions through the developing coma as the comet approaches the Sun. It will be possible to monitor compositional changes in the coma, the development of the contact surface, the bow shock region, the plasma and dust tails, the active areas on the nucleus, extended sources of gas production from the dust, plasma properties, and the decay of dust grains. The erosion of the nucleus and possibly some evolution of the crust and development of crater-like rims can be explored. Also an excursion into the tail will give information about plasma acceleration and dust composition and particle size distribution. A penetrator, deployed from the spacecraft, will pierce the surface of the nucleus and measure its mechanical strength and chemical properties.

1.4 The Oort Cloud Compared to the Planetary Orbits

Figure 1.3 relates the Oort cloud of comets to the outer planetary system. The radially logarithmic scale of the projection on a plane is in astronomical units. On that scale the Earth's orbit (= 1 AU) is the point in the center. The view is a projection on a plane perpendicular to the ecliptic; the orbital motion of the planets is represented by the straight line. The outer radius of the Oort sphere is about one-third to one-half the distance to the nearest star. The Oort cloud is observationally well established but an inner, trans-Neptunian cloud (not shown) along the ecliptic is speculative, although a very active topic of current discussion.

1.5 Summary of Spacecraft Data from Comets

The first *in situ* comet data were obtained by the International Cometary Explorer (ICE) on 11 September 1985. Although the spacecraft was originally not instrumented or intended for a comet mission, the flyby was extremely successful in providing details, unobtainable with ground-based observations, of the plasma structure in the transition region from the coma to the tail. Many of the global features predicted by models of coma–solar wind interaction, such as cometary ion pick-up by the solar wind, cooling, mass-loading, and deceleration of the solar wind ions, could be verified, while many new features, such as a bow shock region (rather than a sharp bow shock) and plasma instabilities in transition regions, had not been included in models. This mission also gave the first and only measurements of the density and low energy distribution of the electrons, suggesting that they can be characterized by a two-temperature approximation.

From 6 through 25 March of 1986 a fleet of six robotic spacecraft traveled through space and met and explored Comet P/Halley on its thirtieth known (and 29th recorded) return to the inner Solar System. At a distance of about 600 km, the spacecraft Giotto (named after the Italian painter Giotto di Bondone, who included the 1301 apparition of Halley's comet as the Star of Bethlehem in his fresco "The Adoration of the Magi") passed closest to the nucleus. At 68 km/s it sped through the coma encountering only about 10^{14} molecules/m^3, less than 10^{-12} of the density of air at sea level. In spite of these "shots through a void," the investigations were spectacularly successful.

The ICE spacecraft, which earlier passed through the onset of the tail of Comet P/Giacobini-Zinner, was the last to pass P/Halley. It made important measurements of the unperturbed solar wind upstream of the comet.

The two Japanese spacecraft, Sakigake (Pioneer) and Suisei (Comet), examined the solar wind and UV radiation interaction with the outer coma. The UV camera aboard Suisei detected a pulsation in the Lyman-α images of the coma that has been linked through the atomic hydrogen production (and therefore the water vapor production from the nucleus) to the spin of the comet nucleus. From this, a spin period of the nucleus has been determined to be 52.9 h. At 4.5×10^5 km from the nucleus Suisei detected a 10°–30° deflection of the solar wind from the direction of the undisturbed flow, indicating the crossing of the bow shock by the spacecraft. It also detected shell structure in velocity space of cometary protons and water-group ions at 2.3×10^5 km from the nucleus. At 1.5×10^5 km its ion energy analyzer detected severe perturbations and a slowing of the solar wind flow to about 56 km/s, caused by the ionization of cometary molecules and their assimilation into the solar wind. The fluxgate magnetometer on the Sakigake spacecraft detected electron cyclotron waves and Alfvén waves with frequencies close to the gyrofrequency of the water ions. This measurement suggests that water-group molecules from the comet were ionized at a distance of 10^7 km from the nucleus.

The Giotto and the two Vega (derived from *Venus-Halley*, using the Russian spelling of Halley) spacecraft investigated the nucleus by remote sensing and the gas and dust environment on the subsolar side of the coma. In an excellent international collaboration, the Vega spacecraft acted as a pathfinder for the Giotto spacecraft which was the last before the ICE to arrive at the comet, but passed closest to the nucleus. Most of the global chemico-physical and plasma properties predicted by existing cometary models, such as ion pick-up by the solar wind (mass-loading, deceleration, stagnation flow, and ion pile-up) and the contact surface, were found to be in good agreement with observations. However, the model predictions for dust composition and size distribution were poor compared to observations. It was found that carbonaceous dust, rich in H, C, N, and O, is very abundant, and there is an excess of particles with masses less than 10^{-17} kg. No detailed model existed for the shape of the nucleus. As expected, the imaging of the nucleus brought many surprises: The nucleus is much bigger than anticipated, very dark with a geometric albedo of about 3%, and deviates significantly from a spherical shape by being about twice as long as it is wide or thick. When the much larger volume of the nucleus is combined with the observed nongravitational acceleration and the associated reaction force from the asymmetric vaporization of the frozen gases, it is found that the density of the nucleus is much smaller than estimates had indicated before the Halley encounters. The models for the coma – nucleus interface, although not correct in their global effects, were found to be in agreement for the active surface areas. About 90% of the coma gas and dust are emitted in jet-like features from a few compact, active source regions that comprise only about 20% of the illuminated surface area at the time of the Giotto encounter. Several of the minor gaseous species have their origin, at least in part, in distributed sources linked to the dust. Entirely unpredicted were the suprathermal ions and the heavy ions with masses up to about 120 amu that appear to be associated with the organic component of the dust. These heavy molecules include polymers and appear to be present throughout the entire nucleus, suggesting that they already existed in the nebula in which comets formed.

1.6 Summary of New Data from Remote Sensing

Remote sensing includes observations from ground-based telescopes, aircraft, rockets, and Earth satellites. Aircraft are most successful for infrared observations, while rockets are used for UV observations. The Halley ground-based observations were an unprecedented effort of international cooperation, coordinated by the International Halley Watch. Many new results including confirmations of *in situ* measurements were obtained. The chemical species HCN and H_2CO that had been identified in only a few other comets were identified in P/Halley with radio telescopes. The abundance ratio of ortho- to para-water was determined from observations made with the Kuiper Airborne Observatory. One of the most controversial observations is the rotation period of the nucleus. It was determined from dust structures in the 1910 apparition of P/Halley and from the Suisei spacecraft to be about 53 h, while photometric data from ground observations, the International Ultraviolet Explorer (IUE) satellite, and Pioneer Venus suggest a period of 7.4 d. This controversy and possible explanations of it will be carried through many chapters of this book.

The interaction of the solar wind with comets can be monitored for long time periods from the ground. Also world-round coverage of the motion of plasma tail structures has been very successful. Accurate determinations of ranges of molecular species and their parents has been the major subject of many observations. The existence of jet-like structures of dust in the coma has been known for a long time, but now similar features have also been detected in the coma gas.

Many types of these observations and monitoring of comets are possible only from the ground. During the past few decades techniques for detecting electromagnetic radiation coming from comets have been improved and expanded to cover the radio, infrared, visible, and ultraviolet wavelength ranges. Major observatories have included comets in their observing schedules and special instruments have been built for comet observations.

1.7 Reasons for Planned and Proposed Comet Missions

The Giacobini-Zinner and Halley missions provided "snapshots" of the physical and chemical conditions in two comets. Although this has very much enriched our knowledge, we are still on a steeply rising part of the learning curve; the advances in comet science raise many new questions that can best be answered through at least three much more complex space missions in addition to another possible comet flyby of the Giotto spacecraft. The first of these three missions is a long-duration close-encounter mission to a short-period comet to investigate the development of the coma as a function of heliocentric distance, to observe coma development over active regions on the nucleus, to correlate coma and tail activity with solar flares, and to probe the crustal layer of the nucleus with a penetrator. The second mission, also to a short-period comet, should physically and chemically probe the nucleus at an area that has recently been active as well as at several inactive areas, and return core samples from the active area and surface samples from the inactive areas for laboratory studies. Finally, the third mission, not yet planned, should probe the coma and nucleus of a long-period comet coming directly from the Oort cloud.

As summarized in mission documents, the planned comet missions should answer questions in various areas of research. In the area of the early history of the Solar System questions include: What were the physico-chemical and thermodynamic conditions in the protosolar nebula? What were the physical processes in the agglomeration of solid protoplanetary matter, on scales from sub-micrometers to kilometers?

What were the conditions that formed cometary ice and organic particles? Did they form in interstellar space, in the protosolar nebula, or in a nebula out of which the giant planets formed? If it is a combination of these, how do the components differ? What is the relationship between organic and silicate particles? Are they totally mixed with the organic component as a binder between the silicate grains or does the organic material form a mantle on these grains? To what extent did comets contribute to the volatile inventory of the terrestrial planets? Is the high deuterium-to-hydrogen ratio in oceans caused by comet bombardment? What is the relative importance of processes that have altered cometary material? What fraction of the organic material known to exist in interstellar molecular clouds survives incorporation into the solar nebula? What organic materials not yet detected are present in comets? What are the important processes of interstellar organic chemistry? Could comet infall have been the major source of organic material on the Earth or on Mars? To what extent have energetic particles produced organic materials on comets and can they do the same on the icy moons of the outer Solar System? Some of these questions are relevant to the investigations of prebiotic conditions, but there are also more specific questions for exobiology: How far did the chemical evolution of prebiotic organic compounds proceed in space? Which of the special structural molecules essential to terrestrial life are products of general cosmic, rather than specific terrestrial, synthesis? Is life a cosmic phenomenon, perhaps with major chemical attributes in common wherever it occurs? Which photochemical, ion–neutral, and surface reactions are important in a space environment? How are complex organic molecules produced in space? Some questions specific to space plasma physics include: How does the free energy associated with counter-streaming plasmas drive collective interactions that result in various ionization and energization processes? How do these processes depend on the parameters that define the states of the gas and plasma? How is momentum transferred from the solar wind to the ions in a comet tail? Under what conditions can a magnetic field penetrate into a flowing, collisional plasma? How stable are surfaces such as the cometopause and contact surface? What is the origin of the energetic ions inside of the contact surface? How does the presence of dust influence chemical and plasma dynamics? What fraction of the dust particles are charged? What is their charge? What is the nature of the very complex molecular ions? How do the gas, plasma, and electromagnetic fields influence the motion of the dust grains?

The Comet Rendezvous Asteroid Flyby (CRAF) mission, a joint endeavor between NASA and the Federal Republic of Germany, fulfills the requirements of the first of the three desired space missions. Its measurement objectives are: (1) to characterize the chemical and physical nature of the atmosphere and ionosphere of a comet, and the processes that occur in them; (2) to characterize the development of the coma and its gas, dust, and plasma components as a function of time and orbital position; (3) to determine the processes forming comet tails; (4) to characterize tail dynamics; (5) to characterize the interaction of a comet with the solar wind and radiation; (6) to characterize the physical and geological structure of the nucleus, including its global shape, spin, topography, and morphology; and, (7) to determine the major mineralogical phases including metals, rocky minerals, carbonaceous materials, and ices on the surface of the nucleus and the spatial distribution

Table 1.1. CRAF instruments

1. Imaging science system
2. Visual and infrared mapping spectrometer
3. Thermal infrared radiometer experiment
4. Penetrator
5. Cometary matter analyzer
6. Cometary ice and dust experiment
7. Scanning electron microscope and particle analyzer
8. Comet dust environment monitor
9. Neutral gas and ion mass spectrometer
10. Cometary retarding ion mass spectrometer
11. Suprathermal plasma investigation of cometary environment
12. Magnetometer
13. Coordinated radio, electrons, and waves experiment
14. Radio science

of compositional units on the surface. To accomplish the measurements, fourteen instruments have been selected for the CRAF mission as shown in Table 1.1.

After the leap upward on the learning curve as a result of the CRAF mission, we still find ourselves on a steeply rising slope. Many new hypotheses will need to be confirmed. Just as the return of Moon samples back to Earth for study in laboratories brought about revolutionary changes in the understanding of the Moon's internal and surface processes and chronology of evolution, so the Rosetta Comet Nucleus Sample Return (CNSR) mission will advance our understanding of the origin and early evolution of the Solar System, put it on a sounder footing, and reveal many important new avenues of inquiry. We will know more about the raw materials of our Solar System that went into short-period comets. Instruments for the CNSR mission are listed in Table 1.2. A similar investigation of Pluto and its satellite Charon might end speculations about their origin. Are they captured giant comets or are they a planet and moon in their own right with a consistent history for *in situ* formation? In the CNSR mission it is planned to return three samples: a core sample taken from a depth of one to three meters, in order to provide the interrelation between volatile and refractory compounds in the comet; a volatile sample, possibly from the bottom of the core sample hole, for analysis of the most volatile components of the comet;

Table 1.2. CNSR science model payload

Essential instruments:
1. Imaging system
2. Temperature probe
Highly desirable instruments:
Orbital instruments:
3. Radar altimeter and sounder
4. Infrared spectral mapper
5. Dust counter
6. Neutral gas analyzer
7. Plasma analyzer
Surface science package:
8. Neutral gas analyzer
9. Borehole logging package

and a non-volatile sample, designated for study of the refractory carbonaceous and inorganic compounds.

Finally, the question whether short-period comets have a different origin from Oort cloud comets must be answered. The evidence that some comets come directly from the Oort cloud is clearly established, but the origin of the short-period comets is unclear. They might come from the Oort cloud, being captured into short-period orbits by Jupiter or Saturn, or they could have a separate origin in the trans-Neptunian region. If these comet groups do indeed have separate origins, one can expect their chemical composition (i.e., the relative abundances of the highly volatile ices) to reflect the different conditions of the nebula regions where they formed. At present, the scatter in the data from remote sensing is too big to show a difference. A space mission with a spacecraft in orbit, ready to intercept an Oort cloud comet, might be feasible early in the twenty-first century.

1.8 Laboratory and Space Experiments and Simulations

Laboratory studies of physical, thermodynamic, and chemical properties of mixtures of frozen gases, refractory organics, and silicate compounds at very low temperatures supply some of the basic data needed to understand the structure of comets and to model their formation 4.5 Gy ago and their subsequent evolution. Natural analogues to cometary material, such as interplanetary dust particles and meteorites, and artificial analogues, such as UV and particle-irradiated ices and ice–dust mixtures, can be studied in the laboratory.

Laboratory experiments should provide data to help answer the following questions: What processes cause low albedo? How does selective devolatilization influence mantle growth on the comet nucleus? What are the visible and infrared reflectance spectra and emission properties for various surface compounds? What are the effects from cosmic radiation, charged particles, high energy particle chemistry, and UV radiation? What is the thermal conductivity and heat capacity of various cometary materials under different conditions of porosity? How does porosity influence mobility of volatiles? How efficient are disproportionation reactions at very low temperatures? What are catalysts for such reactions? What is the equation of state for bulk cometary material? At what relative speeds will colliding aggregates stick together, penetrate one another, or shatter? For the CNSR mission it would also be important to know the response of ice to microwave and radar frequencies.

Many of the laboratory experiments may have to be repeated in a space environment. For these, simulated interplanetary and cometary materials can be stored in a frozen binder material, such as carbon dioxide, released into free space from a satellite or a space shuttle, and irradiated by the full solar spectrum. In more sophisticated experiments one can introduce, at first separately and later simultaneously, refractory grains, refractory organic compounds, ices, plasma-forming materials, and particle clusters of these ingredients to investigate the properties of dusty plasmas in free space. Data can be obtained, for example, by laser interrogation techniques from the shuttle. Characterization of the physical and chemical properties can include elemental analysis, states of excitation of neutrals and ions, light scattering as a function of angle, albedo measurements, particle size and cluster distributions, and motion of charged grains and plasma. In another series of experiments one can investigate the bulk properties, such as conductivities, physical strength at very low temperature, and surface phenomena of fluffy, icy materials. Slow collisions of porous ice – dust mixtures, simulating aggregation to form comet nuclei, can be investigated. Under what conditions will they stick and at which relative speeds will they shatter? Such experiments may give the ultimate answer to some of the questions that can only be partially ascertained in the laboratory.

2. The Nucleus

H. Uwe Keller

2.1 Introduction

Starting with a brief historical sketch, the traditional observing techniques relevant to the extraction of data and knowledge about the nucleus are discussed. Next, in Sect. 2.3, the theories and models for the interpretation of the data prior to the spacecraft encounters are reviewed. The emphasis of the other sections in this chapter is placed on the new results.

The concurrent development of detectors over the entire electromagnetic spectrum, of mass analyzing and timing instruments, of impact sensors and other control devices, and of computers for data analysis has been most active in recent decades, influencing comet science in an indirect way. Some of this was kindled by the enthusiasm for the space missions; an effect that is often overlooked if one considers the "usefulness" of expensive missions. The interpretation of the spacecraft data has not yet reached its peak. The flood of incoming data has to be digested. The phase lag between analysis and interpretation places us in the middle of this exciting period. The images of Comet Halley's nucleus contain most of our present knowledge of comet nuclei. The camera observations, discussed in Sect. 2.4, not only reveal size, volume, structure, shape, albedo, and morphology of the nucleus, but also give insights into the activity of the nucleus, the size of its active regions, and the distribution of dust near the nucleus. The images link the distribution of the sources on the surface to the jet-like dust features observed simultaneously from the ground.

The acceleration of the dust by the subliming molecules just above the nucleus surface has been understood qualitatively since the pioneering work of Probstein (1969). Based on the new results and improved computational means, extensive progress is being made. The status is described in Sect. 2.5.

The physical parameters derived from the new sources of information are summarized in Sect. 2.6. The composition of the nucleus, which can be inferred from the composition of the coma gas, dust, and plasma, is discussed in Chaps. 3, 4, and 5. The conclusions on the composition and physical structure of the nucleus are closely related to its formation and subsequent evolution.

Since most of the new results are derived from the observations of Comet P/Halley, the question of generalization to other comets has to be considered. Aspects directly relevant to the interpretation of the observations of a comet nucleus are briefly reviewed in Sect. 2.7; more details are provided in Chaps. 6 and 7. A summary concludes this chapter, touching also on open questions and the prospects for their solutions. More details will be given in Chap. 8.

2.1.1 Early History

The sudden appearance of a bright object, sometimes spanning the entire sky, and its rapid fading must have been an awesome experience for our ancestors; a great riddle beyond human understanding. In the Middle Ages, after the basis for understanding of our Solar System had been established, the phenomenon of comets was subject to many, mostly superstitious, interpretations. Following Aristotle, the size, appearance, and temporal behavior suggested a connection with the Earth's atmosphere, possibly even an origin in it.

The question about the nature and origin of comets led to extensive observations with the onset of astronomy in Europe in the Middle Ages. In 1577 Tycho Brahe determined from the parallax that comets are not phenomena within the Earth's atmosphere but are farther away than the Moon, i.e., comets are interplanetary bodies. His pupil Kepler thought that comets follow straight lines and concluded in 1619 that they originate outside the Solar System. Hevelius suggested parabolic orbits in 1668. Newton argued against a parabolic orbit of the Comet of 1681 (calculated by Flamsteed); he preferred the hypothesis of two independent comets, one for the inbound and one for the outbound leg (see, e.g., Westfall, 1980). It was Halley's prediction that the comets of 1531, 1607, and 1682 are apparitions of the same comet and that it would return in 1758. This prediction established that comets are part of the Solar System.

2.1.2 The Nucleus as Source of the Ephemeral Activity

While Kant (1755) included comets just like the planets in his nebula hypothesis for the formation of the Solar System, Laplace considered comets to be extra-solar in his cosmogony of the protosolar nebula. The question about the origin and formation of comets has been debated ever since. Öpik (1932) was the first to suggest that a cloud of comets surrounds the Solar System. This idea was corroborated by Oort (1950). He postulated a comet cloud with dimensions of about 10^5 AU and more than 2×10^{11} nuclei, based on an analysis of the almost parabolic orbits of observed comets. These comets were supposedly formed in the planetary nebula and scattered by the planets into the Oort cloud. Passing stars randomize the orbits and eventually perturb some comets so that they pass again through the inner Solar System. This depicts the traditional model of the source of new comets. The details of the hypothesis are still under discussion (see Chaps. 6 and 7). More information on the development of ideas and hypotheses from past to present can be found in the comprehensive review article by Bailey et al. (1986).

Still, the question remained: What turns comets on and off? Why are they so large? There had to be a part of a comet that was stable enough to orbit the Sun for many hundreds of years. A mechanical model (Bredikhin, 1903), describing the differently shaped comet tails, invoked a repulsive force to drive particles away from the central condensation. Later spectroscopic investigations revealed reflecting dust grains and simple molecules, radicals, and ions as constituents of the coma and tail. What is the nature of the central condensation? A "sand bank" model of comets developed from the concept of desorption of gases from dust particles with large surfaces (Levin, 1943a) and from Lyttleton's (1953) theory of solar accretion of interstellar material by a gravitational lens effect. The connection of meteor showers with comet orbits (Öpik, 1966a) further strengthened this model. But how could a loose aggregation survive close passes around the Sun as happens with Sun-grazing comets? On the other hand, splitting of comets (e.g., Comet Biela) was observed several times. The central condensation had to be fragile. Several passages of comets in front of the solar disk were used to determine an upper limit of less than 100 km for the size of the optically thick central condensation.

About the same time that Oort developed his theory of the comet cloud, Whipple (1950, 1951) described the nucleus of a comet as a "dirty snowball," i.e., as a solid body containing large amounts of frozen gases such as CH_4, CO_2, CO, NH_3, and H_2O and refractory grains with sizes ranging from dust to boulders. Kuiper (1951) proposed comet nuclei as building blocks of the Solar System and suggested a belt of comets beyond the region of Neptune and Pluto. Variations and competing hypotheses were developed by Öpik (1973), Safronov (1972, 1977), and Cameron (1962, 1973). Comet nuclei could have formed outside of the planetary system, even outside of the presolar nebula in a collapsing companion cloud that is too small to form a star. They could have formed between the stars that were clustered together (T Tauri association) during the early stages of collapse of the large, molecular cloud. When the young stars separate, many of the comets would be lost in interstellar space.

Only during the last 50 years has the nature of comets become clearer. The inference that almost all of the observed gaseous species are rather short-lived radicals that cannot directly emanate from the central condensation of the comet but must be dissociation products of larger, stable parent molecules such as $(CN)_2$, H_2O, and CH_4 (Wurm, 1934, 1935, 1943) opened the way to our present understanding of comets. No longer was it necessary to store the large amount of chemically very active radicals on or in dust particles; they could form ice mantles on the particles and cause them to stick together. It was Whipple (1950) who combined the astrometrical observations of increasing or decreasing orbital periods with the nature of the comet ices. Comets had to have a solid nucleus that rotates and produces the observed quantities of gas and dust by sublimation (volatilization) of its surface layers. The subliming gases cause a reaction on the nucleus, a nongravitational force that changes the orbital period of a comet depending on its sense of rotation.

The evidence supporting this description of the central body of comets became more and more compelling. The derived production rates of gas into the coma (Biermann and Trefftz, 1964; Huebner, 1965; Keller, 1976) and of dust moving into the tail (Finson and Probstein, 1968a) dismissed the idea of gas desorption from grains; only large quantities of ice could provide the necessary production rates. The sand bank model did not survive. The true nature of comets, their formation and origin could now be explored.

The comet nucleus is the comet proper. It is the body that formed during the formation of our Solar System. It is the body that survived 4.5 Gy. It is the body that orbits the inner Solar System when a comet becomes short-periodic. It is the body whose activity decays by its dwindling size or by formation of a devolatilized crust. However, the misproportion in size of the large phenomenon "comet," visible in the sky, relative to its insignificantly small source, the nucleus, has led to the use of the word "comet" to designate the phenomenon rather than the actual body.

An active comet nucleus cannot be observed with ground-based or Earth-orbiting telescopes. The dust and gas emissions mask the radiation reflected from the nucleus. Estimates of its mass can be derived from Newton's second law; the nongravitational force is obtained from the reaction on the nucleus by the gas and the nongravitational acceleration from the change of the orbital period. From the absence of detectable perturbations of small celestial bodies during close passage of a comet a crude upper limit of 10^{21} kg was derived for its mass (Laplace, 1805).

The reflected light from a nucleus, even at large heliocentric distances, is most often contaminated by scattered light from dust particles in its immediate vicinity. This has led to a systematic overestimate of the size derived from this kind of observation. It was realized that the "nuclear brightness" had to be an upper limit. Deriving the nucleus size from the required surface to produce the observed quantities of gas (mostly water) during the active phase of the orbit yields a lower limit, since not all of its surface, for example the night side, needs to be active. There are other indirect observations providing information on the nature of comet nuclei, such as their splitting, their decay rates, their sometimes erratic behaviors, etc. The chemistry and physics of the coma has been extensively studied to infer information about the nucleus itself.

2.1.3 Space Missions

The advent of the space missions to Comet P/Halley have changed the situation quite drastically. For the first time it was possible to observe the nucleus directly and to resolve it with high resolution cameras. The existence of a solid, single nucleus, as predicted by Whipple (1950, 1951), has been proven beyond any doubt. Here the most obvious quantum jump in our knowledge of comets was achieved. However, the snapshots of the flybys pose a large number of new and more detailed questions. Care has to be taken to interpret the spacecraft observations within the framework of the traditional ground-based observations and model calculations.

2.2 Observational Techniques

Naked eye comets, those that become brighter than 5^m, are usually long-period comets with the exception of a few, notably Comet P/Halley. Comets reach their peak activity when they are close to the Sun. During their perihelion passages their elongation angles from the Sun are often so small that they are not visible during the night. Many new comets have been detected near or after their perihelion passages (Lüst, 1985). To the naked eye they fade within a few weeks. While these bright objects receive substantial public and scientific attention as long as they are spectacular, their decline with increasing heliocentric distance has hardly been observed systematically for two reasons: Bright comets come unannounced, making it difficult to apply for telescope time within the normal scheduling rhythm, and secondly, at larger heliocentric distances, comets are objects of low brightness, and large telescopes have to be used in competition with other observing programs. Therefore our knowledge of the transition from the active to the inactive phase is based on sporadic observations, often by a single, dedicated astronomer. The first worldwide campaign during the 1910 apparition of P/Halley led to a large amount of data while the comet was bright. The collection of the data was difficult, and the assessment was not completed until 20 years after the perihelion passage (Bobrovnikoff, 1930).

2.2.1 Size Determination from Photometry at Large Heliocentric Distances

More recently Roemer (1966a,b) systematically observed comets at large heliocentric distances in order to determine their "nuclear" magnitudes before and after their phase of activity around perihelion. She used telescopes with long focal lengths to minimize the contribution from any potentially still existing coma. Her database comprises about 800 observations of more than 50 different comets, taken over almost two decades from 1957 to 1975. A systematic evaluation of the complete material has not been published. This material could be used not only to determine the relative sizes of nuclei but also to study the decline of the activity when comets recede from the Sun.

The total brightness of a bare comet nucleus should vary proportionally to r^{-2} and Δ^{-2}, where r is its heliocentric and Δ its geocentric distance. The nucleus cannot be resolved and displays a star-like image. Deviations from this distance-dependence have been used to check for contributions from activity in the vicinity of the nucleus. The decrease of activity with increasing r and the decrease of the telescope filling factor of the central condensation with increasing Δ tend to steepen the brightness variation. However, there are some problems even if no activity is present: The unknown phase function of the nucleus corrupts the $r^{-2}\Delta^{-2}$ relationship even though the changes in phase angle are smaller than 14° once the comet is 4 or 5 AU away. Rotation of the nucleus can produce considerable brightness variations caused by its shape or by spots with different albedos. These could be eliminated based on their periodicity if the observations were done frequently enough. Finally, if large dust particles stay with the nucleus because of their low outflow velocities or because they are gravitationally bound, their reflected signals will display the same dependence on the heliocentric distance and can yield a considerable contribution to the total brightness.

Most effects tend to increase the total reflected light, leading to an overestimate of the light scattered from the nucleus. Correlating the magnitudes of comets when they were last observed with their corresponding heliocentric distances led to estimates of their nuclear sizes (Svoreň, 1982). The main result is that the relative size variation of comet nuclei spans about one order of magnitude in linear dimensions and that their radii are usually less than about 10 km. Some of the objects, such as Comet P/Schwassmann-Wachmann 1, seem to be appreciably larger when the same criteria are applied. These comets show other peculiarities indicating that they belong to a different class of objects that may be CO- or CO_2-enriched (Whipple, 1980;

Larson, 1980; Cochran et al., 1980; Cowan and A'Hearn, 1982). Roemer's data show that new and long-period comets on nearly parabolic orbits reflect more light. However, this could be caused by some activity, not related to water sublimation, at large heliocentric distances. An outstanding example was Comet Kohoutek (1973 XII). It was detected at a heliocentric distance of about 5 AU. Its brightness led to an overestimate in the predictions of its activity near perihelion (Roemer, 1973; Delsemme, 1975). There could also be a difference of surface reflectivity rather than a systematic difference in nuclear sizes.

The flux, F_c, of reflected light from the nucleus measured on Earth at a distance Δ in AU from the comet is proportional to the cross section of the nucleus, S, and the geometric albedo, p_v,

$$F_c = r^{-2} \Delta^{-2} \phi(\alpha) p_v S F_\odot \pi^{-1}, \tag{2.1}$$

where $\phi(\alpha)$ is the phase function normalized to the phase angle $\alpha = 0°$ and $F_\odot$ is the solar flux at $r = 1$ AU.

A more suitable form of this equation is found by introducing the magnitude of the nucleus $V(1,1,0)$ at $r = \Delta = 1$ AU and $\alpha = 0°$ and solving for the effective radius of the nucleus, R_N

$$\log R_N = 2.14 - 0.2\, V(1,1,0) - 0.5 \log p_v. \tag{2.2}$$

Nuclear sizes based on magnitude observations require the knowledge of the surface reflectivity. No direct measurements are available. The values found for other Solar System objects span a factor of 25, from about 0.02 to 0.5, and could be considered limiting cases for the nuclear albedo, yielding a range of a factor 5 for the radius. Values between 0.5 to 5 km for short-period comets and 1 to 10 km for long-period comets have been obtained (Roemer, 1966b). This seems to indicate that new comets are bigger than evolved objects in less eccentric elliptical orbits. However, as mentioned in the example of Comet Kohoutek, a higher activity level can distort this conclusion.

2.2.2 Gas and Dust Production

A completely different approach to obtain an estimate on the lower limit for the size of the nuclear surface is based on its observed activity level. The gas production rate, $Z(\theta)$, as a function of the angle θ between the surface normal and the direction to the Sun, is given by the energy balance on the surface considering the attenuated solar radiation input, $F_\odot r^{-2} \exp(-\tau)$, the reflection, the reradiation in the infrared (IR), the energy needed to sublimate the gas, and the heat conduction into the interior

$$\frac{F_\odot \exp(-\tau)\,(1 - A_v)\,\cos(\theta)}{r^2} = \epsilon \sigma_o [T_N(\theta)]^4 + \frac{Z(\theta)\, L(T_N)}{N_o} + \kappa \frac{dT_N}{dR}\bigg|_{R_N^-}, \tag{2.3}$$

where τ is the mean optical depth of the coma, A_v the effective Bond albedo, ϵ the IR emissivity, σ_o the Stefan-Boltzmann constant, $T_N(\theta)$ the radiation temperature of the nucleus surface, $L(T_N)$ the latent heat for sublimation per mol, N_o the Avogadro number, and κ the thermal conductivity of the nucleus at the surface. The quantity R_N^- indicates the interior of the nucleus – coma boundary. The term on the left side is the absorbed solar energy, the first term on the right side is the reradiated energy, the next term is the energy used for sublimation, and the last term is the conduction into the interior. Equation (2.3) is solved in conjunction with the Clausius-Clapeyron equation of state

$$p_N(\theta) = p_r \exp \left[\frac{L(T_N)}{kN_o} \left(\frac{1}{T_r} - \frac{1}{T_N(\theta)} \right) \right],$$ (2.4)

the ideal gas law

$$p_N(\theta) = n_N(\theta)kT_N(\theta),$$ (2.5)

and the gas dynamic relationship of gas escaping from a surface into vacuum

$$Z(\theta) = \frac{1}{4}n_N(\theta)v_N(\theta) = n_N(\theta)\left[\frac{kT_N(\theta)}{2\pi m_u M} \right]^{1/2},$$ (2.6)

where p_r and T_r are pressure and temperature at some reference point, $n_N(\theta)$, $v_N(\theta)$, and $p_N(\theta)$ are the gas number density, velocity, and pressure at the nucleus, k is the Boltzmann constant, m_u is the unit atomic mass, and M is the molecular weight of the gas. Usually the last term in (2.3) is ignored and angular integration is replaced by one of two simplifications: (1) A nonrotating nucleus, or one with spin axis pointing to the Sun, for which the solar energy absorbed by the nuclear cross section πR_N^2 is uniformly redistributed over the sunlit hemisphere $2\pi R_N^2$, or (2) a fast rotator for which the incident energy is uniformly redistributed over the entire surface $4\pi R_N^2$. These simplifications give very similar results at small heliocentric distances, but at large distances both lead to poor estimates of the average surface temperature. The gas production depends on the temperature in a nonlinear way (Clausius-Clapeyron equation of state); an accurate determination of the angular distribution of the temperature is required in order to achieve reliable results. For example, Table 2.1 presents some results for a composition similar to that of P/Halley, containing about 85% water ice, at $r = 1$ AU. In this table $Q = f\pi R_N^2 Z$, where f is a geometry factor; $f = 1$ for a disk, $f = 2$ for a hemisphere, and $f = 4$ for a sphere. In the last row, Z and T_N are mean values, where the production rate and the temperature are obtained from equations like (2.7a), i.e., the exact solution for a hemisphere with uniform surface properties.

For a nonrotating nucleus, the quantity

$$Q = 2\pi R_N^2 \int_0^{\pi/2} Z(\theta) \sin(\theta)\, d\theta$$ (2.7a)

is the total production rate of gas molecules. For a fast rotator θ has no meaning and

$$Q = 4\pi R_N^2 Z.$$ (2.7b)

Table 2.1. Average gas production and temperature at two heliocentric distances

Distribution of energy input	$Z[\mathrm{m^{-2}\,s^{-1}}]$	$fZ\,[\mathrm{m^{-2}\,s^{-1}}]$	$T_N[\mathrm{K}]$
	$r = 1\,\mathrm{AU}$		
Subsolar point (disk)	$1.5\ \ 10^{22}$	$1.5\ 10^{22}$	206
Uniform over hemisphere	$0.67\ 10^{22}$	$1.3\ 10^{22}$	205
Uniform over total sphere	$0.30\ 10^{22}$	$1.2\ 10^{22}$	194
$\cos\theta$ over hemisphere	$0.69\ 10^{22}$	$1.4\ 10^{22}$	196
	$r = 5\,\mathrm{AU}$		
Subsolar point (disk)	$4.7\ 10^{15}$	$4.7\ 10^{15}$	168
Uniform over hemisphere	$1.2\ 10^{13}$	$2.4\ 10^{13}$	144
Uniform over total sphere	$2.4\ 10^{10}$	$9.1\ 10^{10}$	121
$\cos\theta$ over hemisphere	$6.5\ 10^{14}$	$1.3\ 10^{15}$	137

Since the reradiation term is negligible for an active comet near the Sun, (2.3) can be simplified. Considering the solar flux intercepted by the nuclear cross section and ignoring the correction for the optical depth, gives for the nuclear radius

$$R_N = r\,(QL)^{1/2}\,[\pi\,F_\odot(1 - A_v)\,N_o]^{-1/2}. \tag{2.8}$$

The global activity of the comet near the Sun can be used to derive a lower limit for the nucleus size if the heat of sublimation is known or assumed. For H_2O, $L \simeq 50$ kJ mol^{-1}. For P/Halley during the Giotto flyby at $r = 0.9\,\mathrm{AU}$, with $F_\odot = 1.36$ kW m^{-2}, $A_v = 0.03$, and $Q = 7\ 10^{29}$ molecules/s, (2.8) gives $R_N = 3.4\,\mathrm{km}$. This lower limit for R_N is close to the estimates derived from Roemer's observations.

The recent direct observations of the nucleus of P/Halley clearly show that only a minor part of the nuclear surface is active. Therefore the above estimate leads to the size of the active area rather than to the size of the nucleus. The approximate agreement with the lower limits of Roemer's investigations is a coincidence. In fact, recent observations, including IR observations, indicate that several comet nuclei are active only over part of their surfaces and have very low reflectivity, so that both lower limits are rather unrealistic.

Delsemme and Rud (1973) had the idea to combine the two extremes of cometary states using the gas production rate close to the Sun and the nuclear magnitude far from the Sun to determine both radius and albedo. However, their results for Comets Tago-Sato-Kosaka (1969 IX) and Bennett (1970 II) led to albedo values of more than 0.6, which are much too high. It was pointed out by Keller (1973) that the albedo of the extremely dusty Comet Bennett could not be that high and in particular not higher than that of Comet Tago-Sato-Kosaka. Nevertheless, their paper influenced the perception of comet nuclei for a decade.

Based on the experience with P/Halley, the unrealistically high albedos can now be taken as a clue that the gas productions of both of these long-period comets were limited to some active areas on their surfaces. Since the energy absorbed by the active areas and converted into gas production is smaller than that absorbed by the total cross section of the nucleus, the quantity $\pi R_N^2 A_v$ leads to an overestimate of the albedo for the observations at large heliocentric distances. A reinterpretation using a much lower estimate for A_v than the value determined by Delsemme and Rud (1973) shows that long-period comets, such as Bennett and Tago-Sato-Kosaka, are also active only on parts of their surfaces. If this interpretation is combined with IR observations (Sect. 2.2.5) of the distant nuclei to determine their albedos, the percentage of active areas on comet nuclei can be determined.

The onset and increase of activity when comets approach the Sun and the reverse when they recede (if one can ignore thermal lag) are indicators for the latent heat of sublimation (Huebner, 1965). A large number of investigations have concentrated on this effect to derive indirect evidence of the nucleus composition. The latent heat of water ($L_{H_2O} \simeq 50$ kJ mol^{-1}), being larger than that of CO_2 ($L_{CO_2} \simeq 25$ kJ mol^{-1}) or that of CO ($L_{CO} \simeq 6$ kJ mol^{-1}) and most other frozen gas components, suggests an onset of full activity between 1.5 and 3 AU, depending on the albedo (see, e.g., Marsden et al., 1973 and references therein). The production rate should vary proportionally to r^{-2} once the activity is strong and the reradiation term in (2.3) can be neglected, i.e., for $r < 1$ AU. Direct observations of the production rates of the mother molecules are very difficult. Most investigations concentrated on the UV observations, mainly of hydrogen and OH (Keller, 1976; Feldman et al., 1980, 1984; Festou, 1981a,b).

The dependence of the production rates on the heliocentric distance seems in almost all cases to be steeper than that corresponding to r^{-2}, at least during approach. Most of these observations cover a range of heliocentric distance from $r = 0.5$ to about 2 AU. Again it has to be pointed out that the assumption of uniform insolation and hence the corresponding mean temperature are problematic. The production rate is a complex transcendental function of the sublimation temperature. At large heliocentric distances a small variation of T leads to a large change in the value of Z (see Table 2.1). Also, heat capacity and conduction into the interior must be considered. A further condition may be connected with a refractory mantle that prevents free sublimation from the surface (see, Sect. 2.4). The existence of such a mantle complicates the sublimation process considerably and requires more elaborate models for the interpretation of the data.

If the activity of gas sublimation is restricted, a steepening of the r^{-2} law could be explained by a delayed onset of some of the source regions or a gradual enlargement of the regions during approach, stretching the interval from first activity to full activity. A corresponding effect could act during recession of the comet.

2.2.3 Nongravitational Forces

The activity of the nucleus produces nongravitational forces (the recoil effect) and specifically an acceleration in the radial direction. This acceleration can be determined from the interpretation of orbital positions of comets (Marsden et al., 1973; Whipple, 1977). Combining these data with the nuclear brightness, which is proportional to $A_v R_N^2$, leads to limits for the nuclear radius and albedo. Considering the difficulties in the determination of the nongravitational acceleration (unknown mass, lag angle, variable activity, pre- and post-perihelion asymmetry, etc.), the results have been quite consistent in deriving order of magnitude estimates.

2.2.4 Rotation Parameters from Variability of Brightness

Short term variation and structures of the coma can be used to infer the rotation of the nucleus. Stochastic fluctuations are superimposed on periodic variations of the production rate. In this approach, again emphasized by Whipple (1978a, 1980, 1981a), it is assumed that the dust "jets" and halos, often visible in the coma, are caused by the onset and termination of activity as gas- and dust-producing areas on the nucleus come into sunlight and leave it again with each rotation. The method has been applied to a few cases of well observed comets, yielding not only the spin period but also the orientation of the axis and the long-period, forced precession (P/Encke, Whipple and Sekanina, 1979; P/Halley, Sekanina and Larson, 1986). In general, the solutions are not unique and vary appreciably with observational uncertainties.

A much more direct way to determine the rotation rate of comet nuclei would be from the observations of the brightness variations of the nucleus itself, as successfully employed for asteroids of similar size and heliocentric distance (Burns and Tedesco, 1979). As discussed above, the contribution of the nucleus to the total brightness of the inner region is generally rather small, so that the weak periodic changes caused either by varying projections of the nuclear shape or by the variation of the surface albedo are difficult to discern. Roemer's most suitable observations of distant comets produced no conclusive results. The difficulty of this method was demonstrated during the Comet Halley campaign. In spite of the unprecedented early recovery of this comet at more than 11 AU heliocentric distance and its extensive and complete coverage, interrupted only by intervals when the comet's elongation angle from the Sun was too small, a conclusive spin period could not be derived, even though we now know that the shape of the nucleus is irregular (see Sect. 2.4). Various attempts to determine this important parameter were recorded by Sekanina (1985). Nevertheless, Sekanina and Larson (1986) were able to deduce the spin period and possibly the orientation of the spin axis from dust "jet" observations of the 1910 apparition.

In a few instances comets have come so close to the Earth that the high resolution exposures of the central condensation contain enough signal from the nucleus, or fortuitously some recurring activity, to reveal a periodicity. Comet d'Arrest showed a variation in the nuclear magnitude of about 0.1, from which a rotation period of about 5 h was obtained (Fay and Wisniewski, 1978). More recently the period of Comet IRAS-Araki-Alcock was determined to be 51 h by Sekanina (1988a).

2.2.5 Infrared Observations

Another method that has been reasonably successful for the determination of the nuclear size and the albedo relies on observational techniques developed for investigating asteroids. Observations of the IR emission of comet nuclei permit determination of the thermal flux with a peak wavelength beyond $10\,\mu\mathrm{m}$ for temperatures of 150 K and lower. Combination with the visual magnitude gives again two equations for the quantities of albedo and surface area allowing a determination of both. These observations gave the first reliable indications about the rather large size and dark surface of comets, in particular of P/Halley, even before the spacecraft encounters

(Hartmann et al., 1984; Hartmann and Cruikshank, 1982; Cruikshank et al., 1985; Lebofsky et al., 1986). These observations are useful for correlating comets with asteroids.

2.2.6 Radar Observations

A potentially powerful approach is the use of large radar antennas to measure the reflected signal of the nucleus. However, only a few comets come close enough to the Earth to achieve a sufficient ratio of signal to noise. The Arecibo dish received signals from Comets Encke, Grigg-Skjellerup, IRAS-Araki-Alcock, and Halley (Ostro, 1985). The spectral information is used rather than the amplitude of the signal, for which the scattering albedo would have to be known to infer the size of the nucleus. The signal is broadened by the Doppler shift of the spinning nucleus. This reveals the drawback of the method: The orientation of the spin axis and the period of rotation have to be known. Kamoun et al. (1982a) determined the radius of P/Encke to be 1.5 (+2.3, −1.0) km based on the orientation of the axis as determined by Whipple and Sekanina (1979). A second comet observed was P/Grigg-Skjellerup (Kamoun et al., 1982b). Particles comparable to or larger than the radar wavelength of several centimeters may introduce a further complication that makes it necessary to separate the response from the nucleus from that of the large particles in its vicinity. Their signal could actually dominate (Goldstein et al., 1984; Harmon et al., 1989). Not only the signal strength and the Doppler broadening but also the linear and circular polarization can be used to infer properties of the nucleus and the dust surrounding it, if the signals are strong enough.

2.2.7 Direct Observations

The direct observations of the nucleus of P/Halley from the Vega and Giotto spacecraft are leading to a new understanding of the nature of comets. In particular, the Halley Multicolour Camera (HMC) revealed topographic structures, morphology, albedo, and color of the nuclear surface and details of its activity for the first time. In combination with the Vega 1 and 2 observations, it will be possible to determine the shape of the nucleus, its spin axis, and period. The results will be presented and discussed separately in Sect. 2.4, together with a brief summary of relevant data from other *in situ* observations.

In summary, using the detailed determinations of the nuclear properties of P/Halley as a reference, the application of the above methods will yield more reliable results.

2.3 Models of the Nucleus

The existence of comet nuclei as solid bodies and sources of the diffuse and faint dust and gas coma was implied by Newton in his *Principia*. He suggested that solid bodies lose material when they come close to the Sun and are heated up.

2.3.1 The Sand Bank Model

Related to Newton's idea, the sand bank model was promoted by Lyttleton (1948, 1953) and supported by Levin (1943a). In this model, the observed radicals CN, C_2, C_3, CH, etc. were thought to be adsorbed on dust grains. Adsorption could not support the large amounts of gas needed to accelerate the dust into the coma and tail as sublimation of unobservable water ice could (Huebner, 1965). A gas density that is much higher than the emissions of the observed species of CN, C_2, C_3, etc. suggested, had been suspected from the presence of the forbidden oxygen transitions (Biermann and Trefftz, 1964) and by the broadening of the rotational lines of CN (Malaise, 1970). The final confirmation of the high gas density came through the UV observations of the Lyman-α (L-α) transitions of atomic hydrogen (Keller, 1976). The sand bank model could not explain the survival of Sun-grazing comets or the splitting of comets, to mention only some of the most obvious arguments. Intermediate models of an agglomeration of smaller nuclei, gravitationally bound, were also discussed. This kind of a nucleus ensemble could not survive and would collapse already in the Oort cloud (O'Dell, 1973).

2.3.2 Nongravitational Forces

The occurrence of nongravitational forces led Whipple to the quantitative icy conglomerate model for the nucleus of P/Encke. The ice sublimes on the sunward side of the nucleus, the recoil giving it a net acceleration away from the Sun. This interpretation yields the mass of the nucleus as a byproduct. The maximum of the sublimation will generally not lie exactly on the subsolar point of the nucleus because of a thermal lag of a few degrees in the direction of rotation of the nucleus. The main components of the force are in the plane of the comet orbit, causing the deviation from the Keplerian motion. Since this lag angle can only be estimated and depends on material constants, the spin period, and the position of the spin axis relative to the orbital plane of the nucleus, the mass determination of the comet nucleus is rather uncertain.

2.3.3 Icy Conglomerate Nucleus

Whipple's icy conglomerate model introduced the idea that most or at least a large part of the nucleus consists of a volatile, icy component: The unobserved frozen parent molecules H_2O, CH_4, NH_3, and possibly CO, CO_2, and some CN-compounds (Wurm, 1943) of the observed species CN, C_2, C_3, CH, CO^+, etc. It also has meteoritic material from dust to boulder size, like dirt imbedded in a snowball. This concept was conceived to explain the appearance of comets and their relationship to associated meteor streams. It was not derived or based on the ideas of how and where comet nuclei were created. Physical explanations of the formation processes and their influence on the physical structure and composition of comet nuclei have been discussed in a long series of papers (also by Whipple himself) once the basic idea was accepted. More details will be given in Chap. 7. Here we introduce only some fundamental concepts that are required to interpret the observations and recent results of comet nuclei. The arguments in favor of a solid nucleus soon became overwhelming.

2.3.4 Nucleus Formation

In the original concept of the icy conglomerate model, the nucleus contained differentiated refractory matter similar to meteoritic materials. The distribution within the nucleus could be inhomogeneous, a large refractory core seemed to be conceivable (Sekanina, 1972). On the other hand, the concept of the formation of the planetary system from instabilities of the dust component within the molecular cloud suggests that comet nuclei may have been building blocks of the planets in the early stages of the Solar System (Goldreich and Ward, 1973; Safronov, 1972; Cameron, 1973). Comets would have formed on relatively short time scales of 10^5 to 10^6 y from the components of the molecular cloud in the ecliptical dust disk of the contracting nebula. The resulting nuclei should have been homogeneous because large refractory boulders of material probably did not exist. The concept of planetesimal formation by gravitational instabilities has recently been questioned (Weidenschilling, 1988; Weidenschilling et al., 1989), and therefore conclusions drawn on the mass and size distribution of the comets based on this concept (e.g., Biermann and Michel, 1978) may not be valid. Regardless whether the comet nuclei are formed within the solar nebula or in companion clouds (Cameron, 1973; Biermann and Michel, 1978), in their early history they should have consisted of unaltered material homogeneously distributed. The degree of processing that the volatile material underwent before the agglomeration into larger bodies occurred is still unknown. One extreme is the formation directly from interstellar, ice covered grains (Greenberg, 1977).

The comet bodies remain so small that gravitational compaction of the fluffy material does not occur. The formation of bodies larger than 10 km out of a distribution of smaller sizes without major alterations is conceivable only if the relative velocities are small. This concept allows for bodies with densities below $1 \, \mathrm{Mg \, m^{-3}}$. Meteor streams and showers associated with comet orbits consist to a large extent of low density particles (Millman, 1972; Verniani, 1973; Stohl, 1986; Halliday, 1987). Fireballs also indicate extremely porous material (Ceplecha, 1977).

A differentiation of the comet nucleus on its surface as an evolutionary effect has been repeatedly suggested. Cosmic rays induce chemical alterations down to several meters during storage in the Oort cloud, while UV radiation chemically transforms the surface (Donn, 1963; Shul'man, 1972; Whipple, 1977; Johnson et al., 1987). However, this initial layer containing more volatile components and tar-like organic substances may be lost during the first entry of the comet into the inner Solar System.

Another evolutionary effect is the formation of a regolith of dust particles that cannot leave the nucleus. This was already considered by Whipple in one of his early comet papers (Whipple, 1955). The formation of such a mantle and its stability were investigated in several theoretical papers (Mendis and Brin, 1977; Brin and Mendis, 1979). They show that a very thin layer of a few centimeters is sufficient to choke the sublimation of ice. However, unless the material is sintered by the action of cosmic or UV radiation, the physical strength of such a thin layer will not be adequate to withstand the thermal stresses causing expansion and contraction of the surface layer since the thermal gradient can be large because of the good insulation these porous layers provide.

2.4 The Nucleus of Comet P/Halley

Differentiation of the nuclear signals from the scattered light of the coma is very difficult for ground-based observations because of the limited resolution. Only in special cases is it possible, but with an accuracy so low that only very general conclusions can be drawn on the physical properties of the nucleus. The spacecraft missions to Comet P/Halley changed this situation. For the first time it was possible to determine the size, shape, and even details of the surface features (Keller et al., 1987a). These observations of the nucleus have formed a new base for our understanding of comets as a whole.

2.4.1 Expectations

A study group compiled the most probable physical data for P/Halley (Divine et al., 1986) as a baseline for the planning and operation of the flyby missions. The prediction for the nucleus will be summarized here to characterize the state of our knowledge or the scope of our assumptions.

The early recovery of P/Halley at 11.2 AU (Jewitt and Danielson, 1984) and later systematic observations yielded $2.2 \pm 0.2 \, \text{km}^2$ for the product $p_v R_N^2$. It was not possible to infer the spin period, although irregular fluctuations of the signal indicated some level of activity, even at that large heliocentric distance. As a best guess, a radius for the nucleus of $R_N = 3 \, \text{km}$ and a corresponding geometric albedo of $p_v = 0.06$ was suggested. The mass was estimated to be $\sim 10^{14} \, \text{kg}$, assuming a density of $1 \, \text{Mg m}^{-3}$. The spin period showed a long list of "improved" values derived from the timely development of features of the dust coma. The last value before the encounter was 2.17 d (or 52.1 h) given by Sekanina and Larson (1986).

2.4.2 The Giotto Encounter

The first flyby of P/Halley, that of Vega 1, took place on 6 March 1986. The data were transmitted from the spacecraft to Moscow and displayed in real time. The close-up images from a range of $\sim 9000 \, \text{km}$ showed a rather diffuse intensity maximum with a secondary bright spot. Three days later during the encounter of Vega 2, the images looked rather similar. Immediately after both encounters the interpretation of the images was not conclusive; a contour and the shape of the comet nucleus were not obvious. Not until four days later, after the Halley Multicolour Camera (HMC) on ESA's Giotto spacecraft revealed a full contour of the nucleus, could the Vega images be interpreted with confidence. The approach geometry was rather similar for all three spacecraft (see Table 2.2). The visibility of the nucleus during the Vega 1 flyby was

Table 2.2. Vega and Giotto encounter parameters

	Date 1986	Time [UTC]	Distance of closest approach [km]	Phase angle[†] during approach [°]	Azimuth[‡] of target point [°]	Relative velocity [km s⁻¹]	r [AU]	Δ [AU]
Vega 1	6 March	07:20:06	8889	112.2	19.2	79.222	0.7923	1.153
Vega 2	9 March	07:20:00	8030	123.4	15.0	76.785	0.8341	1.073
Giotto	14 March	00:03:01.84±0.20*	596±2*	107.2	−30.6 ± 0.1*	68.373	0.9023	0.960

*Determined by Halley Multicolour Camera
[†] Angle between Sun-comet line and relative velocity vector
[‡] Angle between the spacecraft-comet line at closest approach and the plane defined by the Sun-comet line and the relative velocity vector (Positive: Ecliptic north)

somewhat impaired by the dust in the inner coma. At that time, the dust production was about twice as intense as during the Giotto observations. Unfortunately, the quality of the Vega images was also impaired by technical problems. The Vega 1 camera was out of focus, and the intensity resolution was limited to 5 bits compared to the dynamic range of 12 bits for the HMC images. From the Vega 2 camera all but two images were overexposed in the central part. All images showed strong noise components.

Technical problems of the HMC hardly influenced the quality of the transmitted images. The nucleus of P/Halley became visible from a distance of more than 124,000 km (see Fig. 2.1), about half an hour before the closest approach of the Giotto spacecraft on March 13, 1986 (Reinhard, 1986). The images taken by the HMC revealed a rather elongated and irregular shape of the nucleus, with a 14 km × 7.5 km size in projection on the sky, a very dark surface with albedo less than 4%, and highly concentrated and anisotropic activity (Keller et al., 1988).

A few details about the Giotto encounter with P/Halley and the operation of the HMC are required to understand and interpret the data correctly: The phase angles during the approaches of the spacecraft were rather unfavorable for imaging since the cameras had to look against the Sun. For Giotto the Sun – comet – HMC angle was 107°. Since Giotto was a spinning spacecraft with a period of 4 s the exposure time of the 1 m focal length, high resolution telescope had to be extremely short. Depending on the offset angle, the exposure times decreased from 6 ms down to 64 μs at encounter (Keller et al., 1987b). The images concentrated on the innermost part of the coma, the vicinity of the nucleus out to a few thousand kilometers.

Several hundred images taken by the HMC show the nucleus as a whole or parts of its surface in detail. The maximum field of view of the camera was about 0.5° × 0.4°, corresponding to the size of the CCD detector. During the last five minutes 4 images were taken almost simultaneously during each spin; one image through the full bandwidth and with full spatial resolution, and three through broad-band color filters with the spatial resolution reduced to one half. Because of the limited transmission rate, the field of view was restricted to only 0.1° × 0.1°. The operation of the HMC and the transmission from the entire spacecraft were interrupted by dust impacts about 10 s before closest approach. The last image transmitted was taken from a slant distance of 1680 km.

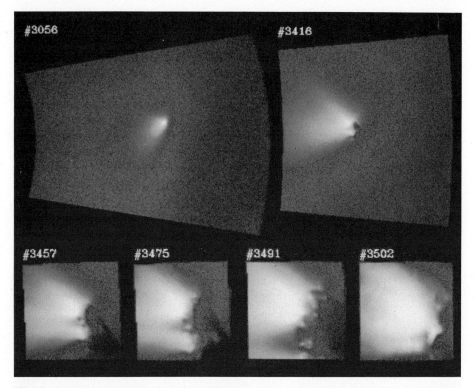

Fig. 2.1. Six examples of HMC images of P/Halley in original frame sizes. Image # 3056 was taken 1814 s (distance to nucleus 124000 km) and image # 3502 was taken 31 s (2200 km) before closest approach.

2.4.3 Imaging Properties and Orientation of the Nucleus

The whole nucleus of P/Halley is visible in the image of Fig. 2.2. It is composed of six images with varying spatial resolution. Images with higher resolution replace the corresponding parts of the images taken from a larger distance. The camera tracked on the brightest point of the comet emission, i.e., the dust from the active area near the northern tip of the nucleus on the sunward side. Therefore the resolution of the composite image improves from 300 m/pixel on the southern limb to 60 m/pixel toward the active area. The aspect angle changes also by up to 11°, particularly for the last two images included in the composite. These changes cause some artificial effects in the composite image that have to be considered when they are interpreted. Celestial north is approximately toward the top.

For most images the Sun is about 17° behind the image plane, slightly changing with the aspect angle, and 25° to 30° above the horizontal. Celestial north is almost aligned with the vertical upward direction of the images; the Sun shines from the northeast (using the astronomical nomenclature with east to the left). This equatorial coordinate system will be used in the following descriptions of the images. Some confusion arises by the use of ecliptic coordinates and a cometocentric system.

Fig. 2.2. A composite of six HMC images ranging in resolution from 320 m to 60 m per pixel. Illumination by the Sun is from the left, about 28° above the horizontal and 12° behind the image plane. Some of the contributing single images are displayed in Fig. 2.1.

This is particularly true here because the spin axis of the nucleus points toward the southeast (see Fig. 2.3). The north pole of the nucleus is put, following IAU convention, almost in the opposite direction of celestial north. Because of the 107° phase angle, most of the nuclear surface facing the observer is not illuminated. The night side part of the surface appears in silhouette against the scattered light from the dust in the background. Approximately 25% of the visible, projected surface is illuminated by the Sun. The terminator is seen from the northern tip down to the southern sunward limb. Strong activity is visible on the sunward part of the surface, concentrated toward the northern tip. The outline of the bright limb is difficult to discern because of the blending of the scattered light from the dust and the reflected signal from the nucleus.

Giotto images alone cannot define the three-dimensional shape of the nucleus and its orientation relative to the Sun and the observer. A combination of images

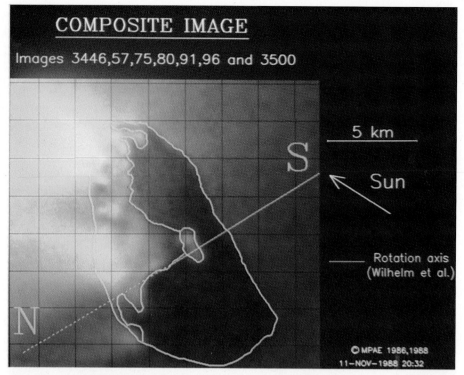

COMPOSITE IMAGE

Images 3446,57,75,80,91,96 and 3500

5 km

S

Sun

Rotation axis
(Wilhelm et al.)

N

© MPAE 1986,1988
11–NOV–1988 20:32

Fig. 2.3. A composite image of the nucleus of P/Halley is overlayed by a mesh with a grid size of 2 km. The outline of the nucleus is shown with the spin axis. The arrow indicates the direction to the Sun.

of the three flybys led to a preliminary deduction of the orientation of the nucleus during the Giotto flyby (Wilhelm et al., 1986; Keller et al., 1987a). The difficulties and uncertainties will be discussed below. It is inferred that the northern tip of the nucleus points about 20° to 30° out of the image plane toward the approaching observer. The spin axis is assumed to be parallel to the maximum moment of inertia, i.e., perpendicular to the long axis of the nucleus. It points about 13° into the image plane. The north pole of the nucleus is near the illuminated limb (see Fig. 2.3). The northern tip rotates away from the observer who looks at the morning terminator.

2.4.4 Shape and Size

The general shape of the nucleus is dominated by its surprisingly distinct elongation of 2:1. Its visible, projected length is 14.2 ± 0.3 km. Image # 1194 from the Vega 2 observations, taken 99 s after encounter, shows a length of 16 ± 1 km, which is considered the maximum extent (Sagdeev et al., 1986a). This is in agreement with the foreshortening derived from the proposed orientation of the nucleus during the Giotto encounter. The overall appearance of the nucleus on the HMC images is characterized by the mild constriction of the dark area in about the middle of the nucleus, giving the impression of two halves separated by a waist (like a peanut).

Indeed, the bright tapering strip from the illuminated limb toward the center of the nucleus is interpreted as a depression, although it is not reflected in a concave limb as might be expected. Thus this depression must be quite shallow. In fact, just south of the depression, the maximum width of the nucleus measured perpendicular to its long axis is 7.8 ± 0.3 km (see Figs. 2.2 and 2.3).

The images of the HMC do not reveal the third dimension of the nucleus. The volume of the nucleus had to be derived from a combination of all spacecraft observations (Wilhelm et al., 1986). This task has not been as straight forward as one might expect. The Vega images do not show any features on the nucleus surface. The outline and terminator are purely inferred and undefined in some cases (see Fig. 2.4). Only the global shape of the nucleus can be compared from image to image, leaving a considerable amount of freedom in orientation. Simultaneously, with the derivation of the three-dimensional shape, the rotation parameters have to be determined; a difficult task, as depicted below.

Information for the determination of the third dimension comes from two sides. The most straightforward would be to derive it directly from the images. During the Vega 1 closest approach, the viewing direction was closely aligned with the long axis of the comet. The projected profile was almost circular, indicating a rotational symmetric ellipsoid for the overall shape of the nucleus. On the other hand, a minimal asymmetry of the nucleus is required from the dynamical behavior of the spin, in particular from the inferred free precession of 7.43 d. Wilhelm (1987) found a self consistent model using 8.2, 8.5 and 16 km for the three axes. The corresponding values for the minor axes derived directly from the Vega images are 7.5 ± 0.8 km and 8.2 ± 0.8 km (Sagdeev et al., 1986a). However, it is not clear whether the orientation of the two sets of body axes are identical to each other.

The southern "hemisphere" seems to be slightly more massive when compared to the northern end. This is in agreement with the Vega observations, although the identification of the "thick" or "thin" end is marginal. Qualitatively the shape of the nucleus can be represented by an ellipsoid with its center close to a point just south and sunward from the "mountain" and with half axes of 7.2 and 3.7 km.

A further interesting property of the projected image of the nucleus is the almost straight limb on the dark side. There are only small deviations visible above the resolution limit of 100 to 200 m. A straight line, with a length of about 10 km and a deviation of only a few percent, is remarkable. Whether this attribute is fortuitous or relates to some physical process is not known.

The straight-line back is in strong contrast to the sharp right angle bend at its southern end. A slight protrusion, the "duck tail" is indicated in the best resolved images (Fig. 2.3 and 2.5). The curvature of the feature is about 300 m, comparable to the resolution limit at that end of the nucleus. The "duck tail" could be a protrusion or mountain with its top visible in projection above the limb. On Vega 2 images, a corresponding feature is seen when the opposite side of the nucleus is visible. This sharp corner is remarkable and contrasts with the appearance of the rest of the nucleus.

The sunward, illuminated limb is less well defined. Careful investigation of the intensity distribution reveals hints of the limb for most of its course. Either the brightness gradient across the limb changes, or a drop of intensity is detectable. The

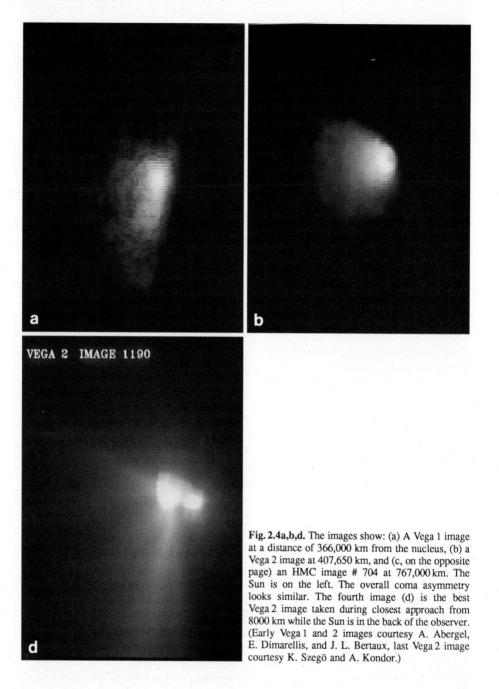

VEGA 2   IMAGE 1190

Fig. 2.4a,b,d. The images show: (a) A Vega 1 image at a distance of 366,000 km from the nucleus, (b) a Vega 2 image at 407,650 km, and (c, on the opposite page) an HMC image # 704 at 767,000 km. The Sun is on the left. The overall coma asymmetry looks similar. The fourth image (d) is the best Vega 2 image taken during closest approach from 8000 km while the Sun is in the back of the observer. (Early Vega 1 and 2 images courtesy A. Abergel, E. Dimarellis, and J. L. Bertaux, last Vega 2 image courtesy K. Szegö and A. Kondor.)

Fig. 2.5. The southern "large" end of the nucleus stretched for high contrast at low intensities to make the "ducktail" visible. The image is presented as a negative to show the undulating structures of the nucleus more clearly. The arrow indicates the direction to the Sun.

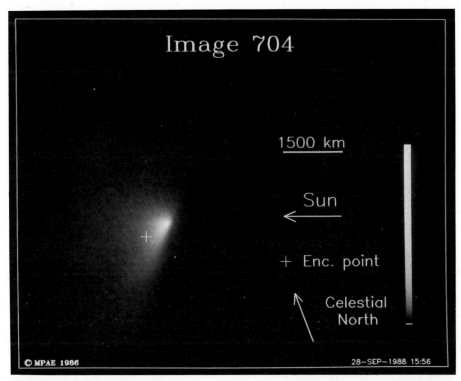

Fig. 2.4c. Caption see opposite page.

▼**Fig. 2.5.** Caption see opposite page.

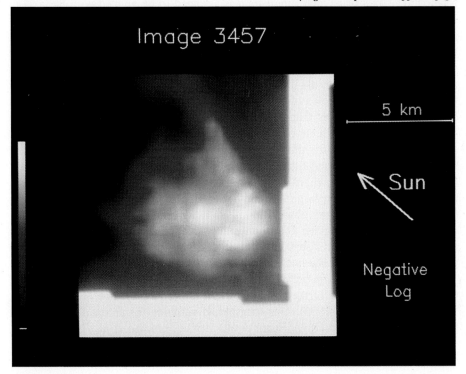

limb has to lie close to the maximum of brightness since the brightest emission has to come either from the illuminated limb itself or from the dust just above its outline, where the dust column density ought to be near its maximum. The situation is a little more complicated at the northeastern active area. The multitude of small "jets" and the apparent topography of the surface blur the outline. The edge of the "crater" coincides with the limb since the wedge-like dark feature toward the Sun is completely featureless.

The surface area and the volume of the nucleus can be estimated from the approximate ellipsoid to be $400 \pm 80 \, \text{km}^2$ and $550 \pm 165 \, \text{km}^3$, respectively. The projected surface during the Giotto approach is about $90 \, \text{km}^2$ (Keller et al., 1987a).

So far it has not been possible to identify surface features of the nucleus on images from the different flybys. This is the main reason why neither the orientations of the nucleus nor its outlines could be determined uniquely.

2.4.5 Spin of the Nucleus

It was planned that all three spacecraft were to image the nucleus during their flybys from three sides, each providing high resolution images of the surface. The time intervals between the passages were long compared to the expected spin period of slightly more than 2 d (Sekanina and Larson, 1986). However, because of the various unfortunate reasons discussed above, it has not been possible to determine the rotational position of the nucleus with any accuracy.

Therefore some assumptions have to made to narrow the possibilities for the rotation. The axis of angular momentum is assumed to be aligned with the axis of maximum moment of inertia. This is obviously perpendicular to the long axis of the nucleus. This assumption, made for the first interpretation (Sagdeev et al., 1986d), still holds. The spin axis points about 13° into the image plane (see Fig. 2.3). The various interpretations of the spacecraft images have yielded similar results with some scatter (Wilhelm et al., 1986; Sagdeev et al., 1986c; Smith et al., 1987). All these results exclude angles of free precession of the spin axis around the axis of angular momentum larger than about 10° half-cone angle. The period is found to be between 52 and 54 h in good agreement with the interpretation of the 1910 ground-based observations, although the derived orientation of the spin axis differs considerably (Sekanina and Larson, 1986). In fact, preliminary analysis of the rotation state of the nucleus, based on high resolution, ground-based observations taken as part of the HMC campaign (Cosmovici et al., 1986), shows that the spin axis derived from the spacecraft observation is not in agreement with the spiral structure of the dust "jets" during the Giotto encounter (Keller and Thomas, 1988). *In situ* observations of the dust distribution in "jets" during the Vega flybys also yield a period of about 50 h for the nucleus rotation (Vaisberg et al., 1986a).

The results for the spin period seem to be in disagreement with observations of the brightness variations of the coma of P/Halley. Here a period of 7.3 d dominates (Millis and Schleicher, 1986). This long-period variation for production of various gases (CN, C_2, OH and H) (A'Hearn et al., 1986b; Feldman et al., 1986; Stewart, 1987), as well as for the production of dust (Larson et al., 1987), is well established.

The two-day period is considerably more difficult to derive from ground-based observations, although Leibowitz (1986) and Schleicher et al. (1986) found it from expanding shells of the CN radical (see also Celnik and Schmidt-Kaler, 1987). A two-day period was also inferred from the Lyman-α observations of hydrogen by the Suisei spacecraft (Kaneda et al., 1986a), in contrast to observations by the Pioneer Venus spacecraft (Stewart, 1987). The dynamics of dust "jets" agree with the short period (Larson et al., 1987; Keller and Thomas, 1988), while the long period seems to be related to the variations in the production of gas and dust. Sekanina (1987a), Julian (1987), and Festou et al. (1987) suggested rotational motions based on solutions for a symmetric top, introducing two spin periods. A rotation around the long axis with the 7.3 d period and a free precession around the axis of angular momentum with a period of about 2 d are suggested. The longer period motion exposes a particularly active spot to the Sun, triggering the increase in productivity. A complete rotation around the long axis is not required. The rotation around the long axis seems to be in disagreement with the spacecraft observations and has actually been rejected by Smith et al. (1987). Festou et al. (1987) suggest a nutation period of 14.6 d that yields peaks of brightness two times per period, as one would expect from a single active spot. They showed that this period is in agreement with the ground-based observations. Wilhelm (1987) simulated the rotation of the nucleus by integrating the Euler equations. He showed that the small asymmetry of the nuclear ellipsoid derived from the Vega 1 near-encounter images (see Sect. 2.4.3) required a longer than 7 d period, that it could be reconciled with 14.6 d, and that it is hardly possible to induce a larger than 3° angle of free precession by the reaction moments of the jet activity of the nucleus. However, an inferred large angle of free precession could also have been produced by a random walk-like process, adding the small angles from orbit to orbit. This requires a very small damping constant of the nucleus and therefore a high rigidity.

A further important result of the numerical simulations of Wilhelm (1987) is that the jet activity of the nucleus probably produces forced precession of the spin axis. A change in position of the spin axis by more than 30° within one perihelion passage seems likely. This makes the poles of the nucleus nonstationary and the whole rotation pattern much more difficult to analyze. While the orientation of the spin axis would change considerably from pre- to post-perihelion, any similarity from one orbit to the next should be limited.

The rotational state of P/Halley has still to be understood. If the long-period variability is caused by free precession, the motion is even more difficult to disentangle. This free precession is not caused by the activity level of P/Halley and therefore it can be expected to be common. The interpretations of variability from ground-based observations have to be viewed with particular care.

2.4.6 Surface

The contrast between light scattered from the dust near the nucleus and the light reflected directly from the surface is small. This was immediately apparent after the first encounter of Vega 1, when it was not possible to determine the outline of the nucleus. Later investigations (Sagdeev et al., 1986b; Keller et al., 1987a) showed that the contributions of the scattered light from the dust columns are optically thin; the dust should be transparent. However, it is obvious on the Giotto images that the intensity in the line-of-sight past the nucleus is often comparable to or even higher than that of the illuminated limb. Appreciable contributions from the dust to the signal should therefore be considered when making photometric studies of the illuminated surface, particularly when looking on the surface at small phase angles along the jet-like dust features. Isophotes of constant reflectivity were given for the image # 1190, taken almost exactly at closest approach of Vega 2 (phase angle 9°). The values are very small (the peak is 0.03) and probably include contributions from the dust. The variation across the surface was very modest. The darkest isophote displayed approximately along the limb and terminator of the nucleus was 0.02 (Sagdeev et al., 1986d). The only discernable features were the two well separated, broad maxima. They seem to coincide with centers of dust activity. It can therefore be concluded that the surface of the nucleus is extraordinarily uniform in reflectance. This has also been confirmed by the observations of the HMC (phase angle of 107°). The local surface brightness varies by less than 50% on a somewhat smaller scale because of the higher resolution and also the smaller amount of dust. Much of this variation is presumably caused by variations of the reflection angles from the local topography. A systematic decrease in reflectance is observed varying from 4×10^{-4} at the limb to less than 1×10^{-4} at the tapering end of the depression near the center of the visible surface (compare Fig. 2.6). This is explained by the increase of the local zenith angle of the Sun toward the terminator.

The geometric albedo of the nucleus at a phase near 0° can be calculated from the Vega 2 observations, yielding $p_v = 0.04$ (+0.02, −0.01) (Sagdeev et al., 1986b), or by using the brightness observations of the nucleus at large heliocentric distances. Hughes (1985) found $p_v S = 1.4\pi \, km^2$. Even at large heliocentric distances, contributions of the surrounding dust have to be suspected, converting this constant into an upper limit. Sekanina (1985) determined $p_v S = 1.2\pi \, km^2$, noting the irregular brightness variations from which, however, no spin period could be derived. A critical evaluation (Divine et al., 1986) yields a constant about half the above value (0.7π). A geometric albedo of 0.05 can be derived as an upper limit; a value as low as 0.02 is in agreement with the earlier photometric observations of P/Halley. This albedo is low, but similar to values determined for C-type asteroids, and in agreement with the determinations using the infrared emission of the nucleus (Bowell and Lume, 1979; Hartmann et al., 1982; Cruikshank et al., 1985).

The reflectivity observed at a phase angle of 107° (Delamere et al., 1986) is about 1/10 of the full phase geometric albedo, a value of similar proportions is found in other planetary objects, e.g., the Moon. The relatively strong decrease with phase angle indicates a rough surface. The law of reflectivity cannot be derived from the HMC data, since the phase angle did not vary. The values obtained from Vega 1 observations (Sagdeev et al., 1986d) contain contributions from dust. They are considerably higher at similar phase angles than those measured a few days later by Vega 2. Nevertheless, the decrease of the reflectivity as a function of phase angle at approximately constant zenith angle is steep, similar to that of the rough lunar surface (Sagdeev et al., 1986b).

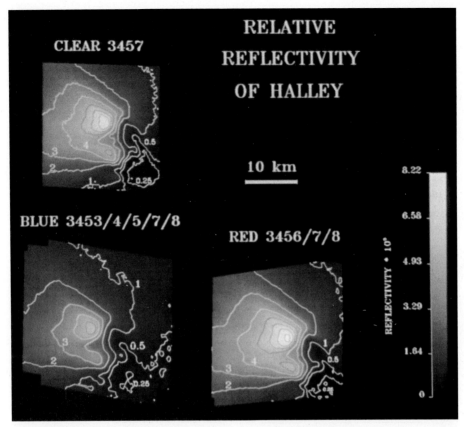

Fig. 2.6. The same area in the image plane is shown for three different spectral passbands: clear, red, and blue broad-band filters. The intensity levels (shown as reflectivity) in each passband would be equal for a solar spectrum.

Variations of the brightness as a function of the solar zenith angle were studied in the area of the central depression from the limb to the center of the nuclear surface. A derivation of a topographic profile from brightness variations has been attempted. Both attempts require the knowledge of the three-dimensional shape of the nucleus and the scattering law. Contributions from dust above the surface strongly influence the results (Schwarz et al., 1987).

The Vega team did not find spectral differences in their observations through three different passbands (0.45 to 0.65 μm, 0.63 to 0.76 μm, and 0.76 to 0.86 μm) within the accuracy of their measurements. The recent calibration of the broad-band color images of P/Halley by the HMC show a red excess of the dust (12%) and a minor excess in the red for the nucleus (Fig. 2.6). The HMC images were taken through filters with effective wavelengths at 653 nm (clear), 813 nm (red), and 440 nm (blue). The scattered light from the dust at phase angle 107° is appreciably redder than that reflected from the nucleus. It is improbable that the reddening is caused by physical scattering properties of the dust grains at the large phase angle. It is more likely caused by the chemical composition of the dust grains as suggested by the high content of organic material. The "color" of the surface is slightly red but the reflected light does not deviate strongly from the solar spectrum.

Summarizing the optical properties of the nuclear surface of Halley's comet: The nucleus belongs to the darkest objects in the Solar System. The surface is probably very rough on a small scale, as indicated by the steep phase angle dependence, and slightly reddish, suggesting organic material.

2.4.7 Surface Features and Morphology

Vega observations have not revealed details or features on the surface of the nucleus, even though it has been repeatedly claimed that linear features could be identified on two different, heavily stretched and processed Vega 2 images (Möhlmann et al., 1986; 1987). It is even difficult to determine the terminator and the limb. Many images published by the Vega team show masks based on outlines derived from a mathematical expression such as the maximum gradient. Any structure on the surface is dominated by noise. As discussed above, the brightness variations due to topographical features are very small. Excellent resolution of contrast is required, considering the interference from the dust.

Several features visible on the HMC images will now be described to characterize the morphology of the nuclear surface (see Fig. 2.7). The northern part of the surface

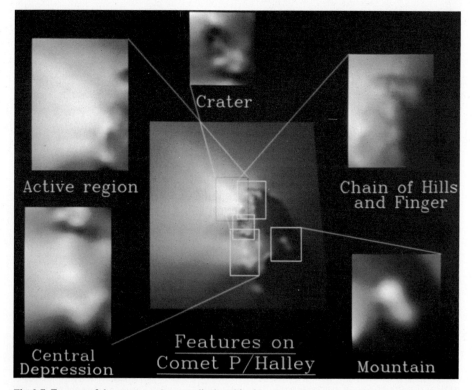

Fig. 2.7. Features of the comet nucleus are displayed in the cutouts that are expanded by a factor of three.

shows the most structure, partly because the resolution is better than in the south, and partly because the illumination of the surface is more favorable. At the southern end the terminator reaches the sunward limb. The features summarized below are illustrated and described in more detail in an article by Keller et al. (1988).

The "finger": The northern tip of the nucleus is not yet illuminated. The morning terminator is interrupted by a bright feature that separates the dark, about 600 m wide and 1.4 km long, finger-like ledge from the main body.

The "crater": A roundish feature at the sunward limb, south of the most active region, looks crater-like. High resolution images show structures within this feature. It is not clear whether the "crater" is really a coherent physical feature, but it is not an impact crater; it could be an area of increased activity when illuminated by the Sun. Its depth is estimated to be 200 m or less and its diameter about 2 km (Schwarz et al., 1987).

The "chain of hills": An undulation of the morning terminator has a scale length of about 1 km. Successive bright and dark patches indicate up and down slopes. The southern end of this feature runs into the area of the "crater." With some imagination the "chain of hills" could be curving around a large plane where all the activity at the northeastern tip is located.

The "mountain": The most spectacular feature on the HMC images of the nucleus is a bright patch in the middle of the nucleus on the night side. It is most probably a hill the peak of which is illuminated by the rising Sun. Its height is estimated to be less than 1 km and depends on the shape of the nucleus and, in particular, on the effective radius of the area of the "central depression."

The "central depression": It seems to lace the nucleus at its center. The terminator recedes toward the center of the outline. However, the limb is not concave, and the limb area appears bright and active where the contrast is very low. A well defined, slightly dark ledge seems to follow the curvature of the terminator at the southern boundary of the "central depression." Interpreting the brightness variations as changes in surface slope yields a terrace-like structure, increasing in height toward the terminator. It is a large-scale feature similar to the "chain of hills."

The "duck tail": This feature is difficult to see on HMC images since the signal to noise is low at the southernmost end of the nucleus. The projected limb of the nucleus displays a sharp right angle turn (see Fig. 2.5). This could be a protrusion on the far side and possibly the same feature that is visible on Vega 2 images.

Shallow, roundish, crater-like features may be more common. At least one more "crater" can be seen just south of the tip of the "finger" on the high resolution images. Its size is about 600 m in diameter (Fig. 2.7).

The surface typically shows a coarse roughness (morphology) in the range of 500 m wherever it can be determined. Images with resolutions down to 50 m per pixel confirm the existence of features of a few hundred meters, in agreement with variations seen on the limb. The altitude of features seen on the surface is very difficult to determine. With the exception of the "duck tail" and possibly the "mountain," they can be very shallow because of the slanted insolation.

The global smoothness of the surface with roughness corresponding to less than 5% of the size of the nucleus, is contrasted by the irregular, 2:1 ellipsoidal shape of the nucleus and the small radius of curvature, the "duck tail," on the southern end. The rugged appearance of this corner and the strong brightness contrast at the "mountain" demonstrate that some sharp protrusions exist on the surface.

It is difficult to discern surface patches of active regions. They hardly show any identifiable properties. Possibly they are somewhat brighter than the surroundings. In two cases dust filaments can be traced to small bright patches with diameters of about 500 m (Thomas and Keller, 1987a). They are so well defined in their perimeters that they must be active areas on the surface rather than dust clouds. Considering the observational geometry of Giotto, most active areas must be near the limb or behind it.

Although the directions of dust "jets" could be reconstructed from the series of Vega 2 observations (Smith et al., 1986), their footprints on the surface were not identified. For the smaller phase angles from Vega 2, the brightness variations on the illuminated surface are weaker than those seen from Giotto.

2.4.8 Activity

Ground-based observations of the inner dust coma of P/Halley have shown some asymmetry toward the Sun. A shift of the optical center, i.e., the maximum of brightness, away from the nucleus in a direction toward the Sun had been expected. Much effort was employed to determine it so that the accuracy of the nuclear ephemeris could be improved for the flyby missions. Nevertheless, the degree of asymmetry of the dust production observed by the spacecraft was still a surprise. Most of the activity is contained within cones of about 120° for Vega and 70° for Giotto on the sunward side but not symmetric to the Sun–comet direction. The strongest emission points south of the Sun–comet line (see Fig. 2.4). This striking similarity over several revolutions of the nucleus suggests that the continuous insolation of the north pole (that points in the southerly direction) could provide the explanation (Keller et al., 1986a).

Images taken with broad-band filters spanning a large distance around the nucleus have to be interpreted with care since the ratio of emission from the dust and that of the gas varies with radial distance. Toward the nucleus, the dust density increases with R^{-2} down to distances comparable with the size of the nucleus. The density of the strongest gas emitters (CN, C_2) shows a much flatter distribution since they are decay products. Outside of about 1000 km gas emission contributes significantly to the brightness.

One of the major surprises immediately revealed from images of the HMC was the strong localization of the dust production. Only between 10 to 20% of the surface showed activity during the Giotto encounter.

Intensity profiles (Keller et al., 1986a, 1987a; Thomas and Keller, 1987a; Tóth et al., 1987) in the radial directions from the nucleus show the R^{-1} variation expected for free outflow of the dust particles. The isophotes indicate a few broad, jet-like features ("jets") with 30° to 40° cone angles overlapping each other. Fine structures produce small kinks that can be followed from isophote to isophote (Keller et al., 1986b). Image processing techniques have been applied to enhance the "jets." These reveal a multitude of subjets (Smith et al., 1986) and with still higher resolution, fine filaments (Thomas and Keller, 1987b). Some of these are only about 500 m across and collimated within a few degrees so that they can still be observed several hundred kilometers from the nucleus. Most of the dust sources are located on the sunward, illuminated side of the nucleus with one exception reported by Smith et al. (1986). Four weak "jets," out of 17 identified on the HMC images, point away from the Sun as projected on the sky plane, however, they may still originate from the day side within 17° of the evening terminator.

Detailed investigations of the intensity distribution of the dust around the nucleus revealed a slight curvature of the main "jet," 50° south of the comet − Sun line on Giotto images, at distances larger than 200 km from the nucleus. The curvature is caused by the rotation of the nucleus. This "jet" is identical with a "jet" seen on ground-based images. It was possible to follow this dust "jet" from the surface of the nucleus out to more than 50,000 km (Keller and Thomas, 1988). Further investigations of the many images taken during and in between the flybys may reveal more information on the dust "jet" distribution and production variation. The potential of this rich source of data has not yet been fully exploited. Correlation studies with ground-based observations are needed.

2.4.9 The Near Source Region

The high-resolution images of the nucleus and its near vicinity reveal new aspects of comet physics that had been investigated only theoretically. This is the region of accelerating the dust through interaction with the gas streaming away from the surface. Ever since the pioneering work of Finson and Probstein (1968a,b), based on the one-dimensional hydrodynamic calculations of Probstein (1969) interpreting successfully the dust tail of Comet Arend-Roland, an acceleration zone of the dust over several kilometers had been expected. Considering conservation of dust particles, $n_d R^2 v_d$ is constant, and the acceleration should result in a steeper slope of the dust density, n_d, than R^{-2}. Instead, a flattening of the intensity profile, when compared to the R^{-1} column density law, was observed (Tóth et al., 1987; Thomas and Keller, 1988; see also Fig. 2.8). However, the spherical or point source approaches are not valid. The flattening of the R^{-1} intensity profile can be understood if it is considered that dust is emitted into cones from source areas that are extended but do not cover the entire nucleus (Thomas et al., 1988; Huebner et al., 1988).

The increase of the source strength, $R \cdot I$, with increasing distance from the nucleus can also be caused by an increase of the scattering cross section that results from fragmentation of the dust particles. In order to discriminate the source enhancements from the geometric effects of the extended source, the total flux of dust particles integrated over a surface enclosing the nucleus (Gauss theorem) can be approximated. From the integration of the column densities along circles centered on the nucleus, an average fragmentation yield of three particles for each original particle and a scale length of about 20 km are obtained. Such a strong source of

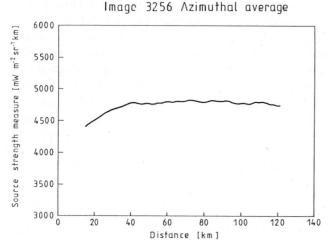

Fig. 2.8. The radial variation of the source strength $R \cdot I$ averaged over the azimuth. For a point source $R \cdot I$ would be constant. The decrease towards the nucleus is obvious.

particles and gas, liberated by the dust fragmentation just above the nucleus, may influence the hydrodynamics of the outflow in that region.

The high-resolution HMC images are difficult to interpret because the contrast is low and "jets" from several sources overlap in the image plane. The strongest visible active area is located near the sunward northern tip of the nucleus (see Fig. 2.2). Part of it is visible, but much of its extension is probably on the rear side of the nucleus. The diameter of the active area is estimated to be about 3.4 km. A similar area of activity is located near the limb at the "central depression." The strongest source supplying about 1.6 times as much dust (as judged from the isophotes at larger distances) has to lie on the hemisphere turned away from the approaching

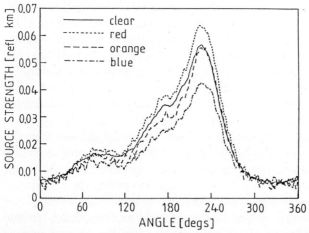

Fig. 2.9. The intensity distribution of the dust near the nucleus averaged over the radial distance as function of azimuth. The Sun is at 180°, the major "jet" is at 240°. Three strong "jets" are visible.

spacecraft. As pointed out by Keller et al. (1986a), part of this source could include the area around the rotation pole that is permanently illuminated by the Sun during the time interval in question. Altogether only about $45\,km^2$ of the nuclear surface are strongly active. This area is sufficient to supply the observed gas production rate if water ice can sublime unobstructedly (Huebner et al., 1986). Solving Eqs. (2.3) to (2.6) yields a production rate of $Z = 1.4 \times 10^{22}$ m^{-2} s^{-1} from a mixture of about 80% water and 20% more volatile frozen gases for a low albedo of 0.04 if a mean insolation corresponding to an average zenith angle of 30° is assumed. Hence the observed gas production of 7×10^{29} s^{-1} during the Giotto encounter can be explained by sublimation from the major source areas alone (see Fig. 2.9 and Reitsema et al., 1989).

2.4.10 Global Dust Distribution

Attempts to construct a three-dimensional distribution for the dust environment of Halley's nucleus during the Giotto encounter have been only partially successful. The lack of a substantial progression of the aspect angles for the HMC makes this task difficult. The *in situ* measurements of the dust distribution by the dust Particulate Impact Analyser (Kissel et al., 1986a) and the dust counting experiments (Mazets et al., 1986a; McDonnell et al., 1986a; Simpson et al., 1986a; Vaisberg et al., 1986a) on the Vega and Giotto spacecraft have revealed dust "jets" – some of them rather narrow – in qualitative agreement with the camera images. A direct correlation has not been established yet. Most important, an onset of increased particle flux was detected before the spacecraft crossed the plane containing the terminator from the night side (McDonnell et al., 1987; Edenhofer et al., 1987). However, measurements indicate that the curvature of the dust particle trajectories is not sufficient to divert large quantities of particles from the day side. Small dust particles have been observed at much larger distances than expected from the fountain model (McDonnell et al., 1986a). The effective radiation pressure on these particles must be small. This will be the case if they are products of the disintegration of big particles at large distances from the nucleus. A ratio of 3:1 is obtained for the column densities in the HMC image plane averaged over the sunward side to that averaged over the antisolar side. This is a much smaller ratio than can be derived from an unrealistically extreme assumption that the dust is ejected uniformly and perpendicular to the insolated nuclear hemisphere, for which the ratio would be 9:1. The background and foreground of particles returning from the up-Sun direction (fountain model) does not suffice to explain the low ratio observed (Thomas and Keller, 1988). The conclusion is that about one quarter of the dust has to have terminal velocities with considerable nonradial components and the grains are emitted in strongly divergent cones. These, probably, relatively small particles spread around the nucleus. There are no conclusive indications that any, even fine dust particles were emitted from the inactive surface.

What are the consequences of the localized dust production and activity? In this connection the spatial distribution of the gas production is also important. Does it have the same sources as the dust? This question cannot be answered unambiguously from the spacecraft camera observations, since the gas emission filters gave only a low signal-to-noise and were strongly contaminated by dust continuum. Spectroscopic observations from Vega 2 seem to show a strong asymmetry for water molecules in the vicinity of the nucleus. An outflow into a 40° cone in direction of the Sun is inferred (Krasnopolsky et al., 1987). These observations are difficult to reconcile with other high resolution observations from the Earth (Larson et al., 1986) that show only a slight sunward – antisunward asymmetry. The creation of isotropic outflow beyond 2000 km seems rather artificial. The *in situ* gas observations show only slight global asymmetries in the case of the Vega flybys (Hsieh et al., 1987) and none for the Giotto Neutral Mass Spectrometer (Krankowsky et al., 1986). The distribution of the primary parent molecule, water, is not conclusively known (see also Chap. 3).

The question of how much gas seeps through the inactive region of the surface cannot be directly inferred from the measurements. However, one can arrive at some conclusions by induction: (a) The size of the active regions is large enough to furnish most of the observed gas production. (b) The "breeze" of dust-laden gas across the nucleus requires a pressure gradient and therefore low gas production above the inert surface. (c) Models for dust "jets" within an isotropic gas flow (Kitamura, 1986) and dust "jets" with localized gas flow (Kitamura, 1987) show that the former model does not agree with the observations (Thomas and Keller, 1987a). All these arguments do not reduce the gas production outside the areas of dust activity to zero but limit them to a very low level. If the permeability of the mantle yields a flux only 10% of that of an active area, about half of the gas would be produced by the inactive areas if the whole surface is considered, and about 25% if only the sunlit hemisphere is considered. Thus, even such a low permeability is too large.

Ground-based images have often shown jet-like features in the coma of comets. This phenomenon is a common feature and has been known for centuries. Old comet drawings demonstrate these structures more clearly than modern-day photography because the observing astronomer tended to substantiate the faint structures he saw during the moments of best seeing (Schmidt, 1863; Whipple, 1981b). The contrast is generally so low that it was not possible to separate the "jets" quantitatively from the dust of the general coma. So the degree of irregularity was unknown. Only when Sekanina and Larson (1984) applied their sophisticated analysis to the development and dynamics of "jets" observed during the 1910 apparition of P/Halley could they conclude that the dust density in "jets" had to be considerably higher than in the surrounding coma. This implies that the highly localized activity may be the rule, since most active comets show jet-like features. However, the concept of linear sources proposed by these authors has not found support from the Giotto observations, although some footprints of "jets" derived from the Vega 2 observations seem to be aligned along a meridian (Smith et al., 1986).

2.5 Modeling and Laboratory Experiments

2.5.1 Sublimation of an Icy Surface

Heating of an exposed icy surface has been assumed to describe the activity of comets (Delsemme and Swings, 1952). The absorbed solar radiation is balanced by the reradiation, a term that is only important at larger heliocentric distances, by the heat conductivity into the ice, which also is quite small and generally neglected as long as pure frozen gases are considered, and by the latent heat of sublimation. The vapor pressure at a given surface temperature can be found from the Clausius-Clapeyron equation describing the phase transition from ice to gas. Solving these equations simultaneously in conjunction with subsidiary equations (see Sect. 2.2.2) yields the production rate of gas molecules, Z, and the temperature of the surface. The gas density is obtained from the ideal gas law and the sound speed from the temperature. The velocity of the gas is approximately one fourth of the sound speed. Typical values for water-dominated ice at $r = 1\,\mathrm{AU}$ are: $T \simeq 200$ K, $Z \sim 10^{22}$ $\mathrm{m}^{-2}\,\mathrm{s}^{-1}$, and $v_s \simeq 300$ m/s. The gas production rate is strongly dependent on the temperature. At a heliocentric distance beyond 2.5 to 3 AU the radiative heat input is no longer sufficient to overcome the latent heat required to sublime the ice. Production of gas therefore subsides very rapidly because reradiation becomes more important than evaporation.

Dust mantles on top of the icy core put emphasis on the heat conduction term in the energy balance equation (Eq. 2.3). Once the input energy is insufficient, the sublimation stops. However, the molecular flux through a thin, porous surface layer is not necessarily reduced; an albedo lower than that of the exposed ice could compensate for the lack of insolation.

In contrast to the regolith type mantles of large dust particles, Horányi et al. (1984) introduced an ice – dust matrix that forms a porous surface layer on top of the icy core after some of the surface ice sublimes. This layer is so friable that the gas percolating through it destroys its outer surface and drags it away as dust particles. The structure of the mantle is similar to that of the core.

2.5.2 Sublimation Through an Inert Layer

While in the past, modeling efforts and interpretations of observations concentrated on the physics of a sublimating icy surface, we will now turn our attention to the possible sublimation through an inert surface layer. As mentioned earlier, the formation of a layer of large dust grains during the aging process of comets had been suggested by Whipple in one of his earlier papers. Dust grains have been treated as impurities of the ice that are useful to decrease the albedo so that more heat could be absorbed (Huebner, 1965). Mendis and Brin (1977) and Brin and Mendis (1979) developed a quantitative model. A formation of even a thin layer of a few millimeters to centimeters tends to choke the sublimation. This dust layer is generally assumed to be very porous and therefore to have poor heat conduction. A strong temperature gradient will result from the equilibrium temperature of water sublimation at about 200 K below the layer to almost the black body temperature of about 400 K at the surface at $r = 1$ AU. The temperature variation during the diurnal cycle will be dampened, depending on the thickness of this layer and its heat capacity. A layer of a few centimeters suffices to suppress the temperature variation of the ice surface (Horányi et al., 1984). The temperature of the ice is then close to the approximation used for a fast rotating nucleus (Huebner, 1965). Fanale and Salvail (1984) considered diffusive flow through a loose dust mantle forming pores and capillaries so that Knudsen gas flow is valid. They applied their model with some refinements to a nucleus on P/Halley's orbit. The feedback of the coma reduces the diurnal temperature differences (Fanale and Salvail, 1986).

The effectiveness of the dust cover to suppress the sublimation of the ice underneath is obvious. The more important question is, how does this layer form and how is the equilibrium of such a dust layer maintained so that it does not choke the sublimation after a very short time but operates on a secular time scale? A possible explanation is that the heat input increases enough so that the layer will be blown away once the comet comes sufficiently close to the Sun. The dust mantle is rejuvenated on each orbit through the inner Solar System. Rather than a slow increase of a ubiquitous dust layer the formation of growing patches seems more plausible.

The physical thickness of a dust layer depends on the mean distance between the grains in the layer, since heat is transported by gas or radiation. Therefore, for predominantly large, centimeter-sized grains (e.g., regolith of dust particles that were not entrained by the gas into the coma) the layer can reach a thickness of about one meter before the sublimation is choked (Mendis and Brin, 1977). Processes that increase the heat conductivity of the surface layer allow the layer to grow in thickness without choking the sublimation. An interesting candidate could be heat transport by inward diffusing gas. The recondensation yields additional heat input. Recent laboratory experiments, simulating physical processes and conditions of the sublimation of a mixture of water ice and dust, lead to a tentative interpretation in support of this gas advection (see Sect. 2.5.8). Gas does not only diffuse inward but it has also to diffuse outward. The dust layer presents a resistance for the gas. The gas pressure at the surface of the subliming ice will increase beyond the value for sublimation into vacuum while the temperature of the ice will increase to the equilibrium value for the enhanced ambient gas pressure. The laboratory experiments show an increase by about 10 K. Under these conditions, a thicker crust can form and still be blown away if some instability occurs. The influence of the dust layer on the process of sublimation itself has not been fully taken into account in model calculations.

2.5.3 Implications from the Restricted Activity

The varnish of a refractory layer is thin compared to the dimension of the nucleus, but also small if compared to the thickness of the layer of icy material lost by sublimation during one orbital revolution. The erosion depends on the density of the material and is estimated to be of the order of 10 m for P/Halley. The concept of a loose dust layer forming on top of an icy interior seems therefore not attractive. The few examples of outcroppings detected on the surface of P/Halley, such as the "mountain" or the sharp corner on the south limb, are not produced by a thin cover on a homogeneous icy nucleus. It is more probable that the outermost layer will be gradually depleted of volatiles during the phase of activity, possibly undergoing some chemical differentiation in addition (Houpis et al., 1985). This process could be facilitated if the structure of the material is porous and is not compacted when the volatile components, the frozen gases, are extracted. All the recent results point into the direction of a low density, porous material, possibly a matrix of refractory material filled with volatiles. The distinction between dust and ice becomes less significant and could even be misleading. The dust consists of semi-refractory organic material and silicates. A core – mantle structure seems probable and is supported by the findings of the PUMA instrument on Vega, (Kissel et al., 1986a) and the Particulate Impact Analyser (PIA) instrument on Giotto (Kissel et al., 1986b). It is plausible that the grains can easily coalesce, even without water ice, during the formation of the nucleus. The water condenses on the grains and, in particular, in the cavities within the grain aggregates.

The localized and restricted activity visible on the surface of P/Halley's nucleus could generate the impression that the comet is about to become extinct. Less than 20% of its surface is active. The persistency and constancy of its recorded apparitions contradict this apparently obvious conclusion. Once only a small fraction of activity is left, the relative decline from one orbit to the next might make itself clearly visible. However, from the constancy (Yeomans, 1985) or the very small decline (Landgraf, 1986) of the nongravitational forces, one can infer that a restricted level of activity is the normal mode and that it has persisted for a long time in the life of P/Halley. The level of activity of P/Halley is very high compared to other short-period comets. It can therefore be assumed that this restricted activity is typical for comets. The level of productivity of most short-period comets is less than 10% of that of P/Halley and often in the range of about 1%. The sizes of these nuclei, such as IRAS-Araki-Alcock (Sekanina, 1988a), Neujmin 1 (Campins et al., 1987), and Arend-Rigaux (Millis et al., 1988), do not seem smaller than that of P/Halley. Therefore their relative activity levels are even smaller in proportion. It has to be concluded that for all short-period comets only minor parts of the surface show activity. Comet P/Halley is not an exception but a typical example.

The physical models for this activity must ensure this persistency. This excludes all models that choke the nucleus by just accumulating debris (regolith) on its surface which reduces solar energy reaching the volatiles. There has to be a process that operates well from underneath an inert surface. The idea of local inhomogeneities, where more volatiles are present and are then released through a vent-like feature whenever the nucleus is heated again during a new perihelion passage, becomes increasingly more attractive. Fresh or recurrent cracks may develop with each revolution. This implication of the observations by the HMC is still not understood.

The dust size distribution from centimeter-sized particles down to molecular clusters (McDonnell et al., 1987), all of low density, fits well into the model of a homogeneous porous matrix. Surface areas depleted of volatiles develop, while the surroundings may still be active. At a later time, the inactive areas may be fragmented by the enormous thermal stresses within an exposed outcropping or by some violent outburst in its vicinity. Considering the low overall density of comet nuclei (Rickman, 1986) and the chemical composition, in particular of the dust (Kissel et al., 1986a), the notion of large boulders within the ice, originally envisaged by Whipple, or of a rubble pile of rocky material bound together by ice (Öpik, 1966b; and recently revived by Weissman, 1986), or even the comet glue model (Gombosi and Houpis, 1986), in which the refractory material shrinks to the size of pebbles, is improbable.

In the model developed by Donn (1963, 1981), Donn and Rahe (1982), and Donn and Hughes (1986) the comet nucleus forms by random accretion of grains into a hierarchy of bodies of different sizes. Differences in the ice to dust ratio are also likely. The accumulation process, at relative velocities less than a few times 10 m/s, will result in a nucleus with a low average density, perhaps about $400 \, \text{kg m}^{-3}$. The collision zone between two aggregates can be a region of density enhancement, where one body penetrates into another, or a region of density lower than the initial value, where the multiply colliding bodies did not quite meet and left cavities. These weak and less dense gaps between the blocks could be the boundaries along which comet nuclei split and can also form pockets of volatiles. The fluffy agglomeration of subnuclei of a large diversity in size would yield a very irregular body. While active, it would rapidly lose its protrusions. In this sense, the shape of P/Halley would be an already strongly evolved nucleus that has shed its sharp features. This model is undergoing development. Numerical simulations of the accumulation of large fluffy aggregates and experiments on impacts of such bodies are in progress.

2.5.4 The Outflow of Dust

The subliming gas entrains the dust from the icy surface into the coma. The terminal outflow velocity of the dust and its density can be derived as a function of dust grain size. The smallest particles ($a < 1 \, \mu\text{m}$) move with speeds of about 500 m/s. Early papers investigated the effusive flux assuming a long mean free path for the molecules (Whipple, 1951; Weigert, 1959; Huebner and Weigert, 1966). A major issue was the initial gas velocity (assumed to be constant, see Wallis, 1982). This issue has recently been revived in the discussion about the nongravitational forces (Sagdeev et al., 1988; Rickman, 1986; and Chap. 7).

The gas production rate, Z, from a subliming surface can be approximated at $r = 1 \, \text{AU}$ to be about $10^{22} \, \text{m}^{-2} \, \text{s}^{-1}$. The cross section for molecular collisions is of the order of $\sigma \simeq 10^{-19} \, \text{m}^2$ and the outflow velocity near the nucleus is $v_g \simeq 100$ m/s. Thus, the mean free path of a molecule above the surface is $\Lambda(R_N) = v_g/(Z\sigma) \simeq 0.1$ m, which is much smaller than the size of the coma, indicating that a fluid dynamic treatment is required for the gas and free molecular flow for the gas – dust interaction. Probstein (1969) provided a one-dimensional treatment. He showed that the gas leaves the surface with a Mach number $M < 1$ because it has to drag the dust. The solution for the outflow is found by a correct description of the subsonic to supersonic transition. This transition point is close above the surface. Probstein applied this gas dynamic approach only for dust particles of a single size. This was generalized to a distribution of particle sizes by Hellmich (1981) whose formalism also allowed consideration of heating of the dust grains by absorption of the solar radiation. Similar numerical steady state solutions were also produced by Gombosi et al. (1983) and Marconi and Mendis (1983). The resulting terminal velocities for dust are about 20% lower (Gombosi et al., 1986) than the corresponding single size solution of Probstein (see also Sect. 4.4 and Fig. 4.5).

More recently Gombosi et al. (1985) developed the formalism of the friable sponge model (Horányi et al., 1984) for the gas seeping through a porous layer into vacuum. The outer surface temperature is higher than the sublimation temperature at the bottom of the layer. Gombosi et al. described the outflow following the concept of a Laval nozzle, with an effective cross section that varies with distance from the nucleus, reaching its minimum at the sonic point. This approach allows for a non-steady state solution describing the onset of outbursts.

The steady state solutions for a dust to gas mass ratios $\chi < 1.5$ show that the sonic point lies between 100 and 400 m above the surface of the nucleus. The dust grain acceleration relative to the gas takes place over several nuclear radii; the small grains reach higher velocities sooner. For gas production rates appropriate for active comets, such as P/Halley, small grains stay coupled to the gas for more than 100 nuclear radii and the solution for the terminal dust velocity is influenced by less than 50% as long as the dust to gas mass ratio $\chi < 1$.

For a more rigorous treatment, the heating and cooling of the gas and dust has to be considered (Marconi and Mendis, 1984). For very large particles the gravitational attraction of the nucleus eventually inhibits the escape. Huebner (1970) and Delsemme and Miller (1971) gave an estimate for the maximum particle size, a_m, that can leave the nucleus (ignoring centripetal force)[1]

$$a_m = \frac{9\,m_u\,M\,Z\,v_g}{16\pi\,G\,\rho_d\,\rho_N\,R_N},$$
(2.9)

where G is the gravitational constant ($6.674 \times 10^{-11}\,\mathrm{N\,m^2\,kg^{-2}}$), m_u the unit atomic mass ($1.66 \times 10^{-27}\,\mathrm{kg}$), and M the molecular weight. Assuming $\rho_d = \rho_N = 500\,\mathrm{kg\,m^{-3}}$ and $R_N = 5\,\mathrm{km}$ leads to $a_m = 10\,\mathrm{cm}$ for typical values of $Z \simeq 10^{22}$ $\mathrm{m^{-2}\,s^{-1}}$ and $v_g \simeq 100$ m/s at $r = 1$ AU. Closer to perihelion and aided by centrifugal force, larger blocks may be lifted and redistributed on the surface. Whether they could form a stable cloud around the nucleus as inferred from radar observations for Comet IRAS-Araki-Alcock (Goldstein et al., 1984) is questionable but certainly an interesting problem to investigate.

However, the above approximation is not valid for large production rates for which the mean free path of the gas is much smaller than the grain size. In this fluid dynamic limit Huebner (1970) obtained (ignoring centrifugal effects)

$$a_m = \left(\frac{27\eta v_g}{8\pi G \rho_d \rho_N R_N}\right)^{1/2},$$
(2.10)

where η is the viscosity

$$\eta \simeq \frac{1.85 \times 10^{-6} T^{1/2}}{1 + 680/T}.$$
(2.11)

---

[1] For a derivation of this equation see Chap. 4, Sect. 4.4, in particular (4.11).

For the same parameters as above and $T \simeq 200$ K, a_m is about 10 cm. The mean free path of the gas, $\Lambda = v_g/(Z\,\sigma)$, is also about 10 cm near the surface if a cross section $\sigma \simeq 10^{-19}\text{m}^2$ is assumed for water – water collisions. However, the viscosity (Eq. 2.11) may be too small.

It is interesting that under fluid dynamic conditions the maximum particle size depends only on the temperature. The surface temperature of subliming ice hardly increases much above 200 K, even at small heliocentric distances. Thus the maximum size of particles that can be lifted from the surface will be limited to the decimeter range.

2.5.5 Two-Dimensional Gas Dynamics

Recently models were developed to calculate the gas dynamic outflow in two dimensions. In this way symmetric "jets" of gas (Kömle and Ip, 1987) and gas loaded with dust (Kitamura, 1986, 1987) could be simulated. Different activity distributions were investigated: (1) with no outgassing from the inactive area and cases (2) to (4) with a uniform outflow from the area surrounding the source region. Cases (2) to (4) are subdivided according to the amount of outflow from the "inactive" area relative to that from the active area. This factor is 10^{-2} for case (2), 10^{-1} for (3), and 1 for case (4). The jet-like structure is followed out to more than 100 km from the nucleus (10 km in case 2). The gas "jet" smooths out and the coma isotropizes. The dust within the "jet" is soon decoupled as discussed in the previous section, i.e., the one-dimensional results have been confirmed.

Small dust grains of high temperature within the "jet" increase the local gas temperature and therefore its lateral velocity component. Near the surface small dust grains attain a sizeable lateral motion that they keep once they are decoupled from the gas flow. However, if the pressure surrounding the active area is sufficiently high (case 3), the lateral motion of the gas is decelerated. The consequence is that the center of the "jet" is depleted and the grain density is enhanced along the interface, a conical surface around the symmetry axis of the "jet." This dust distribution could be called hollow (see Fig. 2.10).

In case (1) the gas expands into vacuum. The pressure gradient drives the gas along the surface more than 90° around the nucleus on each side. The lateral gas density drops quite strongly. The "jet" is clearly pronounced even at distances beyond 100 km, much more than in case (3). If dust is entrained in the "jet," the gas-flow is supersonic everywhere except for the usual zone of a few hundred meters near the surface. Two particle sizes have been considered by Kitamura (1987). Grains with a radius of $a = 0.01$ μm follow the gas flow completely within the hemisphere centered on the "jet" direction (see Fig. 2.10). The larger dust grains with $a = 1$ μm are decoupled earlier and form a more pronounced jet. In all cases the symmetry line of the "jet" is the line of highest density for the gas and the dust.

Kitamura also gives examples for case (4) where the gas outflow from the surface is isotropical but only the active area emits dust, forming a "jet." The gas above the surface within the "jet" is heated by the high temperature dust grains and expands laterally into the surrounding gas. As in case (3) the lateral movement of the gas is

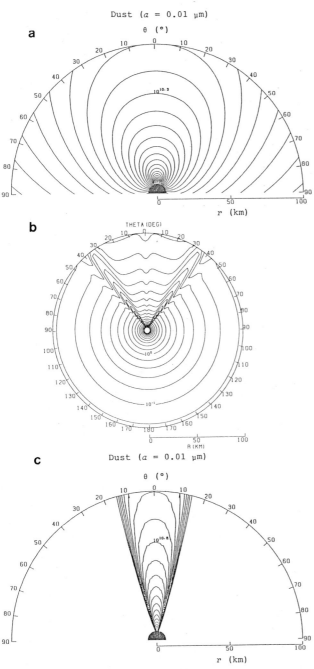

Dust (a = 0.01 μm)

a

θ (°)

r (km)

THETA (DEG)

R (KM)

Dust (a = 0.01 μm)

b

c

θ (°)

r (km)

Fig. 2.10a–c. Two-dimensional fluid dynamic calculations showing the dust distribution near the active surface. (**a**) Particles of size a =0.01 μm are emitted into a cone with a 10° half-angle opening. There is zero gas emission outside of the active region (case 1). (**b**) Particles of size a = 0.65 μm are emitted into a cone with an 8.5° half-angle opening. The background gas emission is 10^{-1} of that in the active region (case 3). (**c**) Particles of size a = 0.01 μm are emitted into a cone with 5° half-angle opening. The gas production in the "inactive" region is the same as that in the active region (case 4). (From Kitamura, 1986, 1987).

quickly stopped, leading to some enhancement of the gas density along the conical interface similar to the dust distribution in case (3). The dust itself is strongly focused within the "jet" and displays strong gradients along the border (see Fig. 2.10).

The intermediate case (2) with a weak background was investigated for pure H_2O gas by Kömle and Ip (1987). They found some depletion of gas in the center of the "jet." As could be expected the formation of an enhancement along the conical interface is more pronounced than in case (3), since the ambient pressure is lower. A ring-like source was also investigated simulating a source region with an inert center. Above about 10 km a maximum of density forms along the center line. The zero boundary condition for the lateral velocity component at the symmetry axis may be unrealistic; it acts like a rigid wall.

All calculations show that the details and some of the more pronounced features, such as the conical interface in case (4) depend on the physical processes of heating and cooling of the gas. In fractions of a second the small dust grains reach temperatures above the black body value and constitute a strong heat source for the gas just above the surface, increasing its lateral flow. At larger distances dissociation of the molecules acts as a heat source, counterbalancing the cooling caused by the gas expansion and losses from IR radiation.

The images taken by the spacecraft during the Comet Halley encounters provide valuable material to which the model calculations have to be compared. A description is given in Sects. 2.5.6 and 2.5.7. The crisscrossing "jets" emanating from the northern active area (Fig. 2.2) indicate that we are not looking at a simple, mathematically defined surface but rather a conglomeration of sources from a rough surface. In fact, it may be easier to learn about the morphology of the surface from an analysis of the observations than about the physical processes governing the gas dynamic outflow.

2.5.6 The Extended Surface Source Model

The first interpretations of the imaging results have been based on more phenomenological models interpreting the dust density distribution based on free outflow. This is certainly valid after the dust has decoupled from the gas at a few nuclear radii. The simplest approach is to assume a point source with a finite opening angle for the dust outflow. Inside the region a change in the density steeper than R^{-2} can be expected because of the acceleration of the grains that start from the surface with zero velocity. This is in contrast to what has been observed (see Sect. 2.4.9). For distances R above the surface comparable to the size of the nucleus, the approximation fails, since it cannot be expected that the dust is flowing directly perpendicular to the surface without any divergence. Even in this simple case the choice of the location of a "virtual" point source somewhere below the active surface would be arbitrary and could not be justified by physical criteria. A different approach is to accept the surface of the nucleus as an extended area of activity where point sources with different source strengths are distributed (Thomas et al., 1988), each emitting into a cone with opening angle equal to that of the entire jet-like feature. At the first stage this model neglects the hydrodynamic complexity near the surface and concentrates on the distribution of the sources and the effects of the morphology of the surface. It is assumed that the "jets" superimpose but do not interact. First calculations show that the model agrees qualitatively with the observations (Huebner et al., 1988).

2.5.7 Comparison of the HMC Observations with Models

The interpretation of the HMC observations is not easy because many "jets" and filaments are seen in superposition and there is almost no information on the third dimension. Some information might be retrieved from earlier HMC observations (18 and 22 h before closest approach) and from ground-based images that may be combined with model calculations of the dust "jet" distribution and effects caused by rotation of the nucleus. Nevertheless some conclusions can be drawn through comparison with the gas dynamic models discussed in Sect. 2.5.5.

Looking globally at the northern active area (Fig. 2.2), the strong focusing of dust caused by isotropic gas emission from the nucleus, case (4) can be rejected. This focusing, also seen in the ring-like source models of case (2), can take place to some degree within an extended source and could explain the filaments. The outlines of the "jets" are not well defined. This blurred appearance could either be caused by the activity distribution within the source region with a smooth transition from maximum activity in the center to the inactive surroundings similar to case (1) or by a physical constellation where the surrounding pressure is low enough so that no enhancement at the conical interface is formed (case 2).

In the case of low ambient pressure the divergence of the flow sweeps gas and dust around the nucleus. This is in qualitative agreement with the strong intensity gradient of emission from the northern tip of the nucleus to its southern end observed above the dark side of the nucleus (Keller et al., 1987a, see Fig. 2.11). However, quantitatively there is a disagreement. The observations require much more dust to be swept around than is shown in Kitamura's model isophotes (see Fig. 2.10). The amount of dust observed by the HMC in the antisunward hemisphere is much larger than predicted by a fountain model for the dust returned from the subsolar hemisphere by solar radiation pressure (Thomas and Keller, 1988). The dust has to originate directly from the nucleus. On the other hand, it cannot be just dust stemming from the enhanced activity of the afternoon hemisphere, pointing away from the observer. The increase of the signal by only a factor of two, when stepping across the dark limb of the nucleus, strongly suggests that the dust column density toward the observer and away from him, as seen from the nucleus, is about the same. The morning hemisphere toward the observer has the same amount of dust. This also argues against a background emission coming from the dark side of the nucleus. One would expect the morning side to be less active. In addition, the nuclear south pole should lie on the visible outline of the nucleus. Since the Sun is supposedly circumpolar on the north pole, the south pole should not be insolated at all.

The explanation that dust is being entrained by a breeze of gas sweeping around the nucleus is most attractive. Effects enhancing the divergence of the dust "jets" of the models in case (1) have to be investigated. The transition across the terminator causes the surface temperature to drop drastically and to reach progressively lower values toward the nonilluminated south pole. Under these circumstances even recondensation of the gas can be expected (Wallis and Macpherson, 1981). Kitamura's calculations should be repeated, introducing the transition from high surface temperature of about 400 K subsolar to the very low values on the night side. Calculations

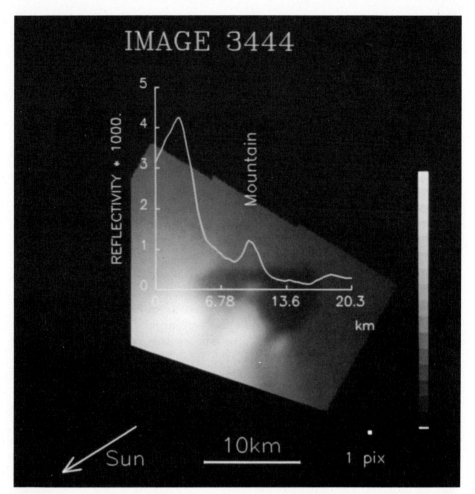

Fig. 2.11. The strong intensity gradient on the dark side of the nucleus caused by the surface breeze sweeping dust around the nucleus from the active area. The secondary maximum corresponds to the "mountain."

by Kömle and Ip (1987) demonstrate that there is a flow to the dark side across the terminator.

Beyond the sonic surface the divergence of the flow is limited. A second possibility to increase the flow around the nucleus exists if the sonic point is further from the surface than the models indicate. This could be the case if heating of the gas is more important than assumed. The pressure increase associated with a higher temperature outside the sonic surface enhances the lateral velocity of the gas but not as effectively as during the subsonic flow.

2.5.8 Laboratory Experiments to Simulate Surface Activity

Laboratory investigations of physical processes relevant to comet nuclei concern the sublimation (condensation) processes, including investigations of phase transitions of low temperature ice, condensation processes, thermal properties of dirty ice, and simulation of the agglomeration to form the nucleus. They also include the aspects of chemical changes, polymerization, crust formation by high energy radiation, and many more. All these investigations are difficult, and their results need careful evaluation because it is hardly possible to simulate the environment of comet formation realistically. The near zero gravity condition, the unrealistic scale lengths within the vacuum chambers, and the high radiation and particle fluxes needed to simulate very long time periods are some of the unrealistic conditions. Only space release experiments can improve on these comet simulations.

Recently a large group in Germany, the KOSI consortium (Grün et al., 1987b), has started a program of laboratory investigations using a cylindrical vacuum chamber with dimensions of 5 m length and 2.4 m diameter. A sample of up to 1 m in diameter can be investigated. Even under these generous conditions the mean free path of molecules sublimating becomes comparable with the dimensions of the chamber. Under these conditions, microscopic effects of the sublimation kinetics could be investigated but not the dust "jets." In the real coma the terminal outflow speed of the dust is controlled by collisions with the gas and the information of the surface properties is masked by the redistribution from the collisions. Scaling from the comet to the laboratory conditions and vice versa is difficult and questionable.

The modifications brought about by illumination of a surface can be studied, and the important questions of the formation of an inert surface layer depleted of volatiles can be investigated. The transport of gas and dust through this porous matrix, the ejection of dust particles from the layer, and the heat transport through the layer and into the icy matrix can be investigated. These are topics of research that seem highly relevant, considering the results from the P/Halley flyby missions. First results show that heat is transported by the vapor not only through the ice-depleted layer but also into the porous icy substrate where the vapor recondenses. The temperature at the top of the ice is about 10° higher than for direct sublimation into vacuum, causing a higher vapor pressure. Other laboratory investigations have concentrated on the behavior of various ice mixtures of soluble and insoluble admixtures. These results confirm that the sublimation of water clathrates (crystalline water enclosing atoms, radicals, or molecules of other substances) does not differ from that of pure water ice. Large clusters with up to several hundred individual grains coming from the disintegration of a thin mantle layer have been found (Dobrovolsky et al., 1986; Ibadinov and Aliev, 1987).

Since water clathrates form only at pressures of 1 bar and higher, their relevance for the composition of comet nuclei is highly questionable. Although clathrates have been introduced into comet modeling (Delsemme and Swings, 1952) they are unlikely to exist in comet nuclei.

2.5.9 Laboratory Experiments for Nucleus Formation

A different line of investigations is directed toward the behavior of the ice at much lower temperatures corresponding to the core of the comet nucleus being heated during its passages through the inner Solar System. If water vapor is condensed onto a cold finger at low temperatures, say 20 to 80 K, it freezes out in an amorphous form.

This aggregate state is not stable. If heated above 137 K, it transforms exothermically into cubic ice crystals, passing several intermediate metastable stages (Laufer et al., 1987). The condensing water vapor can trap other gas molecules depending on the temperature, concentration, and physical properties such as size and affinity of the guest molecules. This can result in fractionation of the gas component at intermediate temperatures above about 50 K. At low temperatures trapping ratios of 1 can be reached. Whether the interpretation of the observed $CO : H_2O$ and $CH_4 : H_2O$ ratios can be used to derive the temperature of formation of comet nuclei based on these experiments is open to question.

The above experiments refer to the formation of comet nuclei out of the gas phase by condensation onto dust grains. A different process is the formation of comet nuclei by coagulation of icy dust grains that are part of the initial protosolar cloud. Their chemical and physical composition may be determined by the history of the interstellar grains before the Solar System formed and by the additional sublimation of gas during the settling of the central dust disk. In interstellar space, long exposure to cosmic radiation changes the chemical composition from simple gas molecules to complex organics. Greenberg and his collaborators (Greenberg, 1977, 1983) have simulated the formation of refractory organics in experiments. Here again the principle can be demonstrated, but quantitative conclusions still need to be evaluated.

2.5.10 Evolution of the Core

Residual nuclear activity in the interior of comets may slightly raise the core temperature while the comets are in the Oort cloud, leading to diffusion of volatiles. Only short-lived radioactive isotopes such as ^{26}Al, if available in sufficient quantities, could cause a considerable temperature increase in the inner core (Wallis, 1980). During passages through the inner Solar System, the ice in a nucleus is gradually heated, and the trapped molecules are released and diffuse according to their mobility through the porous matrix. This could explain the appearance of volatile gases in evolved comets that are periodically heated during their orbital motion (Houpis et al., 1985) in amounts beyond the clathrate hydrate ratio. The observed ratio of volatiles to water during unusual activity of a comet does not necessarily quantitatively reflect the composition of the nucleus. On the other hand, this depletion of very volatile compounds in the outer layers could be the explanation for the low contents of CO, CH_4, etc. observed in P/Halley. It could also explain the often observed enhanced activity of new comets. For these, the near surface layers are not yet depleted, and the very volatile species cause a stronger activity than water ice could produce. It has also been suggested that this enhanced activity is caused by volatile substances in the crust that possibly formed during exposure to interstellar cosmic radiation. Comet Kohoutek (1973 XII) is an example.

2.5.11 Computer Experiments

The formation of nuclei can also be modeled on the computer. Under the assumption of nondestructive cohesive collisions, Donn and Hughes (1986) suggested to

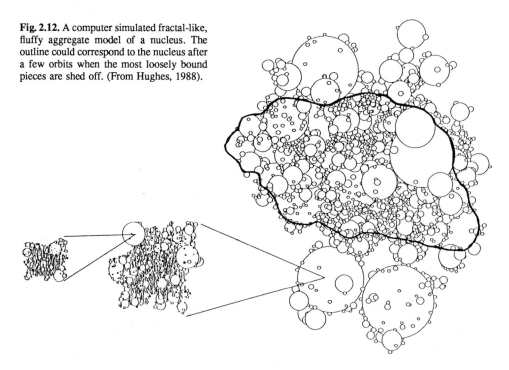

Fig. 2.12. A computer simulated fractal-like, fluffy aggregate model of a nucleus. The outline could correspond to the nucleus after a few orbits when the most loosely bound pieces are shed off. (From Hughes, 1988).

apply the model calculations by Daniels and Hughes (1981) for meteor formation to produce fractal comet nuclei, resulting in structures that are independent of their size. Such a nucleus (Fig. 2.12) can have low density with voids and regions of enhanced density caused by the collision processes. The fractal model breaks down for larger aggregates with significant energy and momentum. They cannot stick where they first touch as with fine grains, but will become compacted until the kinetic energy is dissipated. This will significantly alter the fractal structure, but still produce low-density objects. The numerical simulations are still under investigation.

2.6 Physical Parameters of the Nucleus

2.6.1 Temperature Distribution within the Nucleus

The temperature of the nucleus will adjust to its environment. As long as a nucleus is in the Oort cloud its temperature will be uniform around 10 K. Short-period comets will acquire temperatures that are higher and nonuniform throughout the nucleus varying along their orbits around the Sun. Following Klinger (1985), the mean surface temperature, T_m, of the comet nucleus can be approximated by

$$\epsilon \, \sigma_o \, T_m^4 = \frac{1}{P} \int_0^P [F_\odot(t) - \phi(t)] \, dt, \qquad (2.12)$$

where P is the orbital period and ϕ is the power lost by dissipative processes, such as sublimation. The central temperature depends on the time scale, τ_D, of heat diffusion into the interior. A rough estimate can be found assuming constant diffusivity, $a = \kappa/(\rho c_v)$, which yields $\tau_D = c_v \rho R^2/(\kappa \pi^2)$, where κ is the heat conductivity, c_v the heat capacity, and ρ the density.

If $P \simeq \tau_D$, the temperature will vary within the period (after a few initial periods of settling) with amplitudes that increase the larger P is compared to τ_D. For the condition that the thermal time constant for the whole nucleus is much smaller than the period ($\tau_D \ll P$), the center temperature will oscillate almost in parallel and with the same amplitude as the surface temperature. If $P \ll \tau_D$, the central temperature, T_c, will approach T_m. This last case holds for nuclei of crystalline and even better for amorphous ice.

The material constants are not well known and have to be estimated. For example, the heat conductivity depends inversely on the temperature of the ice, on its compactness, and its physical structure. For a nucleus of 5 km radius and compact ice at a temperature of 30 K, τ_D is as small as several 10^2 y and reaches values of more than 10^5 y for the same size nucleus composed of compact amorphous ice (see Fig. 2.13). If the structure is fluffy, $\rho < 500 \, \mathrm{kg \, m^{-3}}$, these time scales increase. The equilibrium temperature depends on the orbital parameters, such as the eccentricity. Numerical calculations (Herman and Weissman, 1987) show that the analytic approach overestimates T_c. The core temperatures of P/Halley and even of the short-period Comet Tempel 1 are still the same as they were in the Oort cloud.

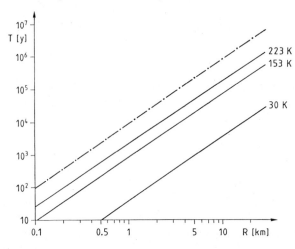

Fig. 2.13. Relaxation times for an icy sphere as function of radius given for different temperatures and diffusivities of the ice. The solid lines are for compact crystalline ice, the dash-dot line is for compact amorphous ice. (From Klinger, 1985).

Another change in the inner nucleus could be caused by the phase transition from amorphous to crystalline ice. If the latent heat is used primarily for additional sublimation, only a thin layer of some 10 m of crystalline ice forms (Herman and Podolak, 1985). If the energy goes into heating the adjacent ice layers, the latent heat of transformation causes the front of the transition to travel inward for tens or hundreds of meters, once the change is triggered on the surface of the amorphous core. Then it takes many revolutions of the comet to lose enough material above the amorphous – crystalline boundary until the temperature rises again above the critical value of 137 K and a new wave is triggered (Prialnik and Bar-Nun, 1987, see Fig. 2.14). If the core temperature rises above about 100 K, the latent heat suffices

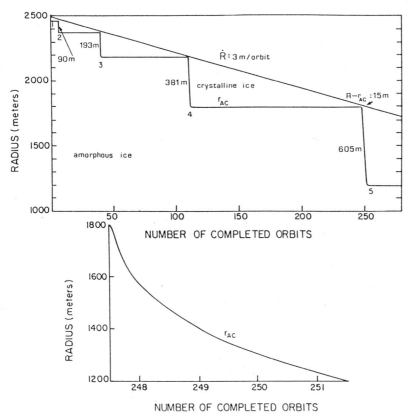

Fig. 2.14. The radius of P/Halley's nucleus, composed of amorphous ice, as a function of the orbit number. Steep decreases occur when phase transitions from amorphous to crystalline ice take place. The curve labeled r_{ac} indicates the transition from amorphous to crystalline ice. The lower diagram shows the decrease of the radius of transition from amorphous to crystalline ice as a function of orbit number. (From Prialnik and Bar-Nun, 1987).

to trigger a runaway, and the whole core is transformed into crystalline ice. The latent heat stored in the amorphous phase is not enough to increase the core temperature of P/Halley above its initial temperature of 10 K during 300 apparitions. The steady state temperature, as defined in (2.12), for a sphere of ice in P/Halley's orbit would be $\lesssim 75$ K. The temperatures of evolved short-period comets are above their initial values, and members such as P/Encke will no longer contain any amorphous ice.

The temperature increase depends on the properties of the surface layer, its albedo, and its heat conduction capabilities. Some effects of the low conductivity of porous material could be balanced by heat transport of the volatile component such as water, or at lower temperatures CO or CO_2. This transport could occur even far below the surface if more volatile substances are present and can diffuse through the pores. This leads to a fractionation of the nucleus far below the sublimation temperature of water ice (Klinger, 1985) and might explain the activity of inbound comets at heliocentric distances far beyond where water could sublime. The outbursts of Comet Schwassmann-Wachmann 1 might be an example. This comet orbits the

Sun outside of Jupiter's orbit and its spectrum during periods of activity shows emissions of CO^+ (Cochran et al., 1980; Larson, 1980).

An unexpected answer to the question of the inner core temperature and the temperature of formation of comets comes from the detection of the ν_3 transition of water in the near infrared at 2 to 3 μm by Mumma et al. (1986) in Comet Halley. The ortho to para ratio of the water molecules in the ice can be determined from the intensity ratio of the corresponding lines. This ratio reflects the spin temperature during condensation of the water before comet formation. A value of 35 K (with a range of +9 to −5 K) was determined by Mumma et al. (1987). This spin temperature is "frozen in" and will not be changed during the warm up or the phase transition from amorphous to crystalline ice. If the observations and the interpretation can be confirmed, this would provide a new tool to look back at physical parameters during the formation of our Solar System.

2.6.2 Surface Temperature

A dust layer forming on top of the subliming ice acts as an excellent insulator. A few centimeters suffice to decrease the temperature of the outer surface, which is close to the black body temperature of about 400 K at $r = 1$ AU, to the sublimation temperature of about 200 K at the ice interface. Heat conduction by the porous dust matrix is very low. Heat transfer by radiation through the pores has also been considered but has no large effect.

The temperature of P/Halley's nucleus, as measured by the IKS infrared spectrometer on Vega 1 at $r = 0.8$ AU, was much higher than the equilibrium temperature of subliming ice of about 200 K (Emerich et al., 1987). The interpretation of the measurements is somewhat model dependent. Interpreting the peak value of scans over the nucleus and coma and assuming a uniformly emitting sunward surface yields a temperature of 320 K. The time variation of the observed signal cannot be explained by this assumption and requires a smaller source region of even higher temperature. For a temperature distribution proportional to $T_{max}(\cos \theta)^{1/4}$, with θ the zenith angle, a maximum temperature of 350 to 400 K has been derived, depending on the details of the temperature distribution and the size of the hot spot. All temperatures are considerably higher than the sublimation temperature of water ice. A minimum of 25% of the projected surface had to be hot.

The IR observations also seem to indicate that the hottest point on the surface was not subsolar. The results of Emerich et al. (1987) represent an independent direct confirmation of the restricted activity on the nucleus and the existence of a thermally insulating crust or mantle covering large parts of the surface.

Even the heat capacity of thin layers (see Sect. 2.5.2) smooths out the diurnal temperature variation of the underlying ice. This should yield a more uniform distribution of the activity with local time. As discussed in Sect. 2.4.8, the observations of P/Halley's nucleus do not show this to be an important effect, since no activity could definitely be detected on the night side of the nucleus. If the surrounding dust coma is dense enough for multiple scattering to occur significantly close to the nucleus, some of the radiation trapped within this region, which is large compared to the nucleus, will add to the insolation of the nucleus (Hellmich, 1981). The diffuse radiation illuminates parts of the nucleus that cannot be reached by direct sunlight, smoothing out lateral temperature gradients. As long as the optical thickness $\tau \simeq 1$, the total energy input onto the nucleus is greater than without dust, however, for larger τ the choking effect dominates.

Many model calculations have been simplified by distributing the insolation of the cross section of a spherical nucleus over its surface (see Sect. 2.2.2). This results in a uniform temperature that is lower than that at the subsolar region. Since at large heliocentric distances the sublimation of frozen gases depends sensitively on the temperature, this leads to a gas production rate for the entire surface that is much lower than the gas production from the subsolar point alone.

2.6.3 Perihelion Asymmetry

If a nucleus rotates very slowly or is insolated pole-on, the temperature on its dark and porous surface can reach values high enough so that water sublimation can still be significant, even at distances out to Jupiter's orbit. This has rarely been considered in discussions about the onset and decline of activity in relation to the volatility of water ice. Comet Bowell (1980 b) was a comet that showed activity beyond 4 AU, including production of water as inferred from the observed OH emission. It is obvious that the symmetry of the activity curve relative to perihelion can be strongly influenced by this seasonal effect (Weissman, 1986). Although the total absorbed energy during a perihelion passage depends only on the orbital parameters, the total amount of activity depends in addition on the orientation of the spin axis of the nucleus and its spin period. The physical reason is the nonlinear dependence of the sublimation rate on the temperature (see Sect. 2.2.2).

A secondary effect that influences the surface temperature is the temperature of the icy layer under the surface. For a given conductivity a larger temperature gradient from the surface into the interior leads to enhanced reradiation, diverting energy from sublimation. This leads to an asymmetry, increasing the sublimation after perihelion (Smoluchowski, 1981a,b). Even in the simplified case of a fast rotator with isotropic insolation, the temperature of the surface is not a simple function of the heliocentric distance.

The asymmetry of an activity curve depends on many effects in an intricate way. Some of them could be compensating for one comet and additive for another. They can be complex even for a single perihelion passage if the orientation of the spin axis changes. In the case of restricted gas production areas, the activity curve can show pronounced seasonal effects, or periodic variations, e.g., the 7.3 d period in the light curve of P/Halley (see Sect. 2.4.5).

2.6.4 Mass

The masses of comet nuclei can only be assessed indirectly. The recent flybys of P/Halley could not directly determine the mass of the nucleus. Even in a close flyby the perturbation on the trajectory of the spacecraft is too small to be measured; it is dominated by deceleration of the probe through impacts with gas and dust particles.

A breakthrough can only be expected from a rendezvous when the spacecraft orbits the nucleus at large heliocentric distances where the activity is sufficiently reduced.

A general estimate is derived from the volume of a nucleus, making an assumption about the density. The error should be smaller than an order of magnitude, considering that the span for estimates of the density lie between $200 \, kg \, m^{-3}$ on the lower end and $2 \, Mg \, m^{-3}$ for a rather unrealistic upper value (see Sect. 2.4.5).

The dimensions of comet nuclei have been determined from observations of the bare nuclei using optical, infrared, and radar observations (see Sect. 2.2). For a typical radius of 5 km the volume is $520 \, km^3$ and therefore the mass is $5.2 \times 10^{14} \, kg$ for a density of $1 \, Mg \, m^{-3}$. The volume of $550 \, km^3$ determined for P/Halley comes close to this typical value although the uncertainty is estimated to be of the order of 30% (Keller et al., 1987a). In principle, the perturbations of orbital parameters of another celestial body closely passing by a comet could yield a determination of the comet mass. However, the nuclear mass is so small that any derived upper limit is too large. The gravitational perturbations of the fragments of a split comet nucleus are also unsuitable, because the relative motion of the subnuclei is dominated by the initial velocity of separation and by nongravitational acceleration from their activity. These parameters cannot be determined with sufficient accuracy from positional observations to extract the weak gravitational forces (Sekanina, 1982).

The mass loss rate per revolution obviously yields a lower limit. Typical values for active comets, e.g., Comets Kohoutek (1973 XII) and West (1976 VI), lie between 10^4 to 10^5 kg/s for the gas production rate at heliocentric distances closer than 1 AU (Keller, 1973, 1976; Feldman and Brune, 1976). Dust production rates have also been determined following the method of Finson and Probstein (1968b), although with larger uncertainties. Considering a period of activity of about 3 months (7.8×10^6 s) and a gas to dust ratio of one, the mass loss per revolution for a typical, active comet is 2×10^{11} to 10^{12} kg. This crude estimate is comparable to the 5×10^{11} kg derived for P/Halley during its last passage (Whipple, 1986). If splitting is neglected, the lifetime of a comet spans at minimum several hundred and possibly many thousand revolutions, yielding a lower limit for the mass of a typical nucleus of 5×10^{13} to 10^{15} kg.

So far the most direct way to determine the mass uses the effects of non-gravitational forces caused by the outgassing of the nucleus. As discussed above (Sect. 2.3.2), the resulting change of orbital parameters depends on the details of the distribution of the "jets" and not only on the mass loss rate as function of heliocentric orbit. About 10^{14} kg have been estimated for P/Halley (Rickman, 1986; see also Chap. 7).

The mass loss rates for bright comets are similar. However, these rates are not a good measure for determining the mass of nuclei, since even bright comets, such as P/Halley, are considerably restricted in their active surface area. The activity level is not a measure for the total surface and therefore for the mass of the nucleus. The sizes of comet nuclei may be considerably larger than indicated from these values.

While it may be obvious that the less active, short-period comets should be smaller in size, this conclusion is not stringent since they may be almost extinct. The nuclei of Comets Encke (Sekanina, 1988b), Neujmin 1 (Campins et al., 1987), IRAS-Araki-Alcock (Sekanina, 1988a) and Arend-Rigaux (Millis et al., 1988) seem to be in the range of a few kilometers. Comet Sugano-Saigusa-Fujikawa (1983 V) could be an exception because its radius of about 0.37 km, determined by IR observations, seems to be too small to support the rather small production rate, even if the whole nucleus would be uniformly active (Hanner et al., 1987).

On the upper end of the mass scale there could be much more massive comets than mentioned above. The Great September Comet of 1882 passed the Sun within 1 Gm, breaking up in about 5 major pieces that were all bright for many months afterwards. Many more Sun-grazing comets have been observed on almost the same retrograde orbit, with an inclination of about 140°. All are members of the Kreutz family of Sun-grazers. Some of them appeared even before the Great September Comet. In fact, two groups with slightly different orbital elements exist. It was shown that they might come from two large parent bodies that split earlier from one giant comet nucleus. The detection of three small comets within about half a year by the solar coronagraph onboard the Solwind satellite (Michels et al., 1982; Sheeley et al., 1982) shows that there may be hundreds of members of the Kreutz group, too small to be visible without a coronagraph. There have been many much brighter comets than P/Halley. While the size of its nucleus could support about 10 times its present production rate if its surface were not covered by a crust, the size and mass of P/Halley is certainly no upper limit.

2.6.5 Comets, Meteoroids, and Meteor Showers

Schiaparelli, in 1866, was the first to connect a meteor shower, the Perseids, to the orbit of a comet, in this case Comet Swift-Tuttle. The ϵ Aquarids and Orionids, appearing in May and October of each year, are correlated with the orbital plane of P/Halley. About 15 comets have been identified as sources of meteor showers. There are three reasons why this connection is important for the physics of comets: (1) The nature of the dust particles, producing the meteors when they enter the Earth's atmosphere, yields information on the composition of comets. (2) The orbital evolution of the meteor streams yields information on the age of the comet on its present orbit. (3) The mass associated with the meteor showers gives a lower limit for the comet mass.

(1) The brighter meteors can be observed spectroscopically. The spectra are dominated by emission lines of metal ions, including silicon. Only these are excited to emit in the visible wavelength range. The relative abundances of Fe, Si, Mg, Ca, Ni, and Cr follow closely the Solar System abundances and are similar to carbonaceous chondrites. However, the more crucial low-mass, volatile elements cannot be observed (Millman, 1977). Collection of the meteoroids in the upper atmosphere by airplanes (Brownlee, 1978; Fraundorf et al., 1982) has been very successful in providing information not only on the bulk composition of the particles but also on their interior structure. These Brownlee particles show a substructure often of silicates of submicrometer size. They are very fluffy with mean densities below or around 1 Mg m^{-3}. The chemical bulk composition of the cometary component, about one third of all interplanetary particles, is similar to carbonaceous chondrites but these porous particles show some distinction in the mineral aspect by being less processed. Other

collection techniques use large exposed surfaces on satellites. The residues within the impact craters are investigated.

It is obvious that all three methods do not deal with a representative sample of cometary material, since only the sturdier particles will survive the journey to Earth and the entry into the Earth's atmosphere. The volatile component will have been evaporated long before.

(2) The particles released by the comet travel with the nucleus but are influenced by radiation pressure and by the planetary perturbations in a slightly different way than the nucleus itself. This leads to a distribution of the particles within the orbital plane of the comet and in some cases to changes of the orbital inclination (Ceplecha, 1977). The evolution times are long compared with the orbital period. The Taurid showers connected to P/Encke needed more than a thousand years to spread out to the Earth's orbit. Therefore P/Encke has been orbiting the Sun, diving within the orbit of Mercury for many hundreds or thousands of revolutions (Whipple, 1985).

Hajduk (1987) determined the evolutionary age of the meteor streams related to P/Halley to 2300 revolutions! It is, however, questionable whether the stability of P/Halley's orbit is sufficient to support such a long interval. The simultaneously derived initial mass of the nucleus of 3×10^{15} kg is rather high. However, Hughes (1985) finds even 3000 revolutions for the age of the meteoroid swarms.

(3) Hughes (1987) demonstrates that the observations of the relatively large particles with masses larger than 10 mg are needed to supplement the spacecraft data that does not reach beyond 1 mg. He determines the initial mass of P/Halley, 3000 orbital revolutions ago, to be about 5 to 6 times its present mass. The mass in the meteoroid streams is estimated to be 5×10^{14} kg.

It may be interesting to note that the mass in meteor showers had been used to determine lower mass limits for comets like Swift-Tuttle and Tempel-Tuttle (Vorontsov-Velyaminov, 1946) even before Whipple published his icy conglomerate model for the nucleus, which was strongly influenced by his studies of meteor showers.

2.6.6 Density

The density of a comet nucleus is even more indirectly assessed because it is derived from an uncertain volume and an uncertain mass (see Chap. 7). A different approach is to infer the nuclear density from observed components, i.e., dust particles. It has been mentioned that the Brownlee particles have a density around 1 Mg m^{-3} and that they represent the sturdier component that survived the process of collection.

Large blocks of cometary material may be observed as fireball meteors of class III that have densities between 600 and 200 kg m^{-3} and masses between 0.1 and 1000 kg (for a summary see Ceplecha, 1977). The tensile strength of the particles of the Giacobinids, a stream of particularly low density, was estimated to be 10^2 Pa (Ceplecha and McCrosky, 1976). Their chemical composition, derived from the spectra that also show CN emissions, confirms a carbonaceous chondrite type material rich in carbon.

The dust counting experiments on the Vega and on Giotto spacecraft showed a strong excess of small particles below 10^{-14} g down to the detection limit of 10^{-16} g (McDonnell et al., 1987; Mazets et al., 1987). Even more surprising is that between 30 to 60% of the tiny particles have a relatively low density below 1 Mg m^{-3}, possibly as low as 0.2 Mg m^{-3} (Kissel and Krueger, 1987; Clark et al., 1987; Jessberger et al., 1988). A minority of grains have densities between 1 and 3 Mg m^{-3}. The densities have been derived from the ratio of silver (the target material) to the total mass of secondary ions from the vaporized dust in the spectra of the Particulate Impact Analyser (PIA). Taking further into account that the ice to dust ratio of the nucleus is about one, it can be concluded that the density of the nucleus should be smaller than 1 Mg m^{-3}. However, values as low as 0.1 to 0.2 Mg m^{-3} are difficult to understand. It should be noted that the PIA observations showed some variations of particle composition, number density, and density as function of the location within the coma, indicating compositional heterogeneity of the source region and hence of the nucleus (Clark et al., 1987).

2.6.7 Splitting and Decay

The splitting of Comet West (1976 VI) was the last spectacular event of a major comet disruption. About 25 such events have been observed (Whipple, 1985) and about 15 have been well recorded (Sekanina, 1977). Splitting is so frequent that comets may break up more than once during their lifetime. Often one fragment dominates, and the others decay rather quickly. Comet Biela was one of the examples where a dual nucleus was observable over two passages (1846 II, 1852 III). Splitting can occur anywhere in an orbit, before or after perihelion, close to the Sun, or at $r >$ 3 AU, and at any comet age, although it has been observed more frequently to happen for long-period or "new" comets. The causes for the splitting remain unknown. Thermal stress has been suggested and analytic approximations yield values high enough to crack at least the surface layers (Kührt, 1984). Often some irregular activity of the comet could be observed before the calculated time of separation of the fragmented nuclei. This may point to the sublimation process as the initializing agent. The release of some stored vapor out of a crevice or cavity may trigger enhanced activity at a local point. Spin-up of the nucleus beyond its limits of cohesion seems unlikely, although a strongly elongated shape of the nucleus would support the loss of parts from its tips. The gravitational attraction is so low that the velocity of escape can be reached even for a relatively slow rotator. Taking P/Halley as an example, equating the gravitational attraction with the centrifugal acceleration $GMR^{-2} = (2\pi/P)^2 R$, the mass of the nucleus $M = 10^{14}$ kg and its maximum radius of $R_{max} =$ 8 km yields $P = 5.5 \times 10^4$ s or 0.64 d as the critical period. This value is not so much lower than the assumed spin period of 2.2 d and certainly not below values that can be expected in other cases. Sekanina (1982), in his comprehensive review on the subject of splitting, derived a velocity of separation for several well observed comets. It varies between 2 and 0.3 m/s, being larger for smaller heliocentric distances. All velocities lie above the $(2\pi/P)R_{max} = 0.26$ m/s for P/Halley. The increase of the separation velocity with decreasing heliocentric distance is not easy to understand if the separation velocity is connected to the spin period. Why should faster rotating nuclei break up closer to the Sun? The elongated irregular shape of comets certainly will contribute to their disintegration; in particular, since the tensile strength of the nuclei is very low, within an estimated range of 10^2 to 10^6 Pa (Whipple, 1963; Sekanina, 1982; Whetherill and ReVelle, 1982).

The smaller fragments often decay within days. This is not easy to understand. The fragments have to be large enough to produce the observed activity, requiring a size measured in fractions of a kilometer if water is the subliming material. On the other hand, within 50 or a 100 days the loss of surface layer is only a few meters thick. Either a much more volatile substance such as CO is responsible for the activity of the fragment, or the activity of the fragment decays because the fresh surface becomes depleted of water ice.

Earlier investigations had used the relative positions of the fragments to derive the velocity of separation (e.g., Stefanik, 1966). However, Sekanina (1977) showed convincingly that the acceleration caused by the sublimation of the nuclei dominates and makes the values for the initial velocities uncertain. There is a clear anti-correlation between the acceleration and the lifetime of the fragments. It is therefore almost impossible to infer physical parameters of the nucleus, such as tensile strength and spin period.

The complete disintegration of a comet, observed for "new" comets several times, can be considered an extreme case of splitting. Comet Biela vanished after two apparitions with a dual nucleus, and only spectacular meteor showers were observed for the following anticipated returns in 1872 and 1885. Such a dissipation of a nucleus, or even of two nuclei simultaneously, suggests an irregular, loosely bound, low density body of extremely low tensile strength. Whether enhanced activity is the initial cause for the disruption is not known (Sekanina, 1984). This behavior of comets warrants more investigation. Here a boundary condition is set that must be met by nucleus models. The large range in deduced tensile strength, as well as the diversity of comet behavior, may indicate a large heterogeneity of the nucleus.

2.7 Summary

The data obtained by the spacecraft instruments and in particular by the Giotto HMC influence our perceptions of a comet nucleus more strongly than any other area of comet physics. The first images coined our thinking so much that the nucleus of P/Halley has become synonymous with comet nuclei in general. This, in fact, may be dangerous since it still has to be shown in what respect, if at all, this nucleus is typical of comets. Next to further investigations of the properties of P/Halley's nucleus, the question of generalization is most important. It can only be solved by further spacecraft missions to other comets.

2.7.1 A New Model of Comet Nuclei

Looking at the spacecraft results from a superficial viewpoint, the ellipsoidal shape and large size of the nucleus came as a surprise. The spherical models for the nucleus have become obsolete. The ramifications that the nuclear albedo is very low and that an inert mantle covers the surface makes the physics of the nucleus considerably more complex.

Only a minor part of the surface of the very active P/Halley emits gas and dust. On other short-period comets, the areas must be smaller yet. The active areas may be only in the percentage range of the total surface. Reevaluation of observations of long-period and "new" comets will probably show that their surfaces are also only in part active. Comet nuclei certainly do not look like "snowballs," not even like dirty snowballs. The appearance of a nucleus is dominated by its dark, inert mantle that is spotted with active areas that also look dark. Comet nuclei hide the volatility of their interiors well.

The active areas where the "interior" of the nucleus should be visible do not differ much from the inert surface in their reflectivity. They may be slightly brighter. This may suggest that some properties of the mantle are not significantly different from the interior matrix, e.g., not darkened more by cosmic radiation processing. The interior is of course enriched by frozen gases, mostly water, and the mantle materials may be sintered. The low reflectivity is probably mainly caused by the porosity of the material rather than its chemical composition although the slightly red color of the nucleus indicates an enrichment of organic substances.

2.7.2 Composition and Structure

The other major surprise, the large content of organic CHON material in the dust particles, indicates that semi-refractory organics are part of the original cometary material and are not later produced by high energy radiation during storage in the Oort cloud.

The high percentage of refractory organic material in the dust grains render the term "dust" ambiguous for the description of the nonvolatile component of the nucleus. This implies something more than just silicate, possibly an enrichment with carbon atoms. (It is quite remarkable how the spacecraft encounters with P/Halley have changed our perception of comets although this is sometimes not so obvious because we still use the old terms, being almost unaware that their meanings have changed, broadened, or shifted.) The dust grains contribute to the gas coma, and the differentiation between volatile and refractory components becomes less distinctive. Not only is "refractory" not a sharply discriminating criterion for the dust but neither is the size of the particles. Low density particles are found in all size ranges, from very large down to molecular clusters. There are differences in chemical composition, from organic only (CHON) to chondritic and silicate material. Setting this apart for the moment, a picture of a homogeneous and homologously structured nucleus appears, built from a matrix of grains covered with water ice and some other minor components. This is not to say that there could not be density variations caused by the intrusions of subnuclei during the formation of the nucleus and some variation of composition from subnucleus to subnucleus. Regions of density enhancements and voids in the original matrix and powdered debris could lead to variability in local composition by secondary processes of the evolving comet. A fluffy-aggregate nucleus should look very irregular until it sheds its protruding subnuclei during its first visits into the inner Solar System. Under these assumptions the nucleus of P/Halley is very evolved.

The water ice sublimes on the surface and a very fluffy structure of grains remains. The grains themselves may be of the core – mantle type, with tiny cores of silicates, similar to chondritic material in composition. The ratio of refractory to volatile material seems higher than previously assumed. The refractory material may even dominate over the icy component by mass.

The observed fluffiness and low density, even of small grains, supports the idea that the small-scale structure of the nucleus is based on the dust particles and not on the ice. It is not an agglomeration of dirty snowflakes but rather a porous matrix of refractory material, the pores and surfaces of which are covered with water ice. The dust matrix is so porous that the addition of water ice may even increase the average specific density.

2.7.3 Formation

Time scales of 10^5 to 10^6 y for the formation of the plantesimals in the early history of our Solar System are too short to produce large molecules, such as polyoxymethylene, from condensed simpler compounds. The grains had to be formed before the aggregation of the planetesimals and are therefore of interstellar origin. The volatile (water ice) component condensed onto the grains while they were coalescing in the dusty disk.

The deficiency of very volatile compounds (CH_4, N_2, etc.) may indicate that the temperature during formation was above 50 K, suggesting a formation just outside the planetary region. A later heating is difficult to envisage, considering the time scales required to heat the core of the nucleus. However, the outer zone of the nucleus warmed up by the Sun is probably depleted of the highly volatile compounds that have diffused through the pores.

2.7.4 Sublimation

In view of the above, it is questionable whether sublimation from an icy surface properly describes comet activity. However, the gas production cannot be much lower than that from ice, otherwise the active fractions of the surface would not sustain the observed coma production. No distinctions between active and inactive surfaces could be found on the spacecraft images. From the point of view of the physics of the nucleus, the details of the sublimation process and the question of what makes an area active will provide essential insight into the nature of comet nuclei. Why is only such a small part active? How can this activity be sustained for many revolutions? Seen from close range, the nucleus of a short-period comet will not give the impression of an active body but rather of an "apathetic," almost dead, irregularly shaped body.

The recent observations will spur new concepts and appropriate modeling. The days of spherical, isotropically emitting nuclei have passed. We enter the era of the first direct observations with all its complexities and difficulties. To understand this new reality, we have to know more and, in particular, whether the one example stands for all.

3. The Neutral Coma

Michael F. A'Hearn and Michel C. Festou

3.1 Introduction

Since nearly all comets bright enough to be seen readily with the unaided eye also show prominent tails, and since these tails are much larger than the coma (even though their density is much lower), the tail is often thought of as the identifying characteristic of a comet and nearly all pre-telescopic depictions of comets show them dominated by the tail. On the other hand, as discussed in the previous chapter, the nucleus is the fundamental body of a comet and the source of all material in the coma and the tail. Nevertheless, it is only the transient existence of a coma that observationally distinguishes a comet from an asteroid and as such the coma is the observational essence of a comet. In this chapter and the next we will consider the production of the coma from the nucleus, the physical processes that occur in the coma, and the loss of material to the dust and plasma tails. There are two reasonably distinct components of the coma: The neutral gas and the dust; they will be considered in separate chapters.

Prior to the 1980s, there were virtually no direct observational data available on comet nuclei except some measurements of brightnesses near aphelion that were thought to refer to the nucleus. Thus all inferences about comet nuclei had to be made from observations of the comae and tails and the orbital motion of comets. These inferences depend critically on understanding the physical processes that occur in the coma. With the recent direct observations of the nucleus of Comet P/Halley and with the *in situ* measurements of its coma, we now have far more information on the processes occurring in the coma and on the correctness of inferences about comet nuclei. Since one of the ultimate goals in studying comets is to learn the composition of the least evolved members of the Solar System, many scientists would argue that the primary goal in studying the coma is to understand the nucleus. Although this attitude ignores much interesting physics, it provides a useful context in which one presents the study of the coma.

In this chapter we will begin with the gas, primarily H_2O, as it is released from the nucleus. In the coma, we will examine the dynamics of the flow of the gas, additional distributed sources of the gas, the chemical changes occurring in the gas, basic physical parameters such as the temperature, and the loss of the gas to the ion tail and the interplanetary medium. We will then consider the nature of the available observational evidence in Sect. 3.3 and, since most of the observational data depend on Earth-based spectroscopic results, we will discuss in some detail the emission processes that control the observations. Closely associated with the emission features are a discussion, in Sect. 3.4, about the composition of the coma. We will conclude in Sect. 3.5 with a prospectus for future research of the coma.

3.2 The Coma: A Transient Phenomenon

3.2.1 Cometary Forms

The visual inspection of a comet head, as noted long ago by Bessel while observing Halley's comet in 1835, suggests by the displayed jets, fans, and rays that matter is ejected from the nucleus in the sunward direction and then is bent as though it were pushed in the anti-sunward direction by a repulsive force [see Fig. 3.1 as well as the extensive sets of drawings published by Rahe et al., 1969 and by Donn et al., 1986, noting, e.g., Comet Tebbutt (1861 II) as well as P/Halley]. A comet seems to behave like a fountain that ejects its water upward. Halos may form that indicate a variable strength of the source function [see the envelopes of Comet Donati (1858 VI) in the atlas of Rahe et al., 1969]. Many comets show round heads, with no internal structure and no tails. Some have a sharp stellar photographic center, while

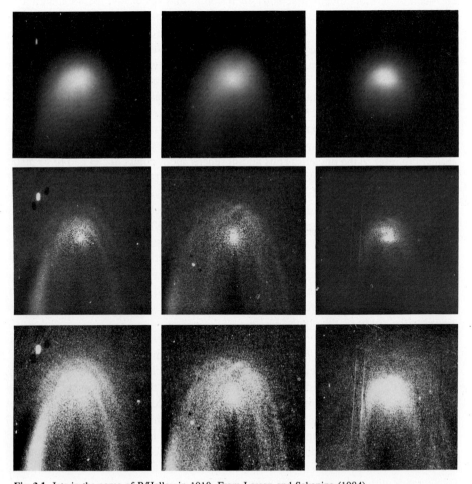

Fig. 3.1. Jets in the coma of P/Halley in 1910. From Larson and Sekanina (1984).

others are very diffuse without a clear central condensation. However, even the best telescopes are unable to show the elusive nucleus that remains hidden in a cloud of solid particles. In the case of Comet IRAS-Araki-Alcock (1983 VII), which passed within 4×10^6 km from Earth in May 1983, the highest spatial resolution images revealed a "planetary" disk of about 25 km diameter showing no phase effect, a behavior typical of an optically thick cloud of solid particles.

The interaction of the solar wind with a comet coma is rather similar to the solar wind interaction with Venus, but because of the lack of gravity in the case of a comet very different spatial distributions for the ions are induced. Most solar wind particles are deviated from their initial straight line trajectories without significant loss of velocity (200 to 400 km/s) and pass around the comet coma relatively far from the nucleus. At large distances upstream from the comet, the solar wind picks up newly ionized cometary atoms and molecules that heat and slow the solar wind until the flow makes a transition from supersonic to subsonic, creating the bow shock or outer shock. After crossing the bow shock, the incoming flux of solar wind particles heading directly at or near the nucleus encounters the cometopause, the region in which cometary ions add significantly to the mass of the solar wind. Inside the cometopause, the solar wind ions pick up additional cometary ions. The flux of solar wind ions is slowed down to a velocity of a few tens of km/s and it deeply penetrates the comet coma to finally deviate from its initial trajectory because of the resistance offered by the cometary ions. The surface inside which solar wind particles never penetrate is called the contact surface. Between the cometopause and the contact surface, solar wind particles can interact directly with cometary molecules with which they undergo charge exchange reactions.

The distances between the nucleus and the various surfaces mentioned depend both on the total gas production rate of the nucleus and on the heliocentric distance of the comet. Close to the Sun, the contact surface is situated at a few thousand km from the nucleus. It was found to be at about 4000 km from the nucleus in P/Halley's coma. Far from the Sun or in extremely weak comets, the comet atmosphere should not present any significant resistance to the solar wind and one may expect some ions to reach the nuclear surface. The bow shock is situated at distances of the order of 10^5 km or more.

Cometary ions produced close to the nucleus do not escape freely: The molecular ions tend to dissociatively recombine with the electrons while atomic ions are expelled in the anti-solar direction with an exceptional strength due to electromagnetic forces induced by the bending of the magnetic field lines embedded in the solar wind around the contact surface. Molecular ions formed further from the nucleus also are expelled in the anti-solar direction. The ion tail is thus formed by a mixture of cometary and solar wind ions that are deviated around the contact surface by the resistance offered by the escaping cometary ions. In the tail, accelerations up to 1000 times that of solar gravity have been measured. In comparison, radiation pressure, which is a very efficient energy transfer mechanism, can only produce accelerations of the order of 1 to 200 times that of solar gravity on light atoms or on fine dust particles. Typical residence times for cometary ions in the tail are measured in hours, or occasionally days in exceptional objects whose tails can reach many tens of millions of kilometers.

Some species that are observed in the coma are chemically very active; with only a few exceptions they cannot exist in the nucleus as such. They arise primarily from the destruction of stable molecules called "mother molecules" or "parent molecules," that have resided frozen in the nucleus since the time comets were formed. In some cases there may be unstable radicals trapped in the frozen ices, but these would be radicals produced by irradiation of stable molecules already trapped in the ices. This is almost certainly the case in the outermost layers of dynamically new comets arriving for the first time from the Oort cloud, since those outermost layers have been bombarded by cosmic rays for 4.5 Gy. Most of the unstable species, however, probably arise from fragmentation of more stable mother molecules.

3.2.2 The Nucleus as the Source of Coma Gas

When comet nuclei are far from the Sun, say beyond Pluto, they are always inert bodies as far as we can determine observationally. Many comets are still inert when they are closer to the Sun than Jupiter's orbit but others are active far beyond the orbit of Saturn. For example, Sekanina (1976a) showed that Comets P/Neujmin 1 and P/Arend-Rigaux, with aphelia of 12.3 and 5.8 AU, are inert over much of their orbits. He subsequently (Sekanina, 1987c) presented a convincing case that Comet P/Tempel 2, with an aphelion of only 4.7 AU, is inert over a significant part of its orbit. It should be remembered that a coma can be present even in a comet that appears stellar, a phenomenon verified by numerous comets that appear to fluctuate in brightness while remaining stellar in appearance. This is particularly exemplified by the disagreement among the many magnitudes published for apparently stellar comets near aphelion. At the other extreme, Meech and Jewitt (1987) observed a significant coma in Comet Bowell (1982 I) beyond 13 AU while West (1988) has measured a larger than expected coma in Comet P/Halley near 8.5 AU from the Sun.

As discussed in Chap. 2, the production of gas from the nucleus is often modeled as the equilibrium vaporization of a surface composed of dirty ice (see Eq. 2.3), but the process must in fact be much more complicated than that since virtually no comet follows the simple heliocentric distance dependence predicted by solutions of that equation except over some portion of its orbit. For example, although the pre-perihelion portion of P/Halley's behavior could be represented by the equilibrium vaporization of water ice from 10 to 20% of the surface as implied by the Halley Multicolour Camera (HMC) images of the active areas on the nucleus, the same simple model does not adequately describe the post-perihelion variation. Increasing the fraction of active area after perihelion can explain the asymmetry about perihelion at small heliocentric distances, but the persistent activity and elevated brightness at large distances post-perihelion relative to pre-perihelion are difficult to explain by sublimation of water ice caused by insolation alone. There are at least four significant complications: (1) Diurnal and seasonal variations when only a portion of the surface is active, (2) development of mantles that may either blow off at some point in the orbit or allow subsurface volatiles to vaporize and percolate through the mantle, (3) gradients in the composition of volatiles with depth within the nucleus, and (4) thermal radiation and scattering by the dust in the coma. All of these complicating factors influence the spatial structures that we observe in the coma in addition to affecting the overall gas production and the chemical composition.

The existence of jets in the coma of P/Halley and other comets has been known for some time. Larson and Sekanina (1984) used a rotational-shift-difference algorithm to computer process photographs of Comet P/Halley from 1910 (see Fig. 3.1) and thus verified the drawings of jets made by visual observers at the time. This in turn validated the drawings of jets and structures by visual observers of other comets. Although these images are predominantly in the sunlight reflected by the dust, that dust is lifted off the nucleus by the gas (see Sect. 3.2.4 and Chap. 4), implying that the gas is also emitted in jets although it disperses more rapidly than the dust does. That the jets in the coma are due to activity from discrete active areas on the nucleus was verified by the HMC images of P/Halley's nucleus as discussed in Chap. 2. Sekanina (1987b,c) has carried out extensive modeling to show that a wide variety of other morphologies observed in comet comae can be explained with such jets simply by varying the geometrical conditions as shown in Fig. 3.2. In particular, he has shown that sunward fans observed in many comets at certain portions of their orbits can be explained as the projection onto the sky of the hollow cone produced by a jet at high latitude in permanent sunlight, and he has successfully applied this model to several comets for which there are extensive sets of observations. The rotation of active areas in and out of sunlight also influences the total gas produc-

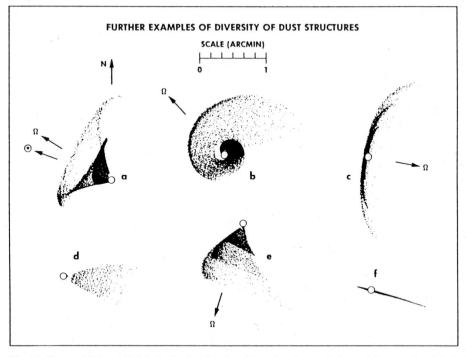

Fig. 3.2. Fans and halos at high latitudes showing the wide variety in morphology of the coma by a single jet. All the differences are caused by variations in the geometry: The cometographic latitude of the jet and the positions of Earth and Sun relative to the comet's axis of rotation. From Sekanina (1987b).

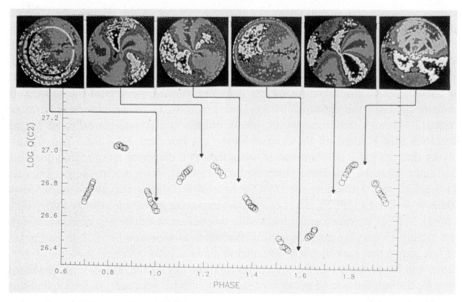

Fig. 3.3. The light curve derived from optical observations (bottom) and the morphology of the gas jets (top) are both periodic at about 7.4 d. The peaks in the light curve are systematically associated with the existence of more jets than are the troughs of the light curve. From Hoban et al. (1988).

tion on short time scales, as was most dramatically exemplified by Comet P/Halley. Since the activity is localized to small areas on the surface, it is clear that the 7.4 d periodicity in the light curve found by Millis and Schleicher (1986) is associated with the motion of active areas into and out of sunlight. This was confirmed by Hoban et al. (1988), who showed a correlation between the number and shape of the CN jets and the phase of the light curve (see Fig. 3.3).

Mantles on comet nuclei and their effect on the gas in the coma have been studied by many modelers. The effect on the total gas production has been studied in the most detail by Fanale and Salvail (1984), who have further developed the ideas previously presented by Brin and Mendis (1979). They show that under many circumstances a mantle will form and then blow off during each perihelion passage. The detailed properties of the mantle and the cometary geometry can lead to asymmetries in either sense with respect to perihelion. They show that these models can qualitatively explain many of the observed asymmetries. The effect on the relative abundances in the coma is somewhat more subtle. If the nucleus contains a mixture of two ices of different volatilities, as opposed to a clathrate structure with the less volatile ice trapping molecules of a more volatile species, then in the outer layers of the nucleus the more volatile ice will diffuse through the less volatile ice over a certain range of heliocentric distances. Houpis et al. (1985) have carried out calculations for a mixture of H_2O, CO_2, and dust and have shown that the ratio of H_2O to CO_2 released at certain heliocentric distances can show an asymmetry about perihelion exceeding a factor of 5. Although there are no observational data available in which to look for such an asymmetry in this particular ratio, large asymmetries about perihelion in other relative abundances have been reported for Comet P/Encke by A'Hearn et al. (1985). Since Comet P/Encke is also thought to have two distinct active areas (Sekanina, 1988b), one dominating the pre-perihelion gas production and the other dominating the post-perihelion gas production, it is impossible at present to decide whether the asymmetry in the abundance ratios is caused by differentiation of the outer layers of the nucleus or by a primordial heterogeneity yielding different compositions for the material under the two active areas. In any case, the process of differentiation must be kept in mind when inferring nucleus composition from coma abundances.

Finally, there is a positive feedback system in which dust entrained into the coma increases the effective insolation and thus increases the gas production. Hellmich (1981) showed that the insolation was increased by the scattered sunlight, while Weissman and Kieffer (1981) also included the thermal radiation by the dust in the coma. The net effect is to increase the gas production by as much as a factor of 2.5 when a comet is active. However, there is no observational evidence for these effects.

Because of the many complicating factors discussed here, it is extremely difficult to use the variation in total gas production to infer properties of comet nuclei. Nevertheless, we must understand these factors if we are to understand the processes in the coma.

3.2.3 Grains as Distributed Sources of Coma Gas

In addition to the gas that is released directly from the nucleus, there is an additional source of gas that has only been realized as significant from studies of Comet P/Halley. It is now clear that a significant amount of gas is released from grains which in turn had been entrained from the nucleus by the gas sublimed directly from the nucleus. It has long been argued that grains of ice should be present in the inner comae of comets and that these should provide additional gaseous H_2O as they evaporate, although in most comets this extra H_2O would be only a small fraction of the total. It is now clear, however, that there are also less volatile grains, thought to be organic (CHON) particles, that release a large fraction of some of the observed gaseous species such as CN and more difficult to observe species such as CO and H_2CO.

Icy grains were first discussed in quantitative detail by Huebner and Weigert (1966) and later by Delsemme and Miller (1970, 1971), who considered subliming icy grains, that had been observed in laboratory experiments, to explain quantitatively both the observed spatial profile of the continuum in Comet Burnham (1960 II) and also the short distance from the nucleus at which certain radicals were observed. For comets at moderate distances from the Sun, the lifetimes of these grains are short, and they therefore totally sublime (evaporate) within the innermost coma where the flow is still hydrodynamic. Longer-lived grains, i.e., larger grains, were invoked by A'Hearn et al. (1984) to explain the OH observed in Comet Bowell (1982 I) beyond 3 AU from the Sun, since the observed OH was far more abundant than could be explained by equilibrium vaporization from a nucleus of plausible size. This gas would have been released in large part in the outer coma, so its flow is different from that of the gas from the nucleus. In special cases such as this, the amount of gas derived from icy grains might be large compared to that derived directly from the nucleus, but this would not be true for comets observed close to the Sun.

Icy grains have also been considered in a somewhat different context by Yamamoto and Ashihara (1985), who showed that the rapid cooling of the coma as it expands from the nucleus should lead to condensation of grains a short distance above the surface. These grains evaporate again, providing gas to the coma. These grains are very small; therefore they sublime very rapidly and are completely gone within a few hundred km from the nucleus. Since the condensation and resublimation is entirely within the inner coma, these grains are not a net source of gas. Their importance is in creating local sources of heating and cooling that would alter the flow of the gas and possibly the chemical reactions in the coma.

One of the most interesting results from the studies of Comet P/Halley was the fact that many gaseous species appear to be released from grains. Although the evidence for this source depends on observations of spatial distributions in the coma, to be discussed below, we present the results here to understand comprehensively the sources of material for the neutral coma. One of the most abundant species coming from grains is CO. Spectroscopic observations from Earth (Woods et al., 1986) implied a relative abundance of CO about 15% of that of H_2O. *In situ* measurements with the Neutral Mass Spectrometer (NMS) on board the Giotto spacecraft (Eberhardt et al., 1987b) showed that the distribution of molecules with $M/q = 28$ was quite different from that of molecules with $M/q = 18$ (primarily H_2O). Whereas the distribution for $M/q = 18$ varied as R^{-2}, consistent with simple radial expansion from the nucleus, modified at larger distances by the dissociation into OH and H, the distribution of $M/q = 28$ required a source for these molecules that was distributed and peaked at a distance of 10^4 km from the nucleus, as shown in Fig. 3.4. Data from the NMS required that at least 1/3 of the $M/q = 28$ peak be due to CO, and since the total abundance required was about 15%, the $M/q = 28$ molecules must have been primarily CO, but less than half of those molecules could be coming from

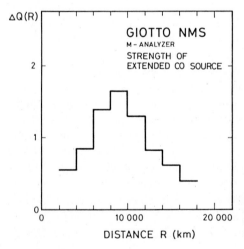

Fig. 3.4. Distribution of source of CO in the coma of Comet Halley as determined from the NMS on the Giotto spacecraft. From Eberhardt et al. (1987b).

the nucleus. Since no potential mother molecules for CO were found in sufficient abundances closer to the nucleus, the source is inferred to be organic grains (CHON particles). Subsequently, further analysis of the data from the NMS showed that formaldehyde (H_2CO) must also originate largely from the CHON particles rather than directly from the nucleus or from the decay of parent molecules (Krankowsky and Eberhardt, 1990). Although alternative explanations are possible – such as passage of Giotto through a jet at a distance near 10^4 km containing enhanced CO and H_2CO, but not H_2O – the evidence favors organic particles as the source.

Evidence from a completely different line of investigation shows that other species must also come from organic particles. A'Hearn et al. (1986a) observed narrow jets of CN (subsequently also detected in other radicals by several groups) that extended beyond $5 \ 10^4$ km from the nucleus. They argued that these could not be produced from a gaseous parent and still maintain the narrow collimation. These jets do not generally overlap with the more obvious dust jets, but they are coincident with jets of various other radicals as shown in Fig. 3.5. Release from CHON particles at large distances from the nucleus is the most likely interpretation of these jets. The fraction of the CN that is produced in this way is probably somewhat less than 0.5.

Quantitative models for the production of stable molecules such as CO or of radicals such as CN from CHON particles are just beginning to be developed (Huebner and Boice, 1989). Such models are difficult because one must assume several physico-chemical properties for the grains, and the nature of the grains has not yet been established beyond the fact that they are probably the smaller CHON particles. The large quantity of CO derived from grains requires that, at least for P/Halley, the mass of grains that is converted to gas be a significant fraction of the total gaseous outflow. In any model it will also be important to take into account the fact that many of these molecules and radicals are released from the grains outside the collision zone. They will therefore not partake of the hydrodynamic flow of the gas

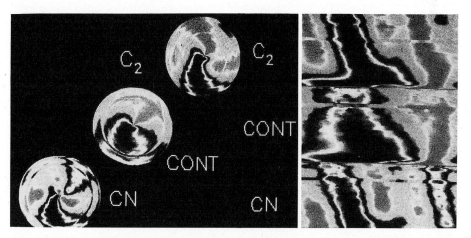

Fig. 3.5. Jet-like features of dust (continuum), C_2, and CN in Comet Halley from ground-based observations. On the left are the regular images processed to enhance the jets. On the right these images have been replotted with radius as ordinate and azimuth as abscissa. The similarities and differences in the jet-like features are more easily seen in the representations on the right.

released from the nucleus. Rather, their overall flow will be determined by the flow properties of the grains, having a lower expansion rate than that of the gas, and by any subsequent processes that act on the individual molecules, such as excess kinetic energy acquired in dissociation processes.

3.2.4 Hydrodynamic Flow

Equation (2.3) in Chap. 2 describes the sublimation (vaporization) of matter from the nucleus in an equilibrium state for dirty ice at the surface. Since the vapor is in equilibrium with the surface, it initially has the temperature of the surface. This temperature will be much below the black body equilibrium temperature because the vaporization carries away much of the heat as latent heat of vaporization. If the vaporization is controlled by CO, that temperature will be between 30 and 45 K for plausible heliocentric distances (0.2 to 10 AU; near 35 K at 1 AU); if it is controlled by CO_2, it will be in the range 85 to 115 K (near 100 K at 1 AU); if it is controlled by H_2O, the range is much larger, varying from 210 K at 0.2 AU through 190 K at 1 AU to only 90 K by 10 AU, as shown in Fig. 3.6. The curve for H_2O has a different shape than the curves for other species because for the volatile species CO and CO_2 nearly all of the incident energy is transferred to the latent heat of vaporization at all these distances, whereas for H_2O this range of heliocentric distances covers the transition between dominance by latent heat of vaporization at smaller distances (the curves are all similar in shape for $r < 2$ AU) and dominance by reradiation at larger distances.

The above results are valid only if the ice is subliming at the surface of the nucleus. Since many models predict that the sublimation frequently occurs below the surface, the temperature of the gas at the surface in these models will not be given by the relationships above. These models predict that the surface of the mantle will be much warmer than the sublimation temperature, typically close to the local black body equilibrium temperature since the nuclear material seems to be very dark. As the subliming gas percolates through the mantle, it will be heated by it and thus the temperature of the gas upon release from the nucleus may be between the sublimation temperature and the black body temperature. Note that this leads to rather different predictions for the variation of the temperature with heliocentric distance since the black body temperature (for a fixed geometry) varies as $r^{-1/2}$, whereas the sublimation temperature in Fig. 3.6 varies in a more complicated way but is nearly constant for $0.5 < r < 2$ AU. This leads to different dependences on heliocentric distance for the initial values of the expansion velocity.

The density of the coma at the surface of the nucleus is calculated as part of the solution for equilibrium vaporization (balancing molecules hitting the surface with those vaporizing). The results depend on which species dominates and controls the vaporization, but the dependence is only weak at small heliocentric distances. At $r = 1$ AU, the density is between 10^{18} and 10^{20} m^{-3}; gross differences among species appear only beyond $r = 2.5$ AU, where an H_2O-dominated nucleus reverts to the mode in which most of the incident radiation is reradiated thermally. At a typical density of 10^{19} m^{-3} and assuming a cross section of 10^{-19} m^2 for water molecules,

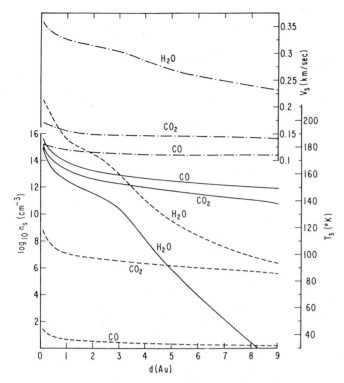

Fig. 3.6. Sublimation temperature, density of the coma at the nucleus, and outflow velocity at the nucleus vs. heliocentric distance r of the comet. From Houpis and Mendis (1981).

the molecular mean free path is just

$$\Lambda = \frac{1}{\sqrt{2}n\sigma} \leq 1 \text{ m.} \tag{3.1}$$

Although this may seem large compared to intermolecular distances at typical laboratory densities, it is small compared to the size of the nucleus and small compared to the scale over which the physical properties change significantly, thus allowing the flow to be treated hydrodynamically in the inner coma.

Since the gas released from the nucleus is essentially expanding into a vacuum, it rapidly reaches sonic velocity in a process that was first modeled by Marconi and Mendis (1982a,b) and was then modeled in detail by Gombosi et al. (1985). The earliest models assumed isothermal expansion either with a velocity derived from $mv^2/2 = 3kT/2$, or at the sonic speed. The first of these assumptions yields $v = \sqrt{3kT/m} = 0.5$ km/s for a water-dominated nucleus at 1 AU. This value is consistent with the expansion speeds found by Whipple (1980) for halos in many comets, but since these halos are probably composed of dust, the relevance of the result is unclear. The next approximation was to use the von Mises solution for the isentropic flow. It allowed for the conversion of internal energy to translational energy and a transition from subsonic to supersonic flow. A proper solution also allows for the acceleration of the gas from the excess energy of the photolytic processes. The solution can be obtained by considering the continuity equations for mass, momentum, and energy. For the case of pure gas without dust (or for perfectly coupled dust), these equations can be written

$$\frac{\partial \rho}{\partial t} + \nabla \cdot (\rho \boldsymbol{v}) = \dot{\rho},$$

$$\frac{\partial}{\partial t}(\rho \boldsymbol{v}) + (\boldsymbol{v} \cdot \nabla)\rho \boldsymbol{v} + \rho \boldsymbol{v}(\nabla \cdot \boldsymbol{v}) + \nabla p = \dot{\boldsymbol{p}}, \qquad (3.2)$$

$$\frac{\partial}{\partial t}\left(\frac{1}{2}\rho v^2 + \rho\epsilon\right) + \nabla \cdot \left[\rho \boldsymbol{v}\left(\frac{v^2}{2} + h\right)\right] = \dot{E},$$

(Schmidt et al., 1988) where ϵ and h are the internal specific energy and enthalpy, and the terms on the right represent sources and sinks of mass, momentum, and energy (e.g., $\dot{\rho}$ might be due to condensation or vaporization of grains, $\dot{\boldsymbol{p}}$ might be due to radiation pressure, and $\dot{E}$ might be due to photolytic heating or radiative cooling). The equations are usually solved in a one-dimensional form with radial variations only. These equations have a singularity at the sonic point, typically within 1 km from the surface of the nucleus. Although there is a singularity at the sonic point in the steady state solution, Gombosi et al. (1985) have pointed out that solving the time-dependent case, in which the flow of dust builds up from zero, eliminates the singularity. One technique for handling the interaction between the dust and the gas is to write equations analogous to (3.2) for the dust, which is thus treated as a separate fluid, and to include the interaction between the dust and the gas in the source and sink terms of the equations. The same procedure is also usually used for components of the gas which do not readily thermalize with the dominant neutral species; species often treated as separate fluids include the hydrogen produced in dissociation of water, the electrons, and the ions. The source and sink terms then include the exchange of matter from one fluid to another. Even with the inclusion of dust, the flow still rapidly becomes supersonic because the sonic speed decreases with distance from the nucleus due to the spherical expansion, but the increase in actual speed of the flow is relatively small. Speeds within a factor 2 of 1 km/s are obtained in all models for a heliocentric distance 1 AU (see Fig. 3.7). Note that the rapid acceleration of the radial flow leads to a decrease in the random motions, i.e., a decrease in the temperature of the gas, as the energy of random motion is converted into radial flow. The gas is cooled even further as collisions excite rotational levels of the H_2O molecules that can then radiate their energy out of the coma (although the coma may be optically thick to this radiation very near the nucleus). Based on numerical simulations, temperatures typically drop to a few tens of degrees in the inner coma.

As the coma gas expands, the density decreases, and eventually the flow becomes a collisionless, free molecular flow rather than a hydrodynamic flow. A convenient although arbitrary boundary is known as the collisional radius, which we will here define as the distance from the nucleus at which an outward moving molecule has a probability 0.5 of escaping to infinity without another collision. Since dissociation and acceleration typically cause the total density to vary no more than a factor 2 from a simple r^{-2} dependence, it is convenient to assume that dependence in which

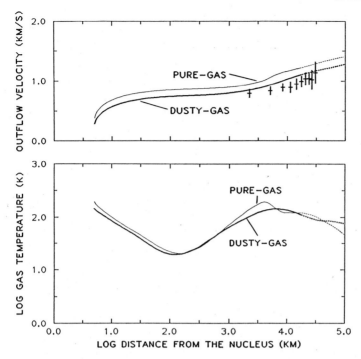

Fig. 3.7. Variation of expansion velocity and temperature with distance from nucleus of P/Halley. From Combi (1989).

case

$$0.5 = \int_{R_c}^{\infty} n(R)\sigma dR = \int_{R_c}^{\infty} \frac{Q\sigma}{4\pi R^2 v} dR, \tag{3.3}$$

which, for constant v, yields

$$R_c = \frac{Q\sigma}{2\pi v}. \tag{3.4}$$

(An alternative definition as the distance at which the mean free path, as defined in (3.1), equals the radial distance yields values $1/\sqrt{2}$ as large.) For an active comet like P/Halley, we can choose $Q = 5 \times 10^{29}\,\mathrm{s}^{-1}$ (at $r = 1\,\mathrm{AU}$), and we can choose again the collisional cross section $\sigma = 10^{-19}\,\mathrm{m}^2$, and a velocity $v = 1\,\mathrm{km/s}$ (see next paragraph), which yields $R_c = 1 \times 10^4\,\mathrm{km}$. As noted, our definition is somewhat arbitrary – measured cross sections for H_2O are larger than the hard-sphere theoretical values assumed here, yielding larger collisional radii, and other definitions of the collisional radius yield smaller values; thus radii of about $10^4\,\mathrm{km}$ seem appropriate for an active comet. A less active comet would have a smaller collisional radius. Crudely speaking, most neutral species (except H and H_2) formed within R_c will be thermalized and participate in the hydrodynamic flow, whereas species formed outside R_c will follow kinematic trajectories determined by the kinematics of their parents at R_c and by any excess velocity imparted in the process of creation. It must be remembered that the collisional radius defined here is the one for the kinematic properties and is not necessarily the effective collisional radius for, e.g., excitation or chemical reactions.

Thus far our discussion has been theoretical, and it is necessary to relate the discussion to actual observations, most of which were obtained on Comet P/Halley. The NMS on Giotto was able to measure the ram energies of the ions and neutral molecules entering the instrument and easily separate the radial motion of the material from the ram motion of the spacecraft through the coma (Lämmerzahl et al., 1987). Those investigators found a velocity of expansion near 0.8 km/s out to a distance $R = 4 \times 10^3$ km with a steady increase to 1.1 km/s at $R = 3 \times 10^4$ km from the nucleus, near the edge of the collision dominated region of the coma. P/Halley was at a heliocentric distance $r = 0.9$ AU at that time. Larson et al. (1987) measured the Doppler profiles and Doppler shifts averaged over most of the coma of the H_2O lines in P/Halley at $r = 1.13$ AU pre-perihelion and $r = 1.03$ AU post-perihelion. The interpretation of the line profiles in terms of the flow velocities is somewhat model dependent, since one has to assume something about the distribution of gas production around the surface of the nucleus as well as assuming something about the distribution of velocities. Nevertheless, the results are in reasonably good agreement with other theoretical and observational results yielding 0.9 km/s and 1.4 km/s respectively for the pre- and post-perihelion expansion velocities. Monte Carlo calculations by Combi (1989), including most of the effects discussed above, have reproduced the flow profile deduced from the Giotto NMS data. Combi has also calculated the flow velocity at the edge of the collision zone and finds values only 10% less than those deduced from the Doppler profiles of the H_2O, even with no attempts to optimize the parameters in the calculation. The pre- and post-perihelion asymmetry is due to the greater total gas production rate post-perihelion, which leads to a larger collision radius, which in turn leads to photolytic heating of the coma (see Sect. 3.2.5) over a larger volume. The models predict that the outflow speed at the edge of the collision zone in Comet P/Halley should vary from 1.8 km/s at perihelion (0.59 AU), through 1.0 km/s at 1 AU, to 0.7 km/s at 3 AU, in good agreement with inferences from the observed Doppler profiles of OH emission. Observations of the flow of species produced outside the collision zone will be discussed later.

3.2.5 Photolytic Processes in the Coma

As the gas expands from the nucleus, it undergoes a number of significant changes in composition. As will be discussed in more detail below, the bulk of the species observable from Earth are unstable species that do not exist in the nucleus. Not only would they be chemically unstable there, but their observed spatial profiles are inconsistent with the hypothesis that they were released from the nucleus as a gas. The primary photolytic processes, and thus the ones considered in even the simplest models, are photodissociation and photoionization by sunlight. In fact, however, more complicated processes can be significant for certain species such as photodissociative ionization, i.e., simultaneous dissociation plus ionization of one of the fragments. The coma itself is basically optically thin except that near the nucleus the lines of very abundant species such as water can become optically thick, as can the ultra-violet continuum, particularly beyond the dissociation limit for H_2O. In very dusty comets part of the inner coma may approach the optically thick state in the continuum even at optical wavelengths. These effects are confined to the region within about 100 km of the nucleus, where most species spend only a tiny fraction of their lifetimes. Thus in most circumstances it is valid to assume that sunlight irradiates all molecules in the coma. Each molecule or radical can thus be considered to have a mean lifetime before dissociation or ionization that varies as r^2 and which may vary with solar activity depending on which wavelengths are most effective in dissociating or ionizing the molecule. The classic description of this process, still widely used by observers to interpret their data, was that of Haser (1957), who considered a steady state, spherically symmetric outflow of parent molecules (not necessarily observable) which photodissociated into some observed species, which in turn continued to expand and eventually decayed into some other species by either ionization or further dissociation. This simple model has advantages because not only is the expression for the volume density, $n_D(R)$, of a given species straightforward but also the solution for the directly observable surface density distribution, $\mathcal{N}_D(\tilde{\rho})$, which can be expressed in terms of Bessel functions:

$$n_D(R) = \frac{Q}{4\pi R^2 v} \cdot \frac{\Lambda_D}{\Lambda_P - \Lambda_D} \left[e^{-R/\Lambda_P} - e^{-R/\Lambda_D} \right] , \qquad (3.5a)$$

$$\mathcal{N}_D(\tilde{\rho}) = \frac{Q}{2\pi v \tilde{\rho}} \cdot \frac{\Lambda_D}{\Lambda_P - \Lambda_D} \left[\int_0^{\tilde{\rho}/\Lambda_D} K_0(y)dy - \int_0^{\tilde{\rho}/\Lambda_P} K_0(y)dy \right] , \qquad (3.5b)$$

where R and $\tilde{\rho}$ are the actual and projected distances from the nucleus, Q is the production rate for the species in question (or the production of its parent multiplied by the fraction that decay in the desired channel), v is the constant expansion velocity, Λ_P and Λ_D are the production and destruction mean distances traveled by the parent and daughter species and K_0 is the modified Bessel function of the first kind. This model is still widely used by observers to interpret their data because it empirically describes many observed distributions in a self-consistent way. However, the model is physically unrealistic for most species because many important physical processes are ignored and the resultant agreement with observed distributions is often fortuitous. Even the earliest applications of this model showed that there were serious problems in identifying suitable parent molecules that could decay into the observed species with the required lifetimes. When the Haser model is used in this empirical way one can be misled by the identical shape but different absolute abundance for the spatial distribution given by (3.5a) or (3.5b) if the parent and daughter scale lengths are interchanged. It is normally assumed that the shorter scale length is that of the parent, but this is not necessarily true. Interchange of the scale lengths will lead to rates of production that differ by the inverse ratio of the scale lengths.

Other significant effects such as radiation pressure and asymmetric outflow of gas and dust from the nucleus have been treated in a different line of models (Eddington, Walace and Miller, Haser 1966). The combination of all effects requires numerical simulations (Combi and Smyth, 1988a). Another important effect is the excess energy which dissociation products receive. Since the dissociating photons can have any energy above the dissociation threshold, the excess energies beyond that required for the dissociation must get distributed among the internal energy states and the kinetic energies of the dissociation fragments. This leads to an additional component of the velocity, v_e, that is isotropically distributed in the rest-frame of the parent molecule and has a mean value and dispersion that varies from species to species. This effect was discussed by Dolginov and Gnedin (1966), although the first calculations and comparisons with observations were carried out by Festou (1981a,b) using what is now usually called the vector model. This model was shown by Combi and Delsemme (1980) to be well approximated by a Haser model with fictitious scale lengths that are related to the true lifetimes by the following relations:

$$\tan \delta \equiv \frac{v_P}{v_e},$$

$$\mu \equiv \frac{\Lambda_P}{\Lambda_D},$$

$$\mu_H \equiv \frac{\Lambda_{PH}}{\Lambda_{DH}}, \tag{3.6}$$

$$\Lambda_D^2 - \Lambda_P^2 = \Lambda_{PH}^2 - \Lambda_{DH}^2,$$

$$\mu_H = \frac{\mu + \sin \delta}{1 + \mu \sin \delta},$$

where subscripts P and D denote parent (production) and daughter (destruction), subscript H denotes the equivalent Haser model, v is the velocity, and Λ is the scale length. This equivalence allows one to solve the vector model in closed form. This model provides a much more realistic description of the spatial distribution of daughter species that are primarily produced well outside of the collision zone. Since this zone is typically less than 10^4 km in size and since many parents, including H_2O, have lifetimes considerably larger than 10^4 s, the model has widespread applicability in the interpretation of observations with large fields of view but is irrelevant for species produced close to the nucleus and observations with very small fields of view.

In current practice, Monte Carlo calculations are becoming most common since they allow inclusion of nearly any desired physical effect, including temporal variations, something which cannot be accommodated in any of the closed form solutions. This approach was followed by Festou in his original use of the vector model mentioned above, and it was used by Combi and Delsemme to show that their equivalent Haser model was valid. In recent years it has become abundantly clear that the temporal variations are extremely important, and it is essential to take these into account when studying many species. If the lifetime of the observed species or its parent is long compared to the time for the gas production from the nucleus to change significantly, these effects will be important. This is often the case for atomic hydrogen because its lifetime is comparable to the time in which the nucleus changes its heliocentric distance by significant amounts (Keller and Thomas, 1975; Keller, 1976). For some comets, e.g., P/Halley, the rotation of the nucleus causes the gas production to change markedly on a time scale of less than a day, i.e., in a time comparable to the lifetime of most observed species.

In order to use any of these models, one needs the lifetime of the species itself and of its parent. These lifetimes depend on photo and collisional processes; photo processes are dominant for a number of species. The rate coefficient for destruction (the reciprocal of the lifetime) of a molecule in sunlight is, in principle, rather easily obtained by taking the integral over wavelength of the product of absorption coefficient and solar flux at the appropriate time and distance:

$$k_i = r^{-2} \int_0^{\lambda_0} \sigma_i(\lambda)\, F_\odot(\lambda)\, d\lambda, \tag{3.7}$$

where $F_\odot$ is the solar flux, $\sigma_i(\lambda)$ is the cross section for dissociation or ionization, and λ_0 is the threshold wavelength of dissociation or ionization of species i. Although it is straightforward in principle to evaluate this integral for any species, the cross sections for many species are not well known over important parts of the spectral range. Furthermore, even for species with well known cross sections, it is important to allow for the variation with the solar cycle of the solar flux at $\lambda < 200$ nm. A suitable set of solar fluxes in the ultraviolet can be found in the work of Mount and Rottman (1985, and references therein), who have used rocket flights at different points of the solar cycle to obtain whole-disk spectra of the Sun at approximately 1 nm resolution. The absorption coefficients are distributed in the chemical physics literature. Note that in addition to evaluating the rate coefficient, one must also evaluate the excess energy received by each species in the dissociation and also the branching ratio among various pathways. A tabulation of lifetimes, energies, and branching ratios based on integrals of this type has been assembled by Levine (1985b).

As an example, the most abundant species in comets is H_2O, which dissociates to the most abundant radicals OH and H; all three of these species are well studied both in the laboratory and astronomically so that the complications are well known. A significant fraction of the OH is dissociated by Lyman-α (L-α), the flux of which varies by a factor 2 from solar minimum to solar maximum. This leads to a variation in the lifetime of OH by 25% (van Dishoeck and Dalgarno, 1984). Furthermore, because some of the dissociation takes place from discrete predissociated levels, the absorptions to those levels are sensitive to the Swings effect, the variation of absorbable solar flux with radial velocity (see Sect. 3.3.1), and this leads also to a 25% variation of the lifetime with heliocentric velocity (Schleicher and A'Hearn, 1988) as shown in Fig. 3.8. Thus for a simple, well studied radical like OH, there are up to 50% variations in the lifetime that are superimposed on the variation with heliocentric distance. The dissociation of its parent, H_2O, is also complicated by the variation with the solar cycle since there are several dissociation branches:

$$H_2O + h\nu \rightarrow \begin{cases} H + OH(X^2\Pi) & 87.6\%,\ 85.3\%, \\ H_2 + O^* & 4.9\%,\ 5.6\%, \\ H + OH(A^2\Sigma^+) & 4.4\%,\ 5.2\%, \\ H_2O^+, H + OH^+, H_2 + O^+, OH + H^+ & 3.2\%,\ 3.9\%. \end{cases} \tag{3.8}$$

The two sets of branching ratios and the corresponding lifetimes of H_2O (7.94×10^4 and 4.35×10^4 s) are for the quiet and the active Sun, respectively (Crovisier,

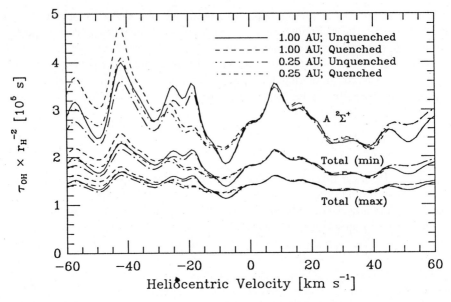

Fig. 3.8. Variation of the lifetime of OH with radial velocity and solar activity. From Schleicher and A'Hearn (1988).

1989). Some of the fragments are formed in excited states of O*, e.g., $O(^1D)$ and perhaps even $O(^1S)$, vibrationally excited $OH^*(X^2\Pi)$ and rotationally excited levels of the electronically excited state $OH^*(A^2\Sigma^+)$, which very rapidly dissociates to O + H. The OH radicals receive an average excess speed of 1 km/s, while the H atoms receive an average excess speed of 18 km/s in the dissociation of water and 7 km/s in the dissociation of OH. Subsequent stages of dissociation lead to additional atomic hydrogen, both from OH as discussed previously and from H_2. The ions are rapidly accelerated out of the coma and do not concern us. Ultimately the OH is dissociated to O + H, primarily by the ultraviolet solar flux, while neutral atoms are ionized, largely by charge exchange with protons in the solar wind (with a lifetime that varies with solar wind conditions between 3×10^5 and 3×10^6 s; Combi et al., 1986), and are then rapidly carried away by it. (The neutralized solar wind proton is carried away by its residual momentum because it is not easily thermalized with the much heavier cometary molecules.) Excess energies are involved in nearly all these steps and must be taken into account properly in model calculations.

Monte Carlo models including these factors have been used by Festou (1981b) to explain the observed brightness distribution of OH, and by many authors, most recently and most completely by Combi and Smyth (1988a,b) to explain the brightness distribution of L-α as shown in Fig. 3.9. The correctness of the basic parameters for the dissociation of water was verified by the *in situ* measurements from the Giotto and Vega spacecraft, although the preliminary results did not allow for the variability of the gas production from P/Halley which turned out to be significant on time scales less than the relevant lifetimes (e.g., Krasnopolsky et al., 1988). Furthermore, the L-α fluxes and the ultraviolet OH fluxes from Comet P/Halley, when interpreted in terms of these models, lead to reasonably consistent (better than a factor of 2)

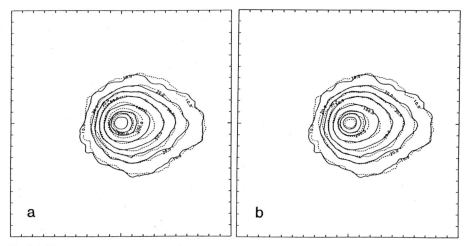

Fig. 3.9. Theoretical and observed isophotes of Lyman-α in Comet Kohoutek. From Combi et al. (1986).

production rates for H_2O, which in turn are consistent with the *in situ* measurements, although direct measurements of the infrared emission by H_2O itself appear to require a higher production rate, a point that will be considered in Sect. 3.3.

It is clear that H_2O, the most abundant molecule from the nucleus, is reasonably well understood in terms of its fragments, although there are some discrepancies in the velocities of the various decay products. Doppler shifts can be measured directly at radio wavelengths (at which, e.g., OH emits) and with suitable equipment on a sufficiently bright comet in the infrared (where H_2O emits). The measurements of H_2O in P/Halley have been discussed above, but more recent measurements of Comet Wilson imply substantially larger velocities. Measurements of OH are generally consistent with the outflow models plus the expected excess energies from photodissociation of H_2O.

For most other species, the association between mother molecules in the nucleus and their fragments is more difficult, sometimes because the relevant reaction rates are not known and in other cases because dominant reaction paths involve unobservable species. Another important constituent of the neutral coma is CO_2; the relative importance of the various photolytic processes that act on it are quite different from those for water. For water, the sum of all the ionizing paths, both pure ionization and dissociative ionization, is only a few percent of the total destruction of H_2O. For CO_2, however, dissociation purely to neutral fragments occurs only half the time; pure ionization occurs more than $1/3$ of the time, and photodissociative ionization occurs 8% of the time, producing CO^+, O^+, and C^+ roughly in the ratio 2:2:1 (rates from Huebner and Giguere, 1980). It is clear, therefore, that the neutral coma cannot be entirely separated from the ion coma.

3.2.6 Chemical Processes in the Coma

Neutral – Neutral Chemistry. Compared to the Earth atmosphere, neutral densities in comet atmospheres are very low. As an example, the density of parent molecules near the surface of a comet nucleus near $r = 0.5$ AU from the Sun will be near 10^{19} m^{-3} (based on an equilibrium vaporization rate of 10^{22} molecules m^{-2} s^{-1} and an expansion velocity of 1 km/s). Similarly, a very bright comet with a total gas production of 10^{30} s^{-1} (comparable to P/Halley) will have a density by number of 10^{16} m^{-3} at a distance only 100 km from the nucleus and a density of only 10^{10} m^{-3} at a distance $R = 10^5$ km from the nucleus, simply because of the radial expansion. By this distance, $R = 10^5$ km, a significant fraction of the original molecules from the nucleus will have been broken down to fragments at the rate of 2 or 3 fragments per original molecule, so the total density by number might be a factor 2 higher. As a consequence of the wide variation in density, chemical reaction rates vary enormously inside a comet atmosphere. In order to understand the role of chemistry, these rates must be compared with the rates of other physical processes. Note that the distance out to which chemical reactions are important is very different from the collisional radius discussed earlier. In the previous context, it was only the mean free path between collisions that mattered, whereas here the relevant comparison is between chemical collision rates and photo rates.

The dominant species released from the nucleus are chemically stable (H_2O, CO, and CO_2), so that rate coefficients for chemical reactions are quite small. For a pair of reasonably reactive species, a rate coefficient might be 10^{-17} m^3 s^{-1}, and these species would be present in the inner coma with abundances of about 1% of that of water, given our current understanding. For a canonically bright comet having $Q = 10^{30}$ s^{-1}, the total density as close as $R = 100$ km is only about 10^{16} m^{-3}. Thus the rate for production or destruction per unit volume for a typical trace species would be of order 10^{11} m^{-3} s^{-1}, whereas for the same 1% abundance, a typical rate for photodissociation per unit volume at $r = 1$ AU would be in the range 10^9 to 10^{12} m^{-3} s^{-1}. Thus photolytic processes, which are independent of density, will always dominate at larger distances, even for chemically reactive species; by $R = 10^3$ km, neutral – neutral reactions will be negligible in virtually all cases. For inert species and even for reactive species with high photo rates, photolytic processes will dominate even as close to the nucleus as several tens of kilometers, where the UV optical depth will become significant for an active comet. For this reason neutral – neutral chemistry is not a dominant reaction mechanism.

Ion – Neutral Chemistry. Ions are created initially by photoionization of neutrals, a relatively slow process when compared to the time that most species spend in the inner coma; but an initial seeding of ions can play an important role in the subsequent chemical evolution of the coma. Because the Coulomb force is a long-range force, the effective cross section of ions is much greater than that of neutrals, and reaction rates are correspondingly higher, rate coefficients in excess of 10^{-15} m^3 s^{-1} being common. The situation in comets is quite different from that in the interstellar medium, even though many of the same species are involved, since the important reactions in the interstellar medium are associative reactions, whereas in comets many of the important reactions are dissociative.

One of the most important types of reactions is dissociative recombination such as

$$CO^+ + e^- \rightarrow \begin{cases} C + O^*, \\ C^* + O, \end{cases} \tag{3.9}$$

which in the coma can be competitive with photodissociation of the ion. Similar reactions are important for the destruction of a number of ions, although in most cases they are of only minor importance for the creation of neutrals.

The main ion – neutral reaction channel is a positive-ion charge-exchange reaction of the type $CO^+ + H_2O \rightarrow CO + H_2O^+$, but positive-ion atom-interchange reactions are also important, particularly $H_2O^+ + H_2O \rightarrow H_3O^+ + OH$. The latter reaction, which is of the associative type (it builds up larger species), leads to the creation of H_3O^+ as the dominant ion in the inner coma, as originally predicted by Aikin (1974). Its presence was not detected, however, until *in situ* measurements with ion mass spectrometers showed that ions with $M/q = 17$, 18, and 19 were all comparably abundant and dominated the ion population. A similar reaction, $H_2O^+ + H_2 \rightarrow H_3O^+ + H$, has an order of magnitude larger rate constant, but since the abundance of H_2 is likely down by two orders of magnitude compared to H_2O, the reaction with H_2O is more important both for creating H_3O^+ and for destroying H_2O^+. Associative reactions of this type are thought to be important for the creation of many other ions that might be abundant, including NH_4^+, CH_2OH^+, and CH_5^+.

Neutral radicals can also be efficiently produced through intermediate ion – molecule reactions. The origin of the C_2 radical has been studied by a number of investigators who have shown that it can result from several processes involving either CH_4 or C_2H_2 as a parent (see Shimizu, 1975; Huebner and Giguere, 1980; Mitchell et al., 1981;). The C_2 is produced in reaction chains that involve either C_2H^+ or $C_2H_2^+$ or both. Thus ion – molecule reactions can play an important role in the production of neutral species in the inner coma because of their higher rate coefficients. In the outer coma, however, even most of the ion – molecule reactions are unimportant compared to photolytic processes.

3.2.7 The Exosphere: Escape from the Coma

Outside the collisional radius, defined previously as the limit of the hydrodynamic flow, the flow would not be a uniform, spherically symmetric radial outflow even if the initial hydrodynamic flow were. Photolytic processes provide their excess energy to fragments that are ejected isotropically in the rest frame of the parent molecule, leading in some cases to radicals that actually move inward toward the nucleus. Furthermore, radiation pressure acts on each species but with an effect that varies significantly from one species to another. These effects are particularly important for understanding the L-α emission, which we will discuss in more detail as an illustrative example. There is no analytic way to combine the hydrodynamic flow inside the collisional radius with the free molecular flow. One must either solve the hydrodynamic equations out to the collisional radius and then switch to the equations of free molecular flow or invoke numerical methods. Full Monte Carlo models can of course cross the boundary, while other numerical techniques have been developed by Huebner and Keady (1984) to gradually leak particles from the collision-dominated fluid to a free molecular flow.

We will first consider the escape of decay products of water; their dynamics have been recently reexamined by Crovisier (1989). The H atoms liberated in the reaction $H_2O + h\nu \rightarrow H + OH$ have an average excess velocity of about 18 km/s. Since the lifetime of H_2O varies between 5 and 8×10^4 s, a significant fraction of these H atoms are liberated inside the collision sphere. They undergo numerous collisions, although they are not thermalized because of the large difference in mass between the H and the dominant species with which it can collide (H_2O and OH). A significant fraction are also released outside the collision zone and will appear as a fast population. In order to conserve momentum, the OH radicals must receive 1/17 as much velocity as the H atoms, i.e., about 1 km/s. The OH radicals, however, are rapidly thermalized if they are formed inside the collision sphere. In the exosphere, H atoms are ionized both by solar photons and by charge exchange reactions with solar wind protons, predominantly by the latter mechanism. Since the charge exchange reactions are sensitive to the properties of the solar wind, there are significant variations in the lifetime of H atoms, from 3×10^5 to 3×10^6 s (Combi et al., 1986). The expansion velocity of 18 km/s and a typical lifetime of 10^6 s immediately imply that the L-α coma should be of order 10^7 km in size. There is a separate population of H atoms released in the reaction $OH + h\nu \rightarrow O + H$; these have an excess velocity averaging 7 km/s. Since much of the OH is produced outside of the collision zone and since its mean lifetime at 1 AU is 1.5×10^5 s, the H atoms from OH will not be be thermalized at all.

Both populations of H atoms are acted on by radiation pressure. In general, the acceleration due to radiation pressure on a molecule or atom is given by

$$a = \frac{h}{mr^2} \sum \frac{g_i}{\lambda_i}, \tag{3.10}$$

where g_i is the fluorescent emission rate or g-factor for transition i and λ_i is the wavelength of the absorbed photons. For atomic hydrogen we can assume that the only important transition is the resonance transition at L-α. In that case we find $a = 3\,\text{mm/s}^2$. With this acceleration acting over 10^6 s, we find a velocity caused by radiation pressure of 3 km/s and a net tailward displacement of 1.5×10^6 km. Clearly both the velocity and the net displacement are significant on the scale of all other processes, and radiation pressure cannot be treated as a minor perturbation. Furthermore, since the gas production of Comet P/Halley was varying by factors of 2 or 3 on time scales of 10^5 s and since in 10^6 s the heliocentric distance typically varies by 10% or more, the variation in gas production is also a significant factor in the problem. Finally, the emission by L-α can be complicated by the optical thickness in an active comet like P/Halley, at least for lines of sight near the nucleus where the coma brightness measured within a small field of view varies as r^{-2} and does not show any asymmetry about perihelion, i.e., it does not mimic the observations with large fields of view or the observations of OH (Feldman, 1990). Extensive modeling of the L-α distribution in comets has been carried out by several investigators with moderate success being achieved only when all these effects are included (see Fig. 3.9).

Since OH radicals are also produced in large part outside of the collision zone and since they can be observed easily in bright comets at radio wavelengths, it is possible to directly measure the Doppler shifts and profiles and thus test the model for the excess dissociation energy acquired by the OH radicals. Theoretical calculations using Monte Carlo methods also reproduce the OH profiles and offsets as observed at various points during the apparition of Comet Halley. We can therefore conclude that the kinematic processes are reasonably well understood in comets, although for species other than water and its fragments the appropriate numerical parameters may not be known well.

3.3 Excitation and Emission Mechanisms

Because so much of our knowledge of comets comes from remote observations, virtually all of which depend on the emission line spectra, it is important to understand the excitation and emission mechanisms in some detail as well as the relationship between the emission mechanisms and observability from Earth. The physically significant method of grouping the lines is by emission mechanisms rather than by wavelength or even by chemical species. In most cases the mechanism is fluorescence, but collisional excitation and prompt emission dominate for some spectral features. In the case of OH, there is even maser emission, i.e., stimulated emission from an inverted population, although the gain is so low that the radiation is not coherent. The population inversion is pumped by fluorescence at much shorter wavelengths and at certain radial velocities can be anti-inverted, i.e., the lower level is populated higher than in Boltzmann statistics, yielding an absorption line. Regardless of the process of emission, the goal is to understand the mechanism well enough to estimate abundances and in addition to infer as many other properties of the comet as possible at the same time. We have already discussed the determination of kinematic properties from line profiles, so these will not concern us here. We will concentrate more on what can be learned from relative intensities, i.e., from the details of the excitation properties.

3.3.1 Fluorescence

Most of the spectral features known in comets are emissions by fluorescence of one sort or another, usually in fluorescent equilibrium. We will use the term fluorescence to mean absorption of a solar photon that excites a molecule or atom followed by spontaneous emission of one or more photons in a single- or multi-step process of decay that may or may not end at the ground state. Figure 3.10 shows a spectrum created by combining ultraviolet observations from the International Ultraviolet Explorer (IUE) satellite with a spectrum taken nearly contemporaneously from the ground. Nearly every spectral feature visible is emission by fluorescence. Some species are in simple resonance fluorescence, such as H (L-α), but most are emitting a more complex pattern.

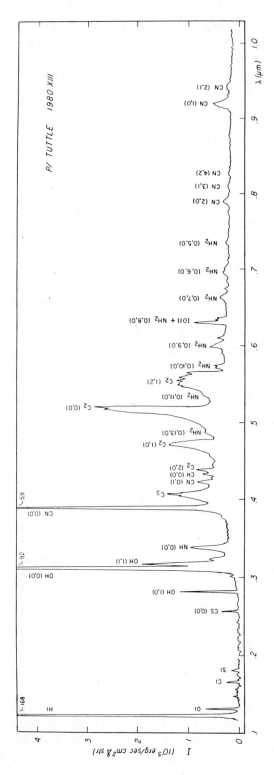

Fig. 3.10. Spectrum of Comet P/Tuttle. Courtesy S. Larson.

The emission in comets is usually characterized by the g-factor or emission rate per molecule, sometimes also called the luminosity per molecule. Calculation of the g-factor and its relationship to both abundances and processes in the coma is an important part of cometary studies. As long as the excitation is simple fluorescence from the ground state of a molecule or atom and nearly the entire population is in the ground state, it is sufficient to calculate the absorption rate of solar photons to the upper level and multiply by the appropriate branching ratio for the transition in question:

$$g = B_{ik}\rho_{\nu_{ik}} \frac{A_{kj}}{\sum\limits_{j<k} A_{kj}}, \tag{3.11}$$

where A_{kj} and B_{ik} are the Einstein coefficients and $\rho_{\nu_{ik}}$ is the solar radiation density at $r = 1$ AU at the frequency of the absorption ν_{ik}.

It is common practice to quote g-factors (fluorescent emission rates) for a given transition since these can be readily related to column densities by

$$B = 10^{-10}\overline{\mathcal{N}}g/r^2, \tag{3.12}$$

where B is the surface brightness in Rayleigh[1], $\overline{\mathcal{N}}$ is the average column density in the volume under consideration, g is the g-factor or fluorescent emission rate per molecule at 1 AU, and the factor r^{-2} accounts for the solar flux variation with r. Equivalently, one can write

$$F = \frac{gN}{4\pi\Delta^2 r^2}, \tag{3.13}$$

where F is the flux, N is the total number of molecules in the volume of the comet under consideration, Δ is the geocentric distance in m (1 AU $= 1.5 \times 10^{11}$ m), and r is in AU.

However, in general, when the excitation is not simply resonance fluorescence, the situation is more complex, and there is a significant population in more than one level. In this case, the usual assumption is that the populations are in equilibrium, although this is not always true. In a steady state, one just sets up a set of balance equations in which, for each level, the rate of transitions into the level is set equal to the rate of transitions out of the level. The mechanism for the transitions is, in principle, completely free, and one can include collisional excitation, photo excitation, absorption of background photons, spontaneous emission, induced emission, collisional de-excitation, etc. In reasonably general form, this can be written

$$x_i \left[\sum_{j\neq i}\left(A_{ij} + B_{ij}\rho_{\nu_{ij}} + \sum_l n_l k_{l,ij} \right) \right]$$

$$= \sum_{j\neq i}\left[x_j \left(A_{ji} + B_{ji}\rho_{\nu_{ji}} + \sum_k n_l k_{l,ji} \right) \right], \tag{3.14}$$

---

[1] The Rayleigh is a unit describing the number of photons emitted per unit time by a column along the line of sight having 10^{10} photons m^{-2} s^{-1}. This corresponds to a surface brightness of $10^{10}/(4\pi)$ photons m^{-2} s^{-1} sr^{-1}, assuming isotropy.

where $A_{mn} = 0$ if $n \geq m$, x_m is the fractional population of level m, n_l is the density of the colliding species l, and k_l is the rate coefficient for collisional excitation or de-excitation, as appropriate. Note that $\rho_{\nu_{mn}}$ can include the radiation density of both solar and background photons. The number of equations equals the number of levels of excitation. One of these equations is redundant and must be replaced by a normalization equation for the fractional populations, $\sum_i x_i = 1$. The equations are solved for the fractional populations in each level and the transition from level i to level j is then given directly by $g_{ij}/r^2 = x_i[A_{ij} + B_{ij}\rho_{\nu_{ij}}]$. In order to estimate whether fluorescent equilibrium is valid, it is important to estimate the time required to reach steady state and compare that with the physically relevant time scales, such as the lifetime of the species in question or its transit time across the field of view. The time to reach steady state can be estimated by calculating the average time between absorption of solar photons and multiplying by the number of cycles that must be completed to populate the levels for steady state. Because of spectroscopic selection rules, this can be very many cycles (approximately equal to the number of levels) for rotational levels in diatomic molecules, although in other cases steady state can be reached in only a few cycles (e.g., steady state for vibrational states of a heteronuclear diatomic molecule). For cometary species, these times range over 10^2 to 10^5 s.

Additional complications that must be kept in mind include the fact that the effective value of ρ_ν is a function of the Doppler shift of the atom or molecule relative to the source (usually the Sun). This has two components – the comet's heliocentric radial velocity, which produces the effect first studied by Swings (1941), and the flow of the gas in the coma relative to the nucleus (Greenstein, 1958). Figure 3.11 shows the solar spectrum in the vicinity of the 0–0 band of the OH

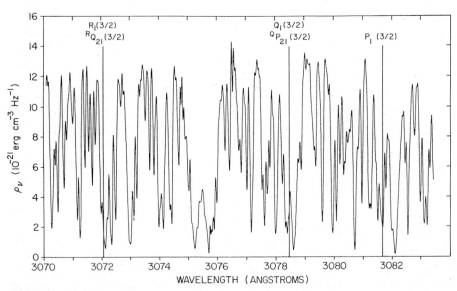

Fig. 3.11. Solar spectrum with the strongest transitions out of the ground state of OH superimposed. From Schleicher and A'Hearn (1982).

$A^2\Sigma^+ - X^2\Pi$ transition. There are only three transitions from the ground state that have significant strength; they are also shown at the positions for zero relative velocity. Because the spectrum of the Sun exhibits absorption bands of OH, these transitions occur near minima in the solar spectrum. Doppler shifts of these lines, however, can lead to significant variations in the solar flux available for excitation. This leads to significant variations in the g-factors.

Furthermore, it should be remembered that, although one normally assumes the rate of emission to vary as g/r^2 (where g is a constant), the variation of heliocentric distance leads to variations in the relative importance of spontaneous decay (independent of r) and of solar pumping, which in turn leads to changes in the steady state populations. In the case of CN, at heliocentric distances $r \leq 0.5\,\mathrm{AU}$, the fluorescent rate of emission becomes comparable to or greater than the rotational decay rate, which leads to a significant deviation from the r^{-2} dependence of the rate of emission for the 0–0 band. Finally we should note that the solar spectrum peaks in the optical and near-infrared, which means that transitions in this range are the ones most strongly pumped by fluorescence. Ultraviolet transitions will generally be weaker because the solar flux is weaker. Transitions at longer wavelengths may be strong in cascade from higher levels, but direct pumping is relatively weak. Typical pumping rates for electronic transitions in the optical are of the order of 10^{-4} to $10^{-2}\,\mathrm{s}^{-1}$.

3.3.2 Prompt Emission

A quite different type of emission can also occur in comets when a particular species is formed in an excited state, since the species can then spontaneously decay to the ground state without having ever absorbed a solar photon. If the excited state is one which is also populated by absorption of solar photons, the prompt emission may be very difficult to separate from the fluorescent emission. If, however, the species is formed in a metastable level, there will usually be no competing fluorescent emission in that transition, and the prompt nature of the emission will be clear. The interpretation of prompt emission is straightforward. If only a single metastable level is involved, the luminosity in photon units is exactly equal to the creation rate of the species in the volume being observed. If one can observe the whole coma, or at least a sufficiently large fraction so that empirical corrections can be made to account for the whole coma, then one has directly the production rate of the observed species. If the branching ratio of the parent's dissociation into the metastable level is known, then the production rate of the parent from the nucleus is also directly known. Furthermore, the distribution of surface brightness is exactly equal to the surface distribution of the decay rate of the parent multiplied by the branching ratio, $B_D = 10^{-10} f \dot{\mathcal{N}}_P$ where B_D is the surface brightness of the forbidden transition of the daughter species in Rayleigh, f is the branching ratio for the parent's destruction by this path, and $\dot{\mathcal{N}}_P$ is the time derivative of the parent's column density. Note that in principle this allows determination of the outflow velocity since the spatial derivative, $d\mathcal{N}_P/d\tilde{\rho}$, is related to $\dot{\mathcal{N}}_P$ by the expansion velocity, although to date no adequate data have been available to attempt this separation.

In general the situation is somewhat more complicated, although not significantly so. The lifetime of the metastable level might be long enough that temporal variability of the comet can be a significant factor, and if the instrumental field of view is relatively small, the motion of the observed species between the time of its formation and the time of its emission of a photon could be significant. In the most important case, production of $O(^1D)$ from H_2O, the lifetime in the metastable level is of order of minutes, and these effects are negligible.

3.3.3 Collisional Effects

As pointed out previously, densities in the coma are generally very low. For this reason, collisions should not generally be important. The time scale between collisions is given by $\tau_{coll} = 1/n\sigma v_{th}$ where v_{th} is the thermal speed, typically $< 100\,\text{m/s}$ for neutral species. A typical cross section for a neutral – neutral collision is $10^{-19}\,\text{m}^{-2}$, leading to collisional rates on the order of $10^2\,\text{s}^{-1}$ near the surface of the nucleus and no more than $10^{-2}\,\text{s}^{-1}$ at a distance $R = 10^3\,\text{km}$ from the nucleus. A typical absorption rate out of the ground state (summed over all upper states) at $r = 1\,\text{AU}$ might be $10^{-3}\,\text{s}^{-1}$, so we see that collisions with neutrals will typically be important only in the inner coma. The energy available in a typical collision is just the kinetic energy of the molecule, which we can take to be water. For a speed of $100\,\text{m/s}$ (less than the sound speed in the inner coma), the available energy is thus $10^{-22}\,\text{J}$ or $< 0.001\,\text{eV}$. This energy is not enough to excite vibrational or electronic levels, but it is enough to excite some rotational transitions or, more generally, transitions at radio wavelengths.

Collisions with ions and electrons are likely to be far more significant. Because they are charged, their effective collisional cross section is typically $10^{-16}\,\text{m}^2$ or greater, i.e., at least 3 orders of magnitude greater than the cross section for neutrals. Measurements at P/Halley showed that in the inner coma the electron density was of order 10^{-3} of that of water, so that the term $n\sigma$ is comparable to that of neutral – neutral collisions. However, the velocity of the electrons is very high, and even the velocity of the ions (which must have the same abundance as the electrons to satisfy charge neutrality) can be somewhat higher than that of the neutrals, at least in rays as they get accelerated into the tail. Thus collisional rates with electrons and ions will be much higher than collisional rates with neutrals, and these can play a significant role in exciting forbidden transitions and in modifying the lowest level populations.

3.3.4 Some Specific Examples

C. Atomic carbon, the energy level diagram of which is shown in Fig. 3.12, exhibits both simple resonance fluorescence and also more complicated processes. Simple resonance fluorescence is exhibited, e.g., by the C I lines at $\lambda = 156.1\,\text{nm}$ and $\lambda = 165.7\,\text{nm}$, which are resonance transitions between the excited $2s2p^3\ ^3D^o$ and $2s^22p3s\ ^3P^o$ states respectively and the ground $2s^22p^2\ ^3P$ state. Since virtually all the atoms are in the ground state and since these are permitted transitions, the column

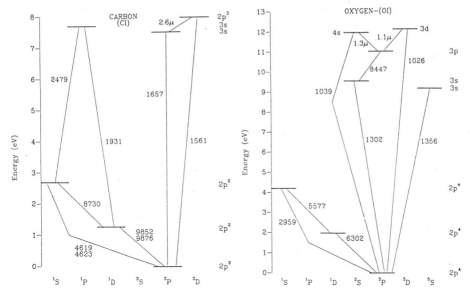

Fig. 3.12. Energy level diagrams for C I and O I.

density of C I can be directly determined by calculating the absorption rate in either transition, which, in this simple case, is the fluorescent emission rate per molecule, g/r^2.

The emission line at $\lambda = 193.1$ nm is also in fluorescence although its lower level is the metastable ^{1}D state. The ^{1}D state, which has a radiative lifetime of 3250 s, is not populated by excitation but may be populated by dissociative recombination of CO$^+$ (Feldman, 1978). Since the lifetime against absorption of a solar photon at 1 AU is comparable to the lifetime against radiative decay, a significant fraction of the carbon atoms will fluoresce at $\lambda = 193.1$ nm before they decay to the ^{3}P state, emitting a near-infrared photon. Since the $\lambda = 193.1$ nm line is produced by newly created carbon atoms, it should have a significantly different spatial distribution than that of the resonance lines. This difference should provide further information on the region in which the atomic carbon is produced and may allow determination of the relative roles of photoionization and electron impact ionization for CO (Woods et al., 1987).

O. Atomic oxygen, the energy levels of which are shown in Fig. 3.12, also exhibits simple resonance fluorescence in the ^{3}S – ^{3}P transition from the ground state at $\lambda = 130.4$ nm. This multiplet, however, is pumped by a narrow solar emission line so that the Swings and Greenstein effects are significant and in the opposite sense from the effects on OH discussed in Sect. 3.3.1. One would expect the emission line to disappear when the Doppler shift exceeds the width of the solar line. The emission does not disappear, however, because there is also pumping in what is known as the Bowen mechanism by solar L-β to the ^{3}D state, which then can decay to the ^{3}S state, which is the upper level for this transition. The two components of the g-factor for this line are shown in Fig. 3.13.

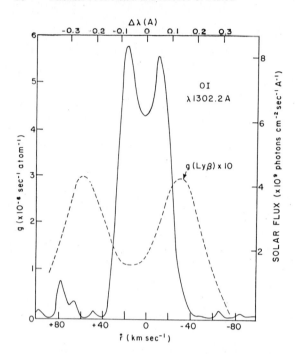

Fig. 3.13. The g-factor for O I showing strong Swings effect in the fluorescence from the solar O I line (solid curve) and Bowen mechanism (dashed line). From Feldman (1982).

Oxygen is also visible in a number of other transitions in prompt emission. Three "forbidden" transitions are observed: $^1S \to {}^1D$ at $\lambda = 557.7$ nm, $^1D \to {}^3P$ at $\lambda = 630.0$ nm and $\lambda = 636.3$ nm, and $^1S \to {}^3P$ at $\lambda = 297.3$ nm. Since these are all LS-forbidden transitions, direct fluorescence is not efficient, nor is there any cascade of permitted transitions from higher levels that could be radiatively pumped. These lines are produced when oxygen atoms are formed directly in the ^1S and ^1D excited states. As noted above, dissociation of water will frequently produce $O(^1D)$ and rarely $O(^1S)$, while dissociation of CO_2 will produce $O(^1S)$. All these lines are difficult to observe: The ultraviolet line is relatively weak (branching ratio only 5% compared to $\lambda = 557.7$ nm), while the $\lambda = 557.7$ nm line is in the midst of the $\Delta v = -1$ sequence of C_2 and the $\lambda = 630.0$ nm line is in the midst of the (0,8,0–0,0,0) band of NH_2. Only the line at $\lambda = 636.3$ nm is relatively free of contamination. Since the lines at $\lambda = 630.0$ nm and $\lambda = 636.3$ nm have a common upper level, their ratio of intensity is well known, and the isolated line at $\lambda = 636.3$ nm can be used to separate the contamination by NH_2 from the 3 times stronger line at $\lambda = 630.0$ nm, which can then be used to map the emission in the coma. The virtue of these lines is that the emission rate of photons is numerically equal to the production rate of $O(^1D)$ in the volume under consideration. By assuming a branching ratio for the dissociation of water, one can then determine the production rate of water directly. In principle this work should allow for the variation of branching ratio with solar cycle and other factors, but this has not yet been done.

OH. The OH radical is interesting because it illustrates yet another example of emission. Its energy level diagram is shown in Fig. 3.14. The principal transitions

ENERGY LEVELS OF OH

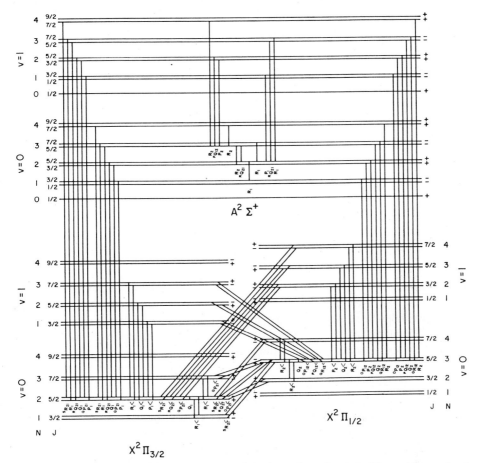

Fig. 3.14. Energy level diagram of OH. Hyperfine splitting is not shown, but Λ-doubling and the rotational, vibrational, and electronic structure are shown. From Schleicher and A'Hearn (1982).

are in the ultraviolet with three vibrational bands of the $A^2\Sigma^+ - X^2\Pi$ transition being strong: The 1–0 band at $\lambda = 282.0$ nm, the 0–0 band at $\lambda = 308.5$ nm, and the 1–1 band at $\lambda = 311.5$ nm. As pointed out in Sect. 3.3.1, the Swings and Greenstein effects are strong, although in the opposite sense from the cases of O and C; the cometary emission is at a minimum rather than at a maximum at $\dot{r} = 0$ as shown in Fig. 3.15. As in all hydrides (except H_2), the asymmetry of OH leads to a large dipole moment and thus to very strongly permitted rotational transitions. As a result, very few rotational levels have significant populations. There is, however, the additional complication that the rotational level of the ground state, $X^2\Pi_{3/2}$ (J = 3/2) undergoes Λ-doubling. The forbidden Λ-doublet transition occurs at a wavelength of 18 cm, with 4 components observable with radio telescopes. Since selection rules apply in the fluorescence process, the Swings effect on the ultraviolet bands leads to variations in the relative populations of the Λ-doubled levels, which would normally

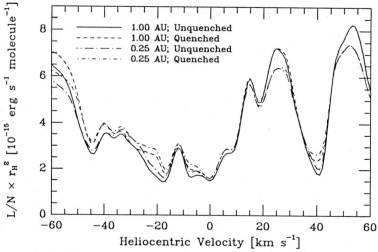

Fig. 3.15. Fluorescence efficiency of the 0–0 band of OH. From Schleicher and A'Hearn (1988).

be populated strictly proportionally to their statistical weights at any temperature above about $0.1\,\mathrm{K}$ (setting $T = h\nu/k$). Since the transition between the levels is highly forbidden, a strong population inversion or even an anti-inversion can build up. The $3\,\mathrm{K}$ black body background (or the galactic background in the Milky Way) can then induce maser action, i.e., stimulated emission, if the population is inverted, or it will be seen in absorption if the population is anti-inverted. As originally discussed by Biraud et al. (1974), this explains the observed variation between emission and absorption as the comet's heliocentric velocity changes. The strength of the emission or absorption is directly proportional to the column density of OH multiplied by the inversion.

Because the Λ-doublet transition is forbidden and the radiation density is low, the inversion of these levels is also sensitive to collisions. At radio wavelengths there appears to be a hole in the OH distribution near the nucleus caused by quenching of the inversion by collisions in the coma. In fact, the hole is quite large, of order several times 10^4 km for P/Halley, and the quenching almost certainly is due to collisions with ions or electrons (e.g., Schloerb, 1988). Maps of the OH emission made with the Very Large Array (VLA) for Comets P/Halley and Wilson showed extensive structures in the OH that have not been reported in the ultraviolet observations. These structures may be tracing the collisional effects rather than the abundance of OH (de Pater et al., 1986), although no explanation for these clumps is fully satisfactory. These effects must be modeled in detail in order to use radio observations of comets to derive abundances.

Prompt emission has been widely discussed in the context of OH but never directly detected. The OH should be formed in both vibrationally and rotationally excited states so that one might detect it in rotational lines of the ultraviolet bands which are not pumped by fluorescence. It is also detectable, in principle, as an excess in the spatial brightness profile of the ultraviolet bands near the comet nucleus. The OH is also observable in the near-infrared (vibrational transitions) and in the submillimeter region (rotational transitions), where one rotational line has already been detected.

C₂. The species with the largest number of different spectral features is C_2. It has not only the Swan system, which dominates the visible spectrum of most comets, but also the Mulliken system in the ultraviolet and the Phillips system in the near-infrared. Because the molecule is homonuclear, it has no dipole moment, and pure vibration – rotation transitions are forbidden. The electronic pumping, on the other hand, is extremely fast with a molecule in the $a^3 \Pi_u$ state absorbing a photon every few seconds at $r = 1\,AU$. The spectrum is therefore very different from that of OH. When fluorescence occurs in OH, leading to population of an excited rotational state, say the $J = 7/2$ level of $^2 \Pi_{3/2}$, pure rotational transitions will cascade down to the $J = 3/2$ level before another solar photon is absorbed. In the case of C_2, however, absorption of another solar photon occurs before the decay, and the subsequent electronic downward transition leads to molecules in even higher rotational levels. Thus C_2 is both vibrationally and rotationally "hot," whereas OH is rotationally and vibrationally "cold." All other things being equal, the vibrational and rotational excitation temperatures would eventually approach the black body temperature of sunlight. This in turn means that many lines are involved in any given band so that the Swings effect, even if it is significant for individual lines, will have virtually no effect on the band intensities for C_2, whereas it has a factor of 7 effect on the intensity of the 0–0 band of OH. The C_2 molecule has been extensively studied because the structure is so complex that it has only recently become reasonably well understood. The dominant Swan system, represented by five band sequences ($\Delta v = 0, \pm 1$, and ± 2), does not terminate in the ground electronic state as shown in Fig. 3.16. The Mulliken and Phillips systems, which do terminate at the ground state, are much weaker.

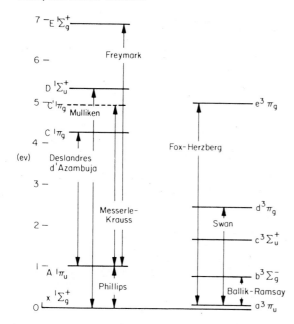

Fig. 3.16. Energy level diagram of C_2. From Cooper and Nicholls (1975).

The problems in interpretation of the C_2 spectrum have historically included an explanation of the rotational and vibrational excitation temperatures, their variation with heliocentric distance, and understanding the transitions between the singlet and the triplet states. It is now clear that intercombination transitions between the singlet and the triplet states play an important role. Observations of Comet P/Halley, both *in situ* and from Earth, suggest that the C_2 triplets are formed vibrationally hot, although it is still not clear whether the C_2 is preferentially formed in the triplet or in the singlet states. Even the intercombination transitions, $a^3 \Pi_u \rightarrow X^1 \Sigma_g^+$, have sufficiently high values of the Einstein A coefficient that triplet – singlet decays occur every few hundred seconds. It therefore turns out that the ratio of singlet to triplet states reaches its equilibrium value of about 1.4 within about 10^3 s. Data are not yet available to determine the ratio of singlet to triplet states with spatial resolution of 10^3 km or better (Vanýsek et al., 1988; O'Dell et al., 1988; Gredel et al., 1989).

3.4 Composition of the Coma

3.4.1 Identification of Species

Although there are very many well identified species in comet comae, many doubtful identifications have also been published. In fact, one of the reasons that results were slow to appear from the NMS on Giotto was the need to eliminate many of the ambiguities in the identification process before announcing the results. There are three methods for identifying species in the coma: photon spectroscopy, mass spectroscopy, and inference. The last of these will not concern us, although it should be pointed out that it was "known" that the dominant volatile was H_2O for many years before it was actually detected. We will be concerned with the spectroscopy of photons at all wavelengths, the range over which spectral identifications in comets have been made extending from 121.6 nm (L-α) to 18 cm. We will consider mass spectroscopy both for ions and for neutrals.

Mass spectroscopy is subject to many ambiguities although the predominant one is due to the coincidence of different chemical species having the same molecular weight. Since only two comets have been studied with mass spectroscopy, we will cite one example from each. In the case of the passage of the International Cometary Explorer (ICE) through the tail of Comet P/Giacobini-Zinner, the Ion Composition Instrument (ICI) detected the expected water-group ions and CO^+, but it also detected an ion with mass-to-charge ratio, $M/q = 24 \pm 1$ amu (Coplan et al., 1987). With the given uncertainty there were a large number of possible identifications of which only one (Mg^+) could be ruled out immediately by contradictory remote spectroscopic data. Many possibilities were ruled out on the grounds of requiring unlikely reaction rates to produce the ion from known neutrals. This still left Na^+ and C_2^+ as possibilities. However, from the ground observations obtained at the time of the ICE encounter with Comet Giacobini-Zinner, Konno and Wyckoff (1989) and

Konno et al. (1990) concluded that the C_2^+ spectral contribution was negligible and that Na^+ could not be the dominant contributor because NaI was not detected. It must therefore be concluded that no species has been identified.

In the case of Comet P/Halley, the most obvious example has been referred to above, the NMS peak at 28 amu which could be due to CO or N_2 or C_2H_4 or even to fragmentation within the NMS of larger species leading to a spurious signal at 28 amu (Eberhardt et al., 1987b). Since the instrument scanned a wide range of masses and there were no peaks at higher masses with sufficient abundance, fragmentation within the NMS could be ruled out. The key question was to resolve the N_2 vs. CO ambiguity. Limits could be set by using two different energies for the ionizing electrons in the mass spectrometer. Electrons at relatively low energy (17 eV in this case) preferentially ionize the N_2 because it has a lower ionization potential than CO, and the ratio between the low-energy mode and the high-energy mode (90 eV) allows an upper limit to be placed on the contribution by N_2 which is sufficient to securely identify CO as being present but not sufficient to determine the abundance of CO precisely. Again, Earth-based spectroscopy, which implied an abundance of CO in the outer coma, was needed to completely resolve the ambiguity.

In both of the cases discussed here, progress was made because the mass peaks under study were relatively strong. In the case of Comet P/Halley, the mass spectrometers detected peaks at nearly every possible mass, and it may turn out to be impossible to identify all of them. The examples illustrate one fundamental difference between ion and neutral mass spectrometry. In ion mass spectrometry, a given species produces a peak at only a single value of M/q. In neutral mass spectrometry, the mechanism for ionization that must be applied often leads to fragmentation as well as ionization. This means that a single species will often produce peaks at several different values of M/q. This can be an advantage if one has relatively few species since it provides a mechanism for resolving ambiguities, but when the mass spectrum is very complex the secondary peaks often add to the confusion rather than clarify the situation. On future missions to comets where there is likely to be more time for the measurements, it may be possible to improve the results for neutral species by including a wide variety of ionization modes in the instrument: Impact ionization at various energies, electron transfer with alkali metal ions, field ionization, etc. Since the different ionization methods have different yields and also different fragmentation patterns, comparison of the peaks for the species in question and for its fragments in the various modes can often yield clean identifications even when species are present at nearly all masses, as they are in the data from P/Halley. Finally we note that the mass spectrometer itself, whether for ions or neutrals, must be extremely well calibrated and studied for the effects that are likely to occur when, e.g., unanticipated temperatures or ram velocities are found.

Photon spectroscopy suffers from different but equally important limitations which can also lead to misidentifications. One of the most common causes of questionable identifications is inadequate signal-to-noise ratio or inadequate spectral resolution. One is naturally limited to identifying those species that emit in the available spectral ranges. This is the reason why there are not yet any measurements of N_2 or the noble gases; their resonance lines are all of shorter wavelengths than L-α. Adequate criteria for an identification vary depending on the type of data available. If one has sufficient spectral resolution that the probability of a chance coincidence of wavelength is negligible, one can in principle identify a species from the wavelength of a single emission or absorption line. This situation, however, almost never occurs in cometary studies, and additional criteria are necessary. Usually one requires several spectral features in physically plausible ratios of intensities and implying physically reasonable abundances, although again one may relax the criterion for understandable intensities if there are a number of spectral features observed with sufficient spectral resolution that a chance coincidence is improbable. In that case, one can admit that the emission mechanism is not understood.

A recent example of a definitive identification is that of S_2 (A'Hearn et al., 1983). Six different ultraviolet spectral features were observed at low (2 nm) spectral resolution during an outburst on the day of closest approach of Comet IRAS-Araki-Alcock. Although this resolution might often be considered too low for identification purposes, the identification was solid from the outset for several reasons. The signal-to-noise ratios of the spectra were high, six features were detected, the relative strengths of the features were consistent (within 50%) with the results of a crude calculation of the fluorescence, all other spectral features within the range of the instrument were predicted to be much weaker, and the required abundance of S_2, although not readily interpreted, was physically plausible in being a small percentage of the abundance of water. Subsequent identification of additional spectral features at longer wavelengths confirmed the identification even more, as did the results of better calculations of the expected fluorescence. Although many questions about S_2 remain, such as the inability to detect it in any other comet to date and the inability to fully explain its presence, the identification is widely accepted.

Formaldehyde, H_2CO, is a species that has been suggested many times and even tentatively identified more than once but which required several pieces of evidence to make the identification convincing. A tentative identification at optical wavelengths by Cosmovici and Ortolani (1984) was based solely on wavelength agreement at low spectral resolution of several features in Comet IRAS-Araki-Alcock. Moroz et al. (1987) identified H_2CO on the basis of a single spectral feature at 3.6 μm, observed at low spectral resolution with the infrared spectrometer (IKS) on the Vega spacecraft. They also determined that a physically reasonable abundance ($\sim$ 5% of H_2O) would be required. Mumma and Reuter (1989) show excellent agreement of this line with a synthesized spectrum. Also a single spectral line of H_2CO was reported at a wavelength of 6 cm by Snyder et al. (1989). Although that feature had extremely low signal-to-noise ratio, it was at precisely the correct wavelength and implied an abundance of the same order as that required by the IKS data. The profile of the feature indicated that H_2CO was not a parent molecule. By this time, the identification of H_2CO in P/Halley was reasonably secure, and the final resolution of ambiguities in the data from the NMS experiment on Giotto (Krankowsky and Eberhardt, 1990) was not needed so much for the identification but rather for showing dramatically that the H_2CO was not a parent molecule. The spatial distribution of H_2CO is very similar to that of CO, suggesting CO is a decay product of H_2CO and H_2CO is a product of a distributed source, possibly containing polyoxymethylene. It still remains to be seen whether all the identifications were correct, but the existence of H_2CO is no longer in doubt.

Recognizing that one must make some arbitrary decisions about which identifications are convincing and also recognizing that continuing analysis of data from P/Halley will uncover many additional species in the near future, Table 3.1 lists the neutral species for which identifications are reasonably sure, based on direct detection either remotely or *in situ*. Some of these species are observed only in Sun-grazing comets (specifically the metals, some of which were also detected in the grains in P/Halley but not in the neutral gas), while others have been observed only in one comet. Species with a question mark are questionable at this time but may be confirmed in the future. Other species can be inferred from their presence as ions, but are also not included here.

Table 3.1. Neutral species directly observed in comets

Metals	Other atoms	Diatomics	Triatomics	Polyatomics
Na	H	CH	H_2O	H_2CO
K	C	CO	CO_2	NH_3 ?
Ca	N	CN	NH_2	CH_3CN ?
V	O	C_2	HCN	
Cr	S	CS	C_3	
Mn		OH	HCO ?	
Fe		NH		
Co		S_2		
Ni		$^{12}C^{13}C$		
Cu		^{13}CN		

3.4.2 Abundances in the Coma

Whether measurements are made remotely or *in situ*, there is always a large uncertainty in relating the measured abundance to a "useful" quantity, such as the abundance in the nucleus or even the total abundance integrated over the entire coma. *In situ* measurements with mass spectrometers can only measure the abundance of a particular species at a single point in the coma; measurements at a variety of points must be made sequentially and, for practical reasons, will probably never be made at enough different positions to determine the total abundance, even if the comet did not vary. Remote measurements could, in principle, determine the abundance of the remotely observable species integrated over the entire coma, but, again for practical reasons, most instruments are capable only of measuring the column density averaged over a relatively small fraction of the coma. Furthermore, since the chemical composition varies dramatically from place to place within the coma, even relative abundances cannot be extrapolated to the whole coma or to the nucleus; any such extrapolation must be model dependent. For that reason, results are often quoted as production rates which were derived from applying a particular model, such as those discussed previously, to observed densities. Recognizing the limitations of the extrapolations, we ask: How can column densities or local volume densities be determined?

In situ determination of volume density is straightforward, provided one is using a well calibrated mass spectrometer. As long as the instrument is well calibrated and the physical conditions of the ambient gas being measured are known, the determination of abundances is straightforward once the identifications are secure. The largest uncertainties are in the models used to relate the local densities to global densities.

Remote observations are somewhat more complicated in practice, although in principle there is no difference from mass spectrometry – as long as the species of interest is well calibrated, i.e., the g-factor is well known, the determination of an average column density is straightforward. Furthermore, one can often observe at least a significant fraction of the coma, which somewhat reduces the model dependence of the results. On the other hand, not all species can be observed remotely. Some, such as N_2, have no emission lines in the observable region of the spectrum, while others, such as S_2, are confined to spatially small regions near the nucleus and therefore cannot be observed except under very unusual geometric circumstances.

As noted above, a prime goal is to understand the chemical abundances in the nucleus, and to this end we must assess not only the abundances in the coma itself, but also, from the spatial and temporal variations, the likely relationships among the various species. In the ultraviolet region of the spectrum there are resonance lines for nearly all the atoms likely to be abundant in comets, except N (which might become accessible with proposed future instruments). Observations with a large field of view, therefore, could yield the relative atomic abundances of all the common atoms in a way that is rather insensitive to models. On the other hand, this does not really tell us the chemical composition of the nucleus. In fact, for only a few species are the parent molecules even known. Until a nucleus can be sampled, only model calculations can provide information on the composition of a nucleus. In Table 3.2 we list the observed species together with the likely sources.

We can estimate the relative abundances of the various species as released from the nucleus. The only species clearly present in P/Halley at more than 1% of H_2O were CO and CO_2, although there have been suggestions that CH_4, NH_3, and H_2CO

Table 3.2. Sources of neutral species observed in comets

Observed species	Dominant source	Other sources
H, OH, O	H_2O	many
CN	HCN, dust	CH_3CN
HCN	parent? dust?	
C_2	C_2H_2?	C_2H_4, C_2H_6, dust?
CO	parent, dust	H_2CO, CO_2
CO_2	parent?	
H_2CO	dust, polyoxymethylene?	parent?
CS	CS_2	
S_2	parent?	
NH, NH_2	NH_3? dust	
C_3	polymers? dust?	C_3H_2
CH	C_nH_m	CH_4
C	CO, dust	CO_2, HCN, CS_2, etc.
N	?	?
metals	dust	

might all be present at the level of a few percent. In the cases of CH_4 and NH_3, the suggestions were based on the ionic composition as measured from Giotto (Allen et al., 1987), but these results are based on measurements of minor branches of the chemistry that are sensitive to the rate coefficients and, at least for NH_3, there was contrary information based on remote observations of the primary decay channels (NH_2 and NH). The high abundances may therefore be regarded as doubtful (Marconi and Mendis, 1988). In the case of H_2CO, the abundance is uncertain and may be variable. The convolution of a production rate variation with lifetime can lead to apparent abundance ratios that are much greater than the time-averaged ratio. There were many species present in P/Halley and in many other comets, at levels less than 1% of H_2O. The measurements of P/Halley are unique in that they provide a more complete inventory than is possible with remote observations. However, it is the remote observations of the few, easily observed species in many comets that permit the interpretion of the P/Halley results.

CO has only been observed in the brightest comets because its emission bands have low g-factors. Nevertheless, it seems to vary greatly from one comet to the next, from less than 1% to perhaps more than 30% relative to H_2O. The only pattern to the variation in the relative abundance of CO is that it is apparently more abundant in comets with a strongly reflecting solar continuum, i.e., in comets normally classed as dusty (cf. Feldman, 1990). CO_2, unfortunately, has not been observed in any comets other than Halley because of a lack of suitable instrumentation. The only observable emission band is at $4.3\,\mu m$, a wavelength at which there are no instruments above the atmosphere and the Earth's atmosphere has strong absorption bands of CO_2, even above the altitudes at which the highest aircraft fly. Although the ion CO_2^+ has been observed in many comets, varying considerably in apparent abundance, the lack of knowledge of ionization processes has inhibited any attempts to deduce the abundance of CO_2 in those comets.

Trace species, such as CN and C_2, have been observed in very many comets. There are some systematic trends with heliocentric distance, such as the ratio of CN/C_2 increasing at large distances, but these may be artifacts introduced by the models used to derive production rates. Compared with the variation of CO, most trace species seem to have a reasonably constant abundance relative to H_2O, as measured either with OH or $O(^1D)$ emission, varying only by small factors in either direction from the average for all comets. Furthermore, much of the variation of CN is strongly correlated with the dust-to-gas ratio of a comet, as shown in Fig. 3.17, implying that the abundance of the CN parent in the ices is very uniform. The variation of C_2 is comparable to that of CN but not well correlated with the dust; this variation might be a reflection of the varying efficiency of ion – molecule reactions in the coma rather than a reflection of the abundance of the parent, whatever it is, in the nuclear ices. As noted above, there is contradictory evidence regarding the abundance in comets of the stable molecules seen abundantly in planetary atmospheres, NH_3 and CH_4, but the preponderance of the evidence seems to favor abundances $< 1\%$, if they are present at all. It is not yet clear whether NH_3 is the parent of the widely observed NH_2 and NH, nor is it clear whether they even share a common parent. Most data are consistent, with both NH and NH_2 coming from NH_3, but the data also seem to suggest a much wider variation in the abundance of NH_2 than in the

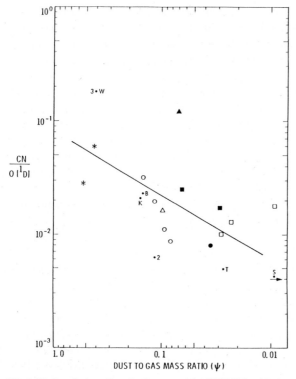

Fig. 3.17. Correlation of production rate ratio CN/O(1D) with the production of dust. From Newburn and Spinrad (1989).

abundance of NH. Since the two abundances tend to come from different data sets, there is very little overlapping data to adequately test the hypothesis.

It is important to ask whether there is any pattern in the relative abundances from comet to comet or whether the variations are random and uncorrelated among the different species. Likewise it is important to learn whether the composition of individual comets is spatially heterogeneous or uniform. Gradients in composition with depth would appear as a variation in relative abundances correlated with the dynamical age of a comet. Clumpy spatial heterogeneity would show up as variations in relative abundances, either with rotation or with orbital position. Because there is considerable current work in progress on the variation of abundances among comets, it is premature to address the results in detail here. We can safely say that thus far no pattern has been detected that is correlated with the dynamical age of a comet, although we must keep in mind that there are no compositional data available for dynamically new comets while still at large distances from the Sun where they would be expected to be outgassing heavily irradiated material.

There are two comets that can be cited as anomalies. P/Giacobini-Zinner is clearly anomalous in the abundances of very many species. Relative to OH, the abundances of both CN and CS seem to be quite typical for all comets. On the other hand, NH_2, NH, C_2, and C_3 are all depleted compared to typical comets by at least a factor 5 (Schleicher et al., 1987; Konno and Wyckoff, 1989; Konno et al., 1990). Among those comets for which a suite of quantitative abundance determinations exists, no other comet stands out in this way. The other anomalous comet is P/Encke, which exhibits a dramatic asymmetry about perihelion in the abundance of CS, CN, and C_2 relative to H_2O. Before perihelion, the abundances are normal, whereas after perihelion these species are all depleted relative to H_2O by at least a factor 3 (A'Hearn et al., 1985). This is the same phenomenon seen at all times in P/Giacobini-Zinner, but the species exhibiting the behavior are not the same: CS and CN are normal in P/Giacobini-Zinner but depleted post-perihelion in P/Encke. In fact all species for which reliable data exist are depleted post-perihelion in P/Encke. Unfortunately, a conclusion in favor of heterogeneity is not clear-cut because mantle effects on the nucleus could play a significant role.

3.4.3 Isotope Ratios

There are only a few species for which isotope ratios are known, but they are important ones since fractionation effects at low temperatures might play an important role during the formation of comets. The potentially most interesting isotope ratio is D/H. Remote observations of OH have provided upper limits near 10^{-3} for the ratio of OD/OH in a number of comets, but the best determination is for Comet P/Halley, where an upper limit of OD/OH $< 4 \times 10^{-4}$ was obtained (Schleicher and A'Hearn, 1986), although inferences regarding the D/H ratio are somewhat constrained by uncertainty in the branching ratio between OD and OH in the dissociation of HDO. Even more important, *in situ* observations of ions in Comet P/Halley allowed independent determinations of upper and lower limits for for the ratio of HDO^+/H_2O^+ and thus $0.6 \ 10^{-4} \leq D/H \leq 4.8 \ 10^{-4}$ (Eberhardt et al., 1987a). This is definitely inconsistent with values of D/H found in the atmospheres of the giant planets Jupiter and Saturn and thus presumably in the gaseous protosolar nebula. It therefore implies either a fractionation of D and H during formation of the cometary ices or a mixing of H_2O from a second reservoir with that in the solar nebula. It should be pointed out that meteorites exhibit several orders of magnitude enhancements of D/H in some compounds but systematically not in H_2O. D/H has not yet been measured in any other compound in comets.

Isotope ratios have also been obtained for oxygen, carbon, and sulfur. The *in situ* measurements at P/Halley were used to derive $^{18}O/^{16}O = 0.0023 \pm .0006$, which is consistent with the terrestrial ratio for these isotopes. The ratio has not been measured in any other comets. The $^{12}C/^{13}C$ ratio has been measured remotely for several comets using the emission by C_2 and by CN. Results appear to vary by nearly a factor of two among the half dozen or so comets studied, but the uncertainties are very large. Most determinations for comets prior to P/Halley implied ratios greater than the terrestrial or Solar System values but typically only by one standard deviation. These were primarily based on the C_2 bands. Determinations using the CN bands in P/Halley imply a ratio two standard deviations below the average Solar System value, $^{12}C/^{13}C = 65 \pm 9$ (Wyckoff et al., 1989). It is still an open question whether the difference is one of technique or if there is a real fractionation of ^{13}C between C_2 and CN. The *in situ* measurements also implied normal, i.e., terrestrial, isotope ratios for sulfur.

Unlike chemical abundance ratios, these isotope abundance ratios in the coma can generally be assumed to reflect isotope abundances in the nucleus, although there are possible fractionation effects that have not yet been evaluated, e.g., in the photodissociation of HDO to OD+H and OH+D. Since the results are so limited, it is premature to interpret them in terms of models for the formation of the Solar System.

3.5 Summary and Prospectus

The study of comet comae and of comets in general is one in which the paucity of existing data and the difficulties of obtaining more data have allowed interpretation to advance to the point that many more effects are incorporated in models than can be constrained by observations. Only the most recent observations have not yet been incorporated in one model or another.

The gas production from the nucleus has been studied in numerous models that include chemical differentiation, diurnal and latitudinal variation of temperature, thermal lag, and insolation from the dust in the coma. Spatial heterogeneity of the nucleus surface has been included in models for certain comets, and models have been calculated to include diurnal variations as active areas of the nucleus move in and out of sunlight, although some of the relevant numerical parameters (conductivity, thermal inertia, mantle thickness, etc.) are still uncertain by at least an order of magnitude. The existence of spatial heterogeneity is unquestionable, although the details of that heterogeneity from one comet to another are not known. Nor is it known whether the spatial variation in gas production also implies a spatial variation in the chemical composition of the material sublimated into the coma. A key goal of future comet observations should be to search for this chemical heterogeneity, something that will be relatively easy to do during the CRAF mission but which should also be pursued in future programs of narrow-band imaging and long-slit spectroscopy of comets. A closely related fundamental question is whether the ostensibly inactive areas of Halley's nucleus are actually releasing significant amounts of gas. This question also will be answered readily when the CRAF mission occurs.

The release of gas into the coma from distributed sources (CHON particles) is another area in which much critical work remains to be done. We do not yet know what fraction of the gas in Halley's coma came directly from the nucleus rather than from dust. Nor do we know the nature of the dust from which the gases are released. Determining the nature of these particles might fundamentally alter our ideas of the origin of comets.

The hydrodynamic flow in the inner coma has also been treated in great detail in current models. At present only observations at radio wavelengths and with infrared Fourier spectrometers have sufficient spectral resolution to measure the velocities directly. As receiver technology improves, it should be possible to make radio measurements of the kinetic properties of many comets at different heliocentric distances to test the models and to search for differences from one comet to another, dependences on total gas production rate, etc. Observations of the Greenstein effect at optical and ultraviolet wavelengths have led to some tantalizing results, but the data are still marginal for inferring kinetic properties. Again, improved detectors with long-slit spectrographs should allow more measurements of this type in the future. The flow in dusty jets of gas has also been studied extensively, and it appears that the models are sufficient to explain most of the extant data.

The photolytic processes in comet comae are quite well understood, although this complete understanding has still not been used widely for interpreting observations. In the future, observers should be including many more effects (such as variations of lifetime with solar activity, variations in excess energy on dissociation, etc.) when interpreting their observations. Chemical processes are understood reasonably well in theory, but the actual models may not be sufficiently complete to properly explain all observable species. They do not include, e.g., charged grains that could play an important chemical role. The models have not yet been applied to the question of variations in relative abundances from one comet to another. To what extent can the variations of abundances in the coma be explained by variations in the chemical rates of formation and destruction and to what extent must we invoke variations in the nucleus abundances? Similarly, excitation processes that allow remote observations are understood in principle, although they have not yet been applied to all known species. It is still not possible to explain the spectral details of some species such as CS.

The fundamental question still not answered, of course, is the complete inventory of species in the coma and their abundances. Observations in new wavelength regions, such as the extreme ultraviolet, will be needed to identify additional species. *In situ* observations will be needed to identify other species. Relative abundances must be determined for a reasonable ensemble of comets and for more than the few most easily measured species. Some of the key questions are:

1. How does the ratio CO/CO_2 vary from comet to comet?
2. Is S_2 present in most comets?
3. Are other sulfur compounds, e.g., SO, SO_2, and H_2S, present in comets in interesting amounts?
4. What is the D/H ratio in a variety of comets?
5. Is the apparent variation in the ratio of CO/H_2O in the coma reflecting a difference in the nuclear ices or a difference in the amount of organic particles? And similarly, what fraction of other species is released from refractory organic particles rather than from the nucleus?
6. Are NH_3 and CH_4 present in most comets at the level of a few percent? Is NH_3 the parent of NH and NH_2?
7. Is N substantially depleted? Or more generally, what is the atomic inventory of C, N, and O relative to H?
8. And finally and most fundamentally, what are the parent species in the nucleus?

Again, the primary lack is observational rather than theoretical.

In the future we can hope for a number of facilities that will address some of these questions. The CRAF mission should yield a wealth of information about the gases in the coma of one comet, answering essentially all the questions above for that single comet. Surveys of many comets will rely on current observing facilities, such as the Hubble Space Telescope (HST, which can address the above questions 2, 3, 4, 5, and 7) and ground-based telescopes (which can address questions 4 and 6), and on proposed facilities, such as the Shuttle Infrared Telescope Facility (SIRTF), the Infrared Space Observatory (ISO, which can answer question 1 and part of question 8), Planetenteleskop (which could answer question 1 for brighter comets and questions 2, 3, and most of 7 for many comets), and the Far-Ultraviolet Spectroscopic Explorer (FUSE/Lyman, which can address questions 4 and 7).

4. Dust

Eberhard Grün and Elmar K. Jessberger

4.1 Introduction

The spectacular display of a bright comet is mostly caused by a cloud of micrometer-sized dust particles of a density that is much less than in the smoke of a cigarette but with the huge dimensions of about 10^5 km. This dust originates from a central source of ices and dust – the nucleus – that had been postulated even before space probes took the first photographs of it in 1986. Upon sublimation of the ices, the dust is entrained in the gas stream leaving the nucleus. Submicrometer-sized particles almost reach the speed of the escaping gases (about 1 km/s), whereas centimeter and decimeter-sized particles hardly reach the gravitational escape velocity from the nucleus (about 1 m/s). At a distance of about 100 km from the nucleus, where the gas density becomes tenuous, the dust decouples from the cometary gas and is left to the forces acting in interplanetary space. These forces cause the dust particles to form a tail about 10^7 km long and generally pointing away from the Sun.

From the similarity of the solar spectrum with that of a dusty comet, it was recognized by Drago in 1820 that the light from a comet is mostly scattered sunlight. Soon after that, in 1836, Bessel developed a mechanical theory to explain the tail and its observed direction away from the Sun; he introduced a repulsive force opposing the force of solar gravity.

With the application of spectroscopy to comet observations, it became evident that there are two types of tails: a plasma tail, that displays only spectral lines of ions and has a long, filamentary structure pointing close to the extended radius vector from the Sun, and a dust tail, that shows mostly the continuum spectrum of scattered sunlight. The dust tail is broader than the plasma tail, is sometimes curved, and lags between the extended radius vector and the path of the comet. Around 1900 Bredichin refined the concept of solar radiation pressure as the repulsive force responsible for the formation of the dust tail. This force originates from the interaction of electromagnetic radiation (sunlight) with matter. It describes the momentum transfer from the radiation field, or photon stream, to the scattering and absorbing dust. The radiation pressure is ineffective for large particles; however, for particle sizes comparable to the wavelength of the radiation, it is very efficient. Therefore in interplanetary space, the repulsive force of solar radiation pressure on submicrometer-sized particles may even exceed the attractive force of solar gravitation.

It was only in 1968 that Finson and Probstein formulated the complete description of the dynamics of dust tails, which also allows comparison with observations. They showed that particles of different sizes moved with different speeds and are therefore separated in the dust tail. With this theory, it is now possible to characterize dust particles according to size and speed of emission from the sphere of influence in the coma.

Comets vary widely in appearance as determined by their composition and observational geometry (distance to the Earth, angle to the Sun, etc.). The compositions range from very dusty comets such as Bennett (1970 II) that are typically very bright, through medium dusty comets like Tago-Sato-Kosaka (1969 IX), to almost dust-free comets like Ikeya (1963 I) that display only an ion tail (Fig. 4.1).

On rare occasions narrow, sunward-directed antitails have been observed, as for example in Comets Arend-Roland (1957 III) and Kohoutek (1973 XII). This phenomenon is explained through edge-on viewing of a relatively thin layer of large cometary grains in the orbital plane of the comet. Because of the projection effects, this dust is observed on the sunward side of the nucleus, displaying a spike in the opposite direction as the ordinary dust tail.

Since comet dust can be observed from the Earth only by its scattered light, the knowledge about the dust was limited to those particles that provide most of the scattering cross section. These particles are typically 1 to 10 μm. Larger grains, up to several 100 μm, are observed only occasionally in antitails. However, the similarity of comet orbits with the orbits of meteor streams provides evidence that still larger grains, typically centimeter-sized, are emitted from comets. Such large grains have been observed in the infrared spectrum with the IRAS satellite (Sykes et al., 1986) and some evidence also exists from radio and radar observations.

Before the Halley encounters, the dust size distribution was determined solely from optical observations of dust tails (see, e.g., Finson and Probstein, 1968a,b) and from infrared observations (see, e.g., Hanner, 1983). A summary of the pre-Halley encounter knowledge on dust is given by Divine et al. (1986). The differential particle density of the dust, n_d, was given in the form

$$n_d = g_0 \left(1 - \frac{a_0}{a}\right)^m \left(\frac{a_0}{a}\right)^n, \qquad (4.1)$$

where the grain radius a had the lower limit $a_0 \simeq 0.1$ μm. In (4.1) g_0 is a normalization constant, and the exponents m, ranging from 4 to 20, and $n = 3.7$ control the slopes at the small and large radii, respectively. This distribution has a peak in n_d for radius

$$a_p = a_0 \frac{(m + n)}{n}. \qquad (4.2)$$

In the coma of a comet the differently sized dust particles are not yet spatially separated enough to be distinguishable. However, any inhomogeneity in the dust emission pattern, together with the rotation of the nucleus, causes structures in the coma brightness. Indeed, structures have been described by visual observers as jets, fans, spirals, and envelopes. Because of the inferior intensity resolution of photo-

---

Fig. 4.1. Spectra of three comets of varying dust emission as evidenced by the strength of the continuum: Bennett (1970 II), Tago-Sato-Kosaka (1969 IX), and Ikeya (1963 I). The spectra were obtained with Coudé spectrographs, in case of Bennett and Tago-Sato-Kosaka with the 5-m telescope of the Hale Observatory on Mount Palomar and in case of Ikeya with the 1.5-m telescope of the Haute-Province Observatory of the French Centre National de la Recherche Scientifique. (Courtesy C. Arpigny, Liège, Belgium).

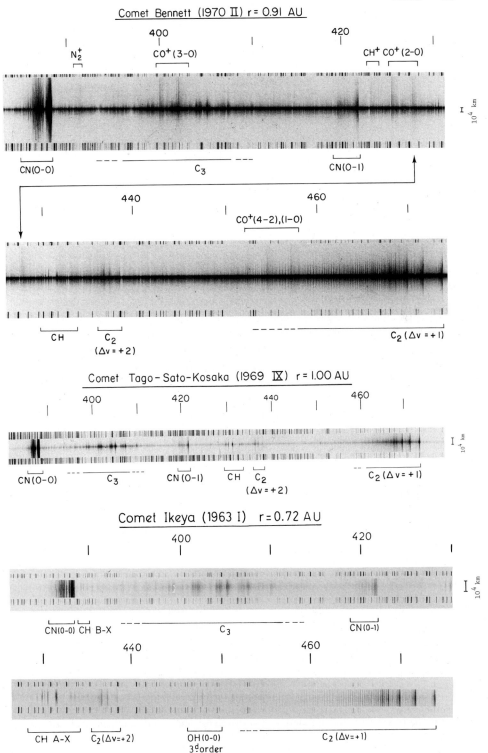

Fig. 4.1. Caption see opposite page.

graphic plates, less structure was noted on photographic records of comets. Only the application of sophisticated contrast enhancing techniques by Larson and Sekanina (1984) revealed structures in photographic comae. From the temporal evolution of the structures they deduced the spin axis, period, and dust emission areas on the nucleus for several comets (Sekanina, 1981; Sekanina and Larson, 1986). Recent observations of Comet P/Halley, however, indicate that the solutions obtained may not be unique; each type of observation, ground-based or *in situ*, results in a different rotational state. No single solution has been found to date that can explain all observations equally well (see, e.g., Chap. 2, Sect. 2.4.5).

What was known about the composition of cometary dust prior to the Halley encounters? Indeed only very little (for a compilation of compositional data see Rahe, 1981). Infrared features at 10 and 18 μm were indicative of the presence of silicates; the metals Na, K, Ca, V, Mn, Fe, Co, Ni, Cu, and possibly Si and Al had been identified in other spectra. Information on their abundances, however, was very sparse. It was concluded that the dust composition may be similar to that of carbonaceous chondrites. From the spectroscopy of meteors and laboratory analyses of captured interplanetary dust particles (both classes of objects may be related to comets), the composition of cometary dust was inferred. However, what was known was mostly based on grain models elaborated to fit spectroscopic observations, such as ice-covered magnetite or olivine grains. The simplicity of the postulated grains reflects a lack of information. More sophisticated dust models inevitably were based on laboratory experiments (Greenberg, 1982) and a general picture of interstellar dust evolution rather than on the direct observation of comet dust.

The P/Halley encounters brought and are still bringing a qualitative and quantitative increase in the knowledge of the composition of comet dust. On board of the three spacecraft Giotto, Vega 1, and Vega 2, impact ionization time-of-flight mass spectrometers facilitated for the first time the *in situ* analysis of the solids in the coma. Three highlights of the results are: (1) The dust is, in greatly varying proportions, composed of a silicate component and a refractory organic component, termed "silicates" and "CHON," respectively. (2) The average abundances of the rock-forming elements are, within a factor of two, the same as their abundances in carbonaceous chondrites and in the Sun. (3) The abundances of the CHON elements are much higher than in any meteorite; especially those of C and N approach solar abundances.

Thus Halley's dust appears to be the least fractionated material assumed to come from the early Solar System that was ever studied in such detail. The analysis of the data from the many thousand grains that hit the dust analyzing instruments is still far from complete. Some unforeseen instrumental peculiarities still have to be studied in the laboratory before it is possible to extract accurate isotopic information, e.g., the important ratio for $^{12}C/^{13}C$. For other isotopes such as ^{26}Mg, the possible daughter of heat-producing ^{26}Al, overabundances larger than a factor of two can already now be ruled out.

Because of possible metamorphism of the dust during its travel from the nucleus to the instruments, the detailed composition of the solid nucleus still has to be modeled, but with far superior boundary conditions than were known before the Halley missions. In the near future space experiments will allow higher precision chemical, molecular, and isotopic analyses close to or even on the nucleus and will also provide basic knowledge of the hitherto unknown structure of the grains. Finally, comet samples returned to Earth will be studied with the full arsenal of sophisticated laboratory techniques and ultimate precision to reveal, with very high confidence, the origin and the detailed evolution of a comet. This, in turn, will further our understanding of the sources of the Solar System and the early processes in it.

4.2 What is Comet Dust?

The classical point of view is that all comet dust is transported from the coma into the tail. Since the color of the dust reflects the solar spectrum, it can immediately be inferred that the particles scattering the sunlight are larger than the wavelength of visible light (400 to 600 nm), typically between 1 and 10 μm in size. In exceptional cases, small, submicrometer and large, millimeter-sized particles appear in the visible display. However, there is no reason to believe that the size range of particulates released from a comet nucleus is restricted to the range that effectively scatters sunlight. Therefore, we will adopt the view that cometary particles range from clusters of a few molecules, several nanometers in size, to chunks of cometary matter, about a meter across, which are eroded from a nucleus but are too small to be considered comets themselves.

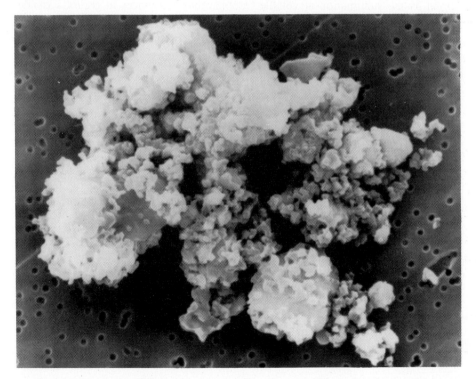

Fig. 4.2. An interplanetary dust particle recovered by a high-flying U-2 aircraft. These particles have a composition similar to the primitive meteorites and are believed to originate from comets. The size of the particle is a few micrometers.

The latter example brings up the next problem: Comet dust is generally assumed to be stable in time or to be modified only very slowly by external interactions, e.g., by solar wind sputtering. Typical examples are interplanetary dust particles (IDPs) that consist of refractory chondritic material, i.e., silicates together with some carbonaceous material (see Fig. 4.2). In extreme situations, as in Sun-grazing comets, even these particles are rapidly modified by evaporation. Evidence for this effect has been shown by Huebner (1970), who analyzed data of observed metal atoms and ions in the comae of comets close to the Sun. However, this picture of relatively stable cometary dust particles also must be modified because of recent observations: Cometary grains can be water ice particles that are quite stable against evaporation at large heliocentric distances. The most prominent example is Comet P/Schwassmann-Wachmann 1, which occasionally displays a spectacular coma (see Fig. 4.3) of ice particles expelled by the drag of more volatile gases such as CO_2 and others. Sekanina (1973) summarized observations of emissions of dust, including ice grains, as far as 15 AU from the Sun. Ice grains can also be emitted in the inner Solar System, however, their lifetimes and hence their ranges from the nucleus before they evaporate are so short (at most a few hundred kilometers, Hanner et al., 1981) that they have not yet been detected.

Besides ice grains that sublimate rapidly, there are indications for other materials that sublimate slowly while the grains are still in the coma. Observations of CN, C_2, and other jet-like features (A'Hearn et al., 1986) led to speculations that they are caused by the sublimation of volatile carbonaceous substances in grains. Extended sources of CO in the coma have been required to explain the measurements of the ion mass spectrometers on Giotto (Balsiger et al., 1986; Krankowsky et al., 1986;

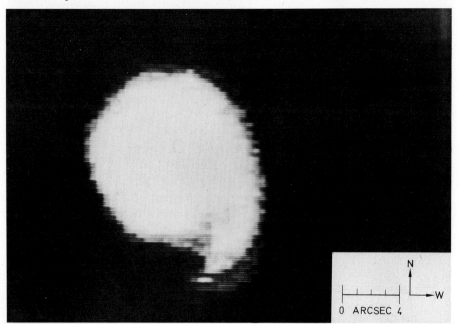

Fig. 4.3. Photograph of Comet P/Schwassmann-Wachmann I taken on 7 February, 1981, by Cochran et al. (1982) during an outburst displaying a one-armed spiral-shaped coma.

Eberhardt et al., 1987). Also, the suspected Polyoxymethylene or POM (Huebner, 1987), which has been deduced from ion analyzer measurements (Mitchell et al., 1986; Gringauz et al., 1987), is believed to originate from cometary dust grains. Evidence that comet dust contains a zoo of organic components has been presented by Kissel and Krueger (1987a).

Sublimation and desorption of volatiles from dust grains is being inferred from the existence of gases in the coma that have not been released from the nucleus itself – at least not according to simple spherical gas emission models. Recent anisotropic emission models (Kitamura, 1986, 1987; Kömle and Ip, 1987) show that the coma is not as isotropic as had been assumed previously. Therefore, deviations of the observed gas density from an R^{-2} dependence (where R is the distance from the nucleus) may be caused by an anisotropic gas emission, and only weak extended sources for the excess gas may be required to explain the observations. Hence, the effects of extended gas sources may not be as extreme, and the importance of dust sublimation for gas production may have been overestimated. Giotto approached the comet from the night side and reached the day side only shortly before closest approach. Therefore, it is quite natural that the gas density does not follow an exact R^{-2} dependence. Only detailed gasdynamic calculations will show whether a source for the excess gas species is needed. Nevertheless, the instability of dust grains was also inferred from *in situ* dust observations. Simpson et al. (1987) and Vaisberg et al. (1987b) report clustering in the detection of small dust grains in the outer coma where certainly it cannot be explained by purely statistical fluctuations of the impact rate. The authors interpret the observations as fragmentation of larger particles during their passage through the coma. Observed changes in the size spectrum of dust particles (Vaisberg et al., 1987b; McDonnell et al., 1987) are also explained in terms of fragmentation effects.

Fragmentation of larger grains or grain clusters are the natural consequence if differential sublimation of grain material occurs. Another possible cause for fragmentation can be the electric charging of grains and their consequent electrostatic disruption. Boehnhardt and Fechtig (1987) recently discussed this effect for the situation during the spacecraft encounters with P/Giacobini-Zinner and P/Halley. They found that electrostatic fragmentation is only effective for submicrometer-sized and very weak (10^2 to 10^3 Pa tensile strength) particles. These tensile strengths are considered typical for fluffy aggregates. Charging of dust particles to high potentials, $\phi \gg 10$ V, occurs mainly because of their interaction with energetic ions and electrons, that do not exist in abundance in the inner part of the coma. Therefore electrostatic disruption of small particles may only be important at distances larger than about 10^4 km from the nucleus. Even in the absence of electrons and ions, dust particles will acquire a positive charge from the photoeffect of solar UV. Wyatt (1969) estimates a flux of

$$F_{pe} = 2.5 \times 10^{10} \, \tilde{\gamma} \, \pi a^2 / r^2 \qquad (4.3)$$

photoelectrons per second, where $\tilde{\gamma}$ is an efficiency factor of the order of 0.1 for dielectrics and r is the heliocentric distance in AU. The surface potential ϕ on grains reaches typically 3 V, thus the grain carries a charge

$$q_d = 4\pi\epsilon_0\phi a, \qquad (4.4)$$

where the permittivity $\epsilon_0 = 8.854 \times 10^{-12} C^2/(N\,m^2)$.

Photoelectron emission from dust grains initially at a rate given by (4.3) and with a total amount determined by Eq. (4.4) is certainly an important factor for charging the cometary environment, especially in regions close to the nucleus where other ionization effects are not efficient.

The electron flux F_e from the ambient plasma onto the unbiased dust grains is given by

$$F_e = n_e \left(\frac{2E_e}{m_e}\right)^{1/2} \pi a^2, \tag{4.5}$$

where n_e is the electron density, m_e is the electron mass, and E_e is the electron energy. If this flux exceeds the photoelectron flux, dust grains are generally charged negatively up to a potential corresponding to the electron energy. The charge on dust grains is reduced if the dust density becomes very high. Some effects of high dust density in a plasma have been discussed by Havnes et al. (1987).

At plasma energies higher than 10 eV, secondary emission from solid particles becomes important and also causes a reduction of the negative equilibrium potential. At plasma energies much higher than 100 eV, the yield of secondary electrons may become larger than 1 for some materials and the charging effect of the impacting electrons reverses because more electrons are released than picked up. The equilibrium potential will then become ~ 10 eV positive, since the energies of the secondary electrons are typically 10 eV. This effect is discussed in more detail by Meyer-Vernet (1982).

4.3 Dust Release from the Nucleus

Dust particles released from a comet nucleus are not necessarily in the same state as they were while residing inside the nucleus or when they were incorporated into the nucleus during comet formation. For example, volatile ices may have filled the cavities inside the particles, especially if particles are fluffy aggregates. But even such a fluffy aggregate structure of a particle may not be original, it could have formed during the sublimation of the ice – dust mixtures. In sublimation experiments of finely dispersed mixtures of clay minerals in water ice at mass ratios of 1:1000, Saunders et al. (1986) found that, in some cases, fluffy filamentary sublimation residues of the minerals were formed, although they made sure that in the original ice mixture the clay particles were not in contact with each other. The residue structure is the result of aggregation during or after sublimation. In these experiments presumably molecular bonds are the binding forces between the particles. In more complex systems that include various organic components, the organic materials could be the binding agents between the individual constituents of the aggregates.

Both theoretical considerations (Shul'man, 1972; Fanale and Salvail, 1984; Brin and Mendis, 1979) and experimental simulations (Ibadinov et al., 1987; Grün et al., 1987b) show that upon sublimation of ice – dust mixtures a mantle of dry dust builds up over the ice (see Chap. 2, Sect. 2.5.2). This mantle can reach thicknesses from millimeters to centimeters before it quenches further sublimation. Pressure build-up under this mantle can lead to disruption of the mantle and to the eruption of particles at speeds of several meters per second (Grün et al., 1989). In this case the released dust grains are fragments of the mantle, which itself is a fluffy aggregate of many individual particles originally dispersed in the ice matrix.

Only in the case of sublimation from a naked nucleus is the ratio of emitted dust to gas the same as within the nucleus, i.e., at least within the surface layer. If a dust mantle forms, this ratio is lower than within the cometary ice. However, if the dust mantle is blown off, e.g., because of an increase in solar heat input during the diurnal or seasonal cycles, the emitted dust to gas mass ratio may exceed the nuclear value by a large amount. With this argument in mind it can be expected that only during phases of high sublimation rates is the dust to gas mass ratio of the emitted material closest to the value within the comet itself.

The loose structure of cometary particles is also demonstrated by the low densities of meteor particles that obviously originate from comets. Densities as low as 0.02 Mg m^{-3} have been observed (Ceplecha, 1977). Ceplecha (1976) reports on a fireball that had about 200 Mg of mass and did not reach deeper into the atmosphere than 55 km height before it was totally ablated. The density of this fireball must have been extremely low.

Another example of fragmentation of larger structures caused by the sublimation of ices has been observed by Bar-Nun et al. (1985). Heating experiments with amorphous water ice in which gas, like CO_2, was trapped demonstrated that large numbers of small ice grains in the range from 0.1 to 1 μm are ejected from the ice upon sublimation of the more volatile components.

Another, quite different process is able to lift dust grains from the nucleus. It is the electrostatic levitation and blow-off of small grains from the surface. Outside of a heliocentric distance of about 5 AU, a water-dominated comet nucleus does not emit enough water molecules in order to have a protective atmosphere. Therefore the nucleus is not shielded from either solar UV radiation or the solar wind. As a result, the nucleus surface gets electrically charged. Because of the predominance of the photoeffect, the central parts of the sunlit side become positively charged and reach a surface potential ϕ_s of about +5 V. Close to the terminator the surface potential is mainly determined by the solar wind electrons. Mendis et al. (1981) discuss the charging process and obtain potentials of a few times 10 V negative at the terminator and up to kilovolts on the dark hemisphere, depending on solar wind conditions. Along with the surface potential, another important parameter is the Debye length, λ_D, in the ambient plasma

$$\lambda_D = \left(\frac{\epsilon_o E_e}{n_e e^2} \right)^{1/2}, \tag{4.6}$$

where e, n_e, and E_e are the electron charge, density, and energy, respectively. The Debye length determines the distance over which an electric field is shielded by the ambient plasma. According to Mendis et al. (1981), $\lambda_D \simeq 10$ m on the sunlit side and $\lambda_D \simeq R_N$ (radius of the nucleus) on the night side of the nucleus. These values are valid for $r \simeq 5$ AU.

The normal component of the surface electric field at the nucleus is $\mathcal{E}_n \simeq \phi_s / \lambda_D$. Dust grains lying on the surface will acquire a charge q_d proportional to its projected surface area

$$q_d = \pi a^2 \sigma_S = \pi a^2 \epsilon_o \mathcal{E}_n \simeq \pi a^2 \epsilon_o \phi_S / \lambda_D, \tag{4.7}$$

where σ_S is the surface charge density. This charge is much smaller than the charge that would be acquired by the grain if it was at potential ϕ in free space (Eq. 4.4).

The electrostatic repulsive force between the nucleus and the dust particle carrying unit charge e is then given by

$$F_{el} \simeq e\,\phi_S / \lambda_D. \tag{4.8}$$

This force is counteracted by the effective gravitational force F_{grav} on a grain lying at the latitude δ on a comet spinning with period P

$$F_{grav} = \frac{4}{3}\pi a^3 \rho_d \cdot \frac{4}{3}\pi \rho_N R_N G \left(1 - \frac{3\pi \cos^2 \delta}{G\,\rho_N P^2} \right), \tag{4.9}$$

where ρ_d and ρ_N are the densities of the dust particle and the nucleus, and $G = 6.67\ 10^{-11}$ N m^2 kg^{-2} is the universal gravitational constant.

Mendis et al. (1981) found that submicrometer-sized dust particles can be levitated over the night side of the nucleus throughout the outer Solar System. Particles that are initially levitated from the surface will eventually escape from the comet because they will acquire the much higher free space charge. This electrostatic dust blow-off mechanism has been proposed by Sekanina (1985) to explain the erratic brightness variation and a possible periodicity of P/Halley at a position in the orbit outside of 7 AU.

4.4 Dust Emission

In this section we describe the dynamics of dust particles from the time that they have been released from the nucleus until they are no longer influenced by either the gas flow or the gravitational attraction from the nucleus. The extension of the gas – dust interaction region has been estimated from modeling to be on the order of a few tens of nuclear radii (see, e.g., Keller, 1983). The gravitational sphere of influence with radius R_{GS} can be estimated according to Öpik (1963)

$$R_{GS} = r \left(\frac{m_N}{2M_\odot} \right)^{1/3}, \tag{4.10}$$

where r is the heliocentric distance of the comet and $M_\odot$ is the solar mass. Introducing typical values one obtains, for $r = 1$ AU $= 1.5 \times 10^8$ km, $R_{GS} \simeq 100\ R_N$.

If, for simplicity, we neglect both solar gravity and radiation pressure inside this region, the equation of motion of an individual dust particle in the radially expanding gas flow is given by

$$\frac{4\pi}{3} a^3 \rho_d \frac{dv_d}{dt} = \pi a^2 \frac{C_D}{2} \rho_g (v_g - v_d)^2 - \frac{4\pi}{3} a^3 \rho_d m_N \frac{G}{R^2}, \tag{4.11}$$

where a, ρ_d, and v_d are the dust particle radius, density, and speed, respectively. The first term on the right-hand side describes the gas drag, where ρ_g and v_g are the gas density and speed and C_D is the drag coefficient (e.g., for free molecular flow $C_D = 2$). The second term describes the gravitational attraction from the comet nucleus of mass m_N at distance R.

In a simple model, one can assume that the gas is emitted isotropically and hence the gas density decreases as $\rho_g = n_g M m_u \sim R^{-2}$, where n_g is the gas number density, M the molecular weight, and m_u the unit atomic mass. In this case, for large particles $v_g \gg v_d$ and the ratio of the gas drag to gravitational pull by the nucleus is constant. Therefore, all particles lifted off the nucleus will leave the comet. However, at least the night side of the nucleus surface and in many cases parts of the day side do not emit gas. Therefore, the gas density over active areas on the nucleus decreases more rapidly than R^{-2}. The largest particles that could just be lifted off the surface will only reach a height of the order of the extent of the active area before they fall back onto the nucleus, most probably outside of the active area. This falling back of large particles with speeds up to 1 m/s can trigger new emission activity, thereby spreading active areas over parts of the sunlit surface that are not protected by a tough and cohesive crust.

If activity persists at one location for several orbital periods then rim-like structures can build up around the active area giving crater-like appearances. The speed at which the build-up occurs is comparable to the speed at which the floor recesses because of sublimation (i.e., meters per orbital period). This fall-back material further enhances the insulation of inactive regions thereby perpetuating any heterogeneity on the surface.

In a simple model of the gas – dust interaction (Whipple, 1951; Weigert, 1959), it is assumed that the gas particles have mean free paths larger than the grain size and collide elastically with the dust. This assumption is equivalent to $C_D = 2$. With the further assumptions of a steady state and a constant gas flow velocity v_g in the dust acceleration region, Gombosi et al. (1986) obtain the expression for the dust velocity v_d

$$\frac{v_d}{v_g - v_d} + \ln\left(1 - \frac{v_d}{v_g}\right) = K\left(1 - \frac{R_N}{R}\right), \tag{4.12}$$

with the coefficient

$$K = \frac{3\,C_D\,Z\,R_N}{16\,a\,\rho_d\,v_g}, \tag{4.13}$$

where Z is the gas production rate per unit area. More realistic models take into account the energy exchange between the gas and the dust (Probstein 1969). This can be described by the energy balance equation for a dust particle

$$\frac{4\pi}{3}\,a^3\,\rho_d\,c_d\,\frac{dT_d}{dt} = 4\pi\,a^2\,q_{gd} + \pi\,a^2\,\alpha_d\,\frac{F_\odot}{r^2} - 4\pi\,a^2\,\epsilon\,\sigma_o\,T_d^4, \tag{4.14}$$

where T_d is the dust particle temperature, q_{gd} is the gas – dust heat transfer rate per unit area, c_d is the specific heat of the dust, α_d and ϵ are the dust absorption in visible light and infrared emissivity, respectively, σ_o is the Stefan-Boltzmann constant, and

$F_\odot = 1370$ W m^{-2} is the radiative energy flux from the Sun at 1 AU. By coupling the gas to the dust, the dust to gas production ratio, χ, becomes important. Using a single characteristic dust size, Probstein (1969) obtained steady state numerical solutions to the coupled gas – dust equations. Figure 4.4 shows the normalized terminal dust velocities for different values of χ as a function of the parameter

$$\beta_d = 8\,a\,\rho_d \frac{(c_p\,T_o)^{1/2}}{3\,Z\,R_N},$$

(4.15)

where c_p is the gas specific heat at constant pressure and T_o is the surface temperature.

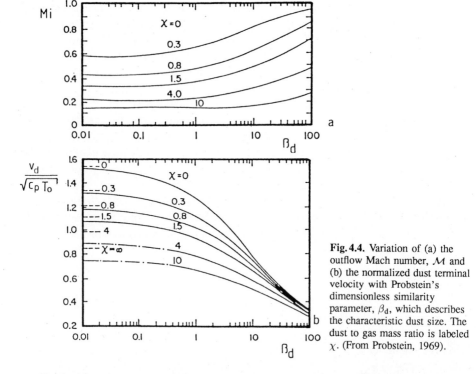

Fig. 4.4. Variation of (a) the outflow Mach number, $\mathcal{M}$ and (b) the normalized dust terminal velocity with Probstein's dimensionless similarity parameter, β_d, which describes the characteristic dust size. The dust to gas mass ratio is labeled χ. (From Probstein, 1969).

Further improvements of the models can be obtained by taking into account a realistic dust particle size distribution. Gombosi et al. (1983) showed that, with an assumed size distribution $\sim a^{-4.2}$, the terminal dust velocities decreased by about 20% with respect to a single size dust distribution (Fig. 4.5; see also Chap. 2, Sect. 2.5.4).

Gombosi et al. (1986) calculated the radial evolution of gas and dust parameters in the dust acceleration region. Some results are shown in Fig. 4.6. In this calculation, $T_o = 200$, $c_d = 8\ 10^2$ J kg^{-1} K^{-1}, the ratio of the specific heats of the gas $\gamma = 4/3$, $\chi = 0.27$, and $r = 1$ AU. It can be seen that, within 2 R_N, the gas approaches 80% of its terminal velocity value, while the dust accelerates somewhat more slowly. Within a distance of about 10 R_N both gas and dust almost reach their terminal velocities.

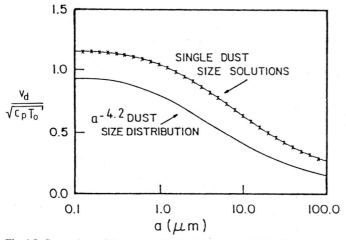

Fig. 4.5. Comparison of dust terminal velocities obtained with single dust size solutions and using $a^{-4.2}$ dust size distribution for a dust to gas mass ratio $\chi = 0.5$. (From Gombosi et al., 1983).

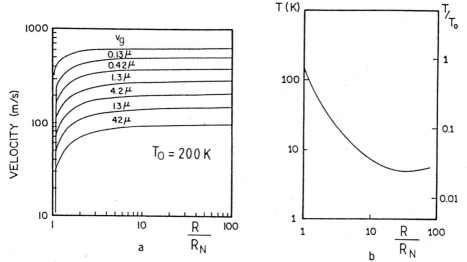

Fig. 4.6. Dependence of (a) steady state gas and dust radial velocity and (b) gas temperature for dust size distribution defined by (4.1). (From Gombosi et al., 1986).

Adiabatic cooling decreases the gas temperature to a few K at about 30 R_N; then heating from photodissociation reverses the trend. The dust temperature (not shown) increases from T_o and rapidly converges to its asymptotic value

$$T_d = \left(\frac{\alpha_d F_\odot}{4\epsilon\sigma_o r^2} \right)^{1/4}.$$

(4.16)

Calculations of the terminal dust velocity, taking into account the recent Halley measurements, have been performed by Gombosi (1986) and Crifo (1987). Figure 4.7 shows the terminal velocities as a function of particle size. The main difference

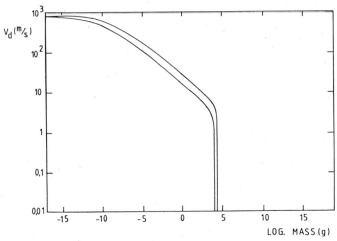

Fig. 4.7. Dust terminal velocities for a dust to gas production ratio $\chi = 2.5$ and for two different dust particle densities $\rho_d = 0.3$ Mg m^{-3} (lower curve) and $\rho_d = \rho_0 - \rho_1 a/(a_1 + a)$ with $\rho_0 = 3$ Mg m^{-3} and $a_1 = 2$ μm (upper curve). (From Crifo, 1987).

in the basic assumption was that Gombosi (1986) used $\chi = 0.3$, whereas Crifo (1987) used $\chi = 2.5$. Both values are not in contradiction with the observations since the largest particles observed were 1 mg and therefore most of the mass can be hidden in the not observed larger particles (see Sect. 4.7). Nevertheless, both results are in good agreement, mainly because the gasdynamics are mostly controlled by the mass load in small grains as noted by both authors. Much larger uncertainties would be introduced by time fluctuations in the gas production rate and by taking into account a spread in the shapes and densities of the grains. The terminal velocity of particles larger than 0.1 mm decreases approximately as $a^{-1/2}$ until the particles become too heavy to be lifted off the nucleus by gas entrainment.

4.5 Dust Coma Structure

4.5.1 Simple Models

The simplest model for the structure of a comet coma is described by a point source with a production rate of Q_d dust particles per unit time. The number density n_d of dust particles at the distance R from the nucleus is given by

$$n_d(R) = \frac{Q_d}{4\pi R^2 v_d},$$ (4.17)

where v_d is the dust speed. The column density is then obtained by integrating along a line of sight, l, at distance $\tilde{\rho}$ from the center,

$$N_d(\tilde{\rho}) = \int_{-\infty}^{\infty} \frac{Q_d}{4\pi R^2 v_d}\, dl = \frac{Q_d}{4\tilde{\rho} v_d},$$ (4.18)

where $R^2 = \bar{\rho}^2 + l^2$. If $N_d(\bar{\rho})$ is multiplied by the scattering efficiency of the dust particles, then the surface brightness distribution of a coma is obtained. Since a coma extends typically out to several times 10^4 km, one can see that the brightness varies by about the same factor from the center – which is not resolved by ground-based observations – to the outer fringes of the coma.

The point source model of the coma has to be modified in order to account for the real situation. Beyond a distance of a few times 1000 km from the nucleus, a distance that can just be resolved by ground-based observations, micrometer-sized particles are increasingly influenced by solar gravitation and radiation pressure forces

$$F = F_{grav} - F_{rad} = (1 - \beta) \frac{4\pi}{3} a^3 \rho_d \, G \, \frac{M_\odot}{r^2}. \tag{4.19}$$

The radiation pressure constant, β, is the ratio of the radiation pressure force F_{rad} over the solar gravity force,

$$\beta = \frac{F_{rad}}{F_{grav}} = \frac{3 F_\odot r_o^2}{4 c \, G \, M_\odot} \cdot \frac{\eta_{pr}}{\rho_d \, a} = \frac{5.78 \times 10^{-4} \, \text{kg m}^{-2} \, \eta_{pr}}{\rho_d \, a}, \tag{4.20}$$

where c is the speed of light and $r_o = 1.5 \times 10^{11}$ m (=1 AU). The constant β does not depend on the distance from the Sun but only on particle properties like radius a, density ρ_d, and the radiation pressure efficiency η_{pr}. The dependence of η_{pr} on particle size and material is shown in Fig. 4.8 (Burns et al., 1979; Hellmich and Schwehm, 1983).

The radiation pressure efficiency reaches a maximum for particle sizes comparable to the wavelength of visible light. For big particles $\eta_{pr} \sim 1$ (for dielectric or only slightly absorbing grains at $a \gtrsim 1$ cm, for absorbing particles at $a \gtrsim 10$ μm), whereas for small particles ($a < 0.1$ μm) it decreases. The decrease is more rapid for nonabsorbing dielectric materials (olivine) than for absorbing materials (magnetite).

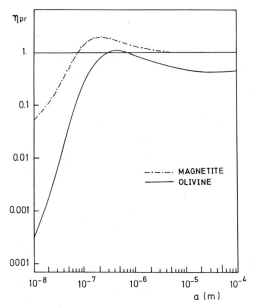

Fig. 4.8. Radiation pressure efficiency η_{pr} for an absorbing (magnetite) and a mostly nonabsorbing dielectric (olivine) particle as a function of grain radius a. (From Hellmich and Schwehm, 1983).

In a first approximation, the dust particle dynamics can be described by the force expressed in (4.19) and the initial conditions, namely the terminal velocity at which dust particles leave the sphere of influence of the comet. In a cometocentric system, the trajectories of micrometer-sized particles are approximately parabolic (see Fig. 4.9). Each parabola has one point tangent to an envelope which is specific to the particle size. This envelope is a paraboloid of revolution with focus in the nucleus. The sunward apex distance R_A is given by

$$R_A = \frac{r^2}{2\,G\,M_\odot} \cdot \frac{v_d^2}{\beta}, \tag{4.21}$$

where r is the heliocentric distance of the comet in AU. Since both v_d and β vary principally with particle properties like mass, density, and type of material, different envelopes are formed for each set of parameters. The variation is not very strong since for large particles ($a > 1\ \mu$m) both v_d^2 and β depend on the cross section of the particle and therefore this dependence cancels for R_A. Because of the low emission speed ($v_d < 100$ m/s) of large particles ($a > 10\ \mu$m), the orbital motion of the comet has to be considered. Especially close to perihelion, the comet position changes significantly during the time interval that the dust particles need to reach the envelope, if they reach it at all. In Fig. 4.10 trajectories of 10 μm and 8 mm particles are shown (Fertig and Schwehm, 1984). The envelope of the 10 μm particles is distorted from the ideal situation, whereas no envelope at all forms for the big, slow ($v_d = 30$ m/s) particles.

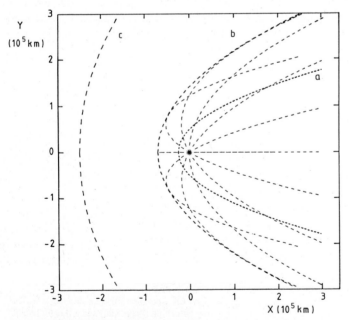

Fig. 4.9. Fountain-like trajectories of dust particles emitted from a comet nucleus. The Sun is to the left. Depending on the initial conditions (emission speed v_d and radiation pressure constant β) the trajectories have different envelopes (a, b, c).

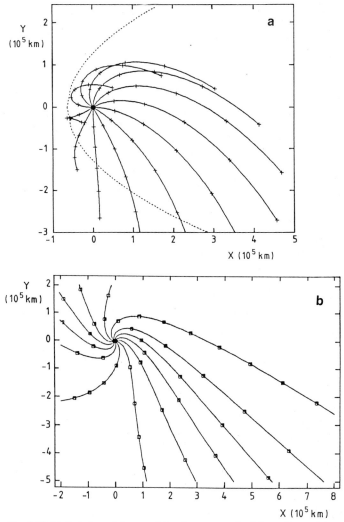

Fig. 4.10. Trajectories of dust particles that were emitted from Comet P/Halley at its perihelion passage (a) 10 μm particles, (b) 0.8 mm particles (Fertig and Schwehm, 1984). The dotted curve shows the ideal envelope.

In this simple model it was assumed that dust particles are emitted isotropically from the nucleus. However, similar envelopes form if emission occurs only from the sunlit side of the nucleus. If the emission is even more asymmetric, only partial envelopes will form. The situation is further complicated by the fact that the nucleus spins about one axis or even shows a more complex rotational behavior (precession, nutation, see Chap. 2, Sect. 2.4.5).

4.5.2 Advanced Models

In the following we discuss some simple cases of the interplay between anisotropic dust emission and nucleus spin. Let us assume there is a single, small, active area on

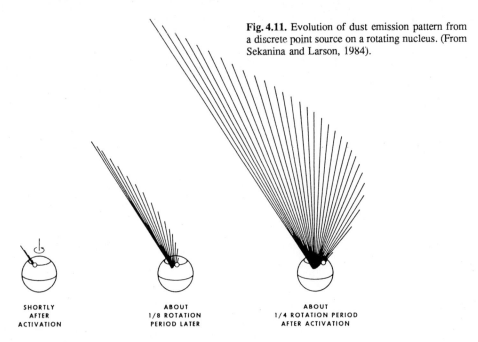

Fig. 4.11. Evolution of dust emission pattern from a discrete point source on a rotating nucleus. (From Sekanina and Larson, 1984).

SHORTLY
AFTER
ACTIVATION

ABOUT
1/8 ROTATION
PERIOD LATER

ABOUT
1/4 ROTATION PERIOD
AFTER ACTIVATION

a spinning nucleus which starts to emit dust for some time in the radial direction at a fixed velocity. Then the particle jet evolves with time in a spiral, like water droplets from a rotating garden hose (see Fig. 4.11). Since not all particles have the same speed because of different sizes and dispersion effects, the particles are spread on a conical surface inside the spiral which represents the maximum emission speed. The picture for all particles approaches that of a conical fan. How such a fan is seen by an external observer depends strongly on the viewing geometry. Figure 4.12 shows three examples of fan images that were computed by Sekanina (1987b). All images have been obtained for the same rotation period of the nucleus and the same emission characteristics, i.e., the same velocity distribution and source position with respect to the spin axis, but for different orientation of the spin axis with respect to the observer and the Sun. The appearance of the fans are quite different just for geometric reasons. In all cases the active dust emitting region was illuminated all the time. If the source is switched off part of the time (on the night side), only a partial fan appears. Figure 4.12c has a spiral-like appearance because the observer is located inside the cone. The edges of the cones are distorted (see Fig. 4.12a) because the dust particles are deflected from the radial direction by solar effects.

Spirals or shell-like structures as seen in Fig. 4.12 are typical for fast rotating nuclei ($P \simeq 0.4$ d). For rotation periods of a few days, radiation pressure effects distort the spiral before it is fully evolved. Figure 4.13 shows a dust jet from a nucleus rotating with a period of 2.2 d (Massonne et al., 1985). In addition to a high dust emission rate from a single active area, low dust activity was assumed from the rest of the sunlit surface. The sunward part of the jet has reached the envelope and is about to be deflected into the tail direction, which is up.

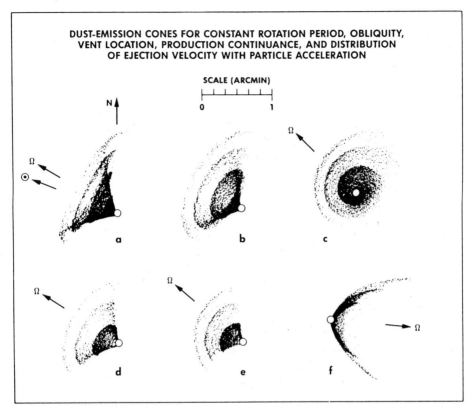

DUST-EMISSION CONES FOR CONSTANT ROTATION PERIOD, OBLIQUITY, VENT LOCATION, PRODUCTION CONTINUANCE, AND DISTRIBUTION OF EJECTION VELOCITY WITH PARTICLE ACCELERATION

Fig. 4.12. Computer-generated images of dust ejecta from a discrete source on a fast rotating nucleus with spin period $P = 0.4$ d. In all cases it has been assumed that there is continuous activity, but the location of the observer changes with respect to the emission cone (Sekanina, 1987b). Directions of the projected spin vector (Ω), north (N), and the Sun ($\odot$), are indicated.

4.5.3 Observations and Measurements

We are now ready to look at some real coma images and to understand the basic features. Visual observations of the dust coma, that were manifested by hand drawings of last-century astronomers, showed manifold structures in the near-nucleus region (see Fig. 4.14). They interpreted these structures as inhomogeneous dust emission patterns. Because of the inferior dynamic range and intensity resolution of photographic plates, less structure has been noted by photographic comet observers; only the application of sophisticated digital image processing techniques (Larson and Sekanina, 1984) revealed structures in the photographed comae (see Fig. 4.15). From the time evolution of these structures, the spin axis and the dust emission pattern on the nucleus can be deduced in principle. Mainly Sekanina and coworkers have used this method with varying success. For example, the analysis of the dust emission of Comet Swift-Tuttle (Sekanina, 1981) is in accordance with all available observational data, whereas the rotational state and the dust emission of Comet Halley is still not understood (see the conflicting results presented by Sekanina and Larson, 1986; Sekanina, 1987a; and Chap. 2; Sect. 2.4.5).

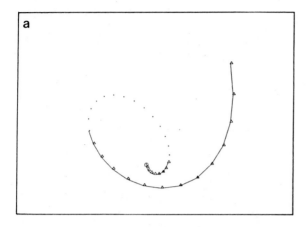

a

Fig. 4.13. Dust jet simulation for Comet P/Halley. (a) Center of jet. Triangles connected by solid lines indicate particles that were emitted from the illuminated hemisphere. (b) Computer-generated image of the same jet as in (a) but with a background dust emission from all the illuminated surface. The white spot marks the position of the nucleus. (From Massonne et al., 1985).

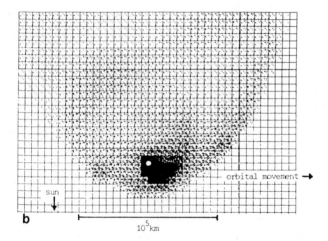

b

10^5 km

Modern Charge-Coupled Device (CCD) observations of comets with their superior dynamic range and resolution allow the easy and quantitative observation of fine structure in the comae of comets. Figure 4.16 shows an image of P/Halley taken in December, 1985. The general brightness decrease with distance from the center was removed from the image by normalizing the intensities to a $\tilde{\rho}^{-\alpha}$ dependence ($\tilde{\rho}$ being the projected distance from the nucleus). For an image only in continuum light, $\alpha = 1$ is the appropriate value (see Eq. 4.18). Because a broad-band filter has been used for this image, $\alpha = 0.83$ takes into account the contribution from the gas. The dark spot and the brighter small ring around the nucleus position are an effect of the normalization procedure on the seeing disk. Two jets are emitted in the sunward (left) hemisphere and bend at some distance into the opposite, tail direction.

Fig. 4.14. Drawings of Comet Swift-Tuttle by Winnecke in 1862. The visual observations were made with the 30 cm refractor at Pulkovo. The direction to the Sun is indicated by the dotted circle. (From Sekanina, 1981).

Fig. 4.15. Digitally processed images of Comet P/Halley from photographs taken on 12 to 15 May, 1910. The dates are in UT. The Sun is at the top and each frame is 200000 km on a side. (From Larson and Sekanina, 1985).

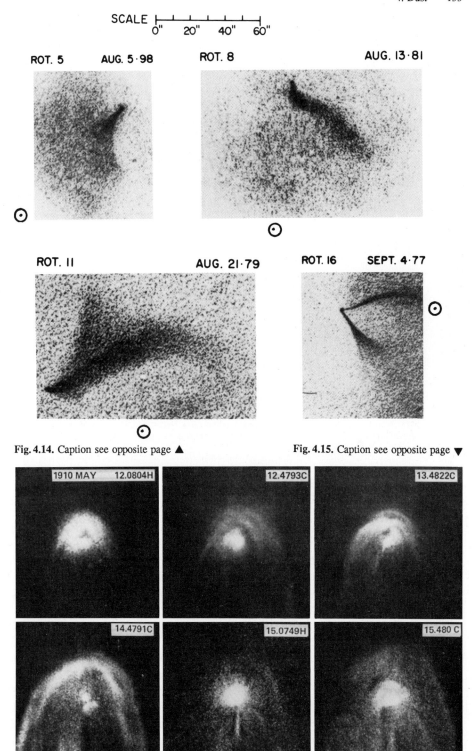

SCALE

0"   20"   40"   60"

ROT. 5   AUG. 5·98

ROT. 8   AUG. 13·81

ROT. 11   AUG. 21·79

ROT. 16   SEPT. 4·77

Fig. 4.14. Caption see opposite page ▲

Fig. 4.15. Caption see opposite page ▼

1910 MAY   12.0804H

12.4793C

13.4822C

14.4791C

15.0749H

15.480 C

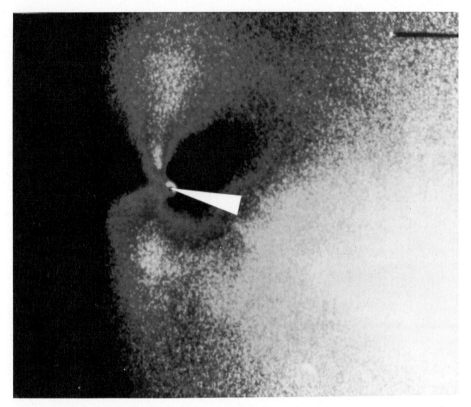

Fig. 4.16. Processed CCD image of Comet P/Halley taken on 7 December, 1985, by Birkle and Hopp. The nucleus position is indicated by the white wedge. The Sun is to the left. The frame is 60000 km on a side. (From Grün et al., 1987a).

A full analysis of the images recorded during the 1985 – 1986 apparition of Comet Halley, together with the observations taken from the Vega 1, Vega 2, and the Giotto spacecraft, should eventually allow disentanglement of the complex rotational state of Halley's nucleus and determination of the distribution of active (dust emitting) areas on it. In order to achieve this goal, a complete sequence of images has to be analyzed spanning several rotational periods (P = 2.2 d or 7.5 d; see Sekanina, 1987a) with a time resolution that includes several images of each jet during its active period (see also Chap. 2, Sect. 2.4.5).

The ground-based observations are complemented by the three spacecraft observations. Images taken by the Vega 1, Vega 2, and Giotto spacecraft show the activity distribution on the nucleus at three distinct times (1986 March 7.3, 9.3, and 14.0 UT; see Fig. 2.4).

Figure 4.17 shows images of Comet Halley taken on March 13, 1986, by the camera on the Giotto spacecraft (Keller and Thomas, 1988) and by a ground-based telescope. Both images were processed removing the strong radial intensity variation. Since the images were taken from different positions in space, there is no one-to-one relationship between the jets on them. Cosmovici et al. (1986) suggest that

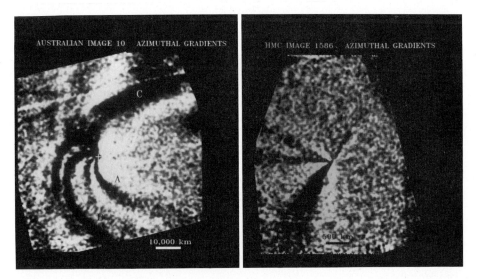

Fig. 4.17. Images of the coma of Comet P/Halley taken 3.5 h apart on 13 March, 1986. (*left*) From the ground by Cosmovici, frame size 10^5 km and (*right*) from the Giotto camera, frame size 6000 km. The Sun is to the left. (From Keller and Thomas, 1988).

the southern jet (A) corresponds to the strongest emission on the Giotto image. Comparisons like this between long term ground-based observations together with highest resolution space observations either from spacecraft or the Hubble Space Telescope will help us to understand the development of coma structures and will finally resolve open questions concerning the rotational state of the nucleus and the distribution of active areas on it.

Three spacecraft probed the coma of Comet Halley with dust detectors on board. The spacecraft penetrated down to a distance of 8890 km, 8030 km, and 596 km for Vega 1, 2, and Giotto, respectively. Figure 4.18 shows impact rate profiles measured by the SP-1 detectors on board of Vega (Vaisberg et al., 1987b). Inside an outer boundary (the outermost envelope between 1 and 2×10^5 km), the impact rate increases roughly with the inverse square of the cometocentric distance R. Superimposed on this slow increase are distinct steps that are interpreted as envelopes of particles of different sizes and materials. We have listed all envelopes recorded by the *in situ* dust experiments in Table 4.1. First and last recorded particles have been included in the table. Closer to the nucleus some of the envelope positions may be confused with timely and spatially varying emission rates at the nucleus.

Figure 4.19 shows the position of the envelopes along the trajectories of the three Halley spacecraft as a function of the parameter v_d^2/β, which is proportional to the envelope distance (see Eq. 4.21). The distances differ for the inward and outward legs of the trajectories. For large emission speeds v_d and small values of v_d^2/β, they reach the values of the fountain model (solid lines). Deviations from the fountain model are only significant for low emission speeds on the inward leg of the trajectories (Fertig and Schwehm, 1984).

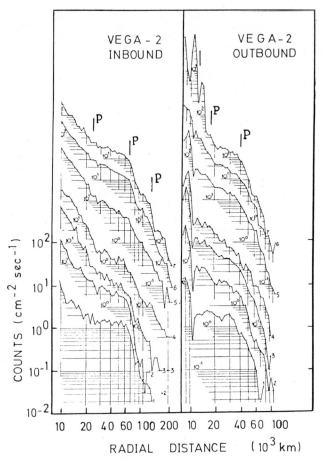

Fig. 4.18. Radial distribution of dust impact rates measured by the Vega 2 spacecraft, both inbound and outbound. The different profiles in each panel correspond to different mass channels of the SP-1 instrument. The positions of different dust envelopes are marked. (From Vaisberg et al., 1987b).

With the help of Fig. 4.19 we now assign values of v_d^2/β to the envelope distances of Table 4.1 and plot them as function of the particle mass (see Fig. 4.20). For comparison, we also show v_d^2/β dependencies calculated from speed (Gombosi, 1986) and β values (Hellmich and Schwehm, 1983) for olivine and magnetite. The latter curves show strong minima between 10^{-14} g and 10^{-13} g. Most of the spacecraft data are compatible with the theoretical curves in that mass region. Only the first particle detected by the DUCMA experiment (at $v_d^2/\beta \simeq 8.5\ 10^6\,\mathrm{m}^2\,\mathrm{s}^{-2}$) requires a factor of 3 higher emission speed or a factor of 8 lower β than the model values. Neither possibility is easy to accept. This particle may have been a stray interplanetary dust particle. The envelope distances for the 10^{-16} g particles point to somewhat absorbing particles, intermediate between olivine and magnetite.

Particles of mass $m \gtrsim 10^{-10}$ g have been observed only well inside the expected envelope distances for both absorbing and dielectric particles. Therefore, structures in the impact rate profiles for these big particles are probably caused by heterogeneous

Table 4.1. Spacecraft observations of potential dust envelopes

No.	Experiment	Distance (10^3 km)	Leg of trajectory	Particle mass (g)	Ref.
Vega 1					
1	SP 1	260*	in	10^{-16}	a
2	SP 1	70	in	10^{-12}	a
3	SP 2	280*	in	no mass given	b
4	SP 2	138	in	10^{-14} to 10^{-10}	c
5	SP 2	64	in	10^{-13}	c
6	SP 2	33	out	10^{-12} to 10^{-8}	c
7	SP 2	64	out	10^{-9}	c
8	DUCMA	637*	in	10^{-13}	d
9	DUCMA	120	in	10^{-13}	e
10	DUCMA	80	out	10^{-13}	e
Vega 2					
11	SP1	320*	in	10^{-16}	a
12	SP1	130	in	10^{-16} to 10^{-14}	f
13	SP1	80	in	10^{-12}	f
14	SP1	24	in	10^{-11}	f
15	SP1	19	out	10^{-11}	f
16	SP1	43	out	10^{-16} to 10^{-14}	f
17	SP2	133	in	10^{-13} to 10^{-10}	c
18	SP2	105	in	10^{-12} to 10^{-10}	c
19	SP2	27	out	10^{-14} to 10^{-12}	c
20	DUCMA	255*	in	10^{-13}	d
21	DUCMA	70	in	10^{-13}	e
22	DUCMA	50	out	10^{-13}	e
Giotto					
23	DID-1	253*	in	10^{-11}	g
24	DID-4	287*	in	$4 \; 10^{-9}$	g
25	DID-4	202*	out	$4 \; 10^{-9}$	g

*First or last particle detected
[a] Vaisberg et al. (1986b)
[b] Mazets et al. (1986a)
[c] Mazets et al. (1986b)
[d] Simpson et al. (1986b)
[e] Simpson et al. (1987)
[f] Vaisberg et al. (1987b)
[g] McDonnell et al. (1986)

emissions from the nucleus. In conclusion, it can be said that the reported envelopes are mostly compatible with absorbing particles. Only some observations of envelopes for particles in the mass range 10^{-14} g to 10^{-12} g can be caused by weakly absorbing dielectric particles.

Particles of different sizes have different envelope distances; the mass distribution of dust particles in the outer coma reflects this situation. In Fig. 4.21, a theoretical mass distribution (Massonne, 1985) is compared with one that has been measured (Mazets et al., 1987). Both show remarkable similarities which demonstrates the qualitative understanding of dust coma dynamics.

Inside the envelopes, the impact rate increases roughly as R^{-2}, as would be expected for a radial dust outflow from the nucleus (see Eq. 4.17). Deviations from

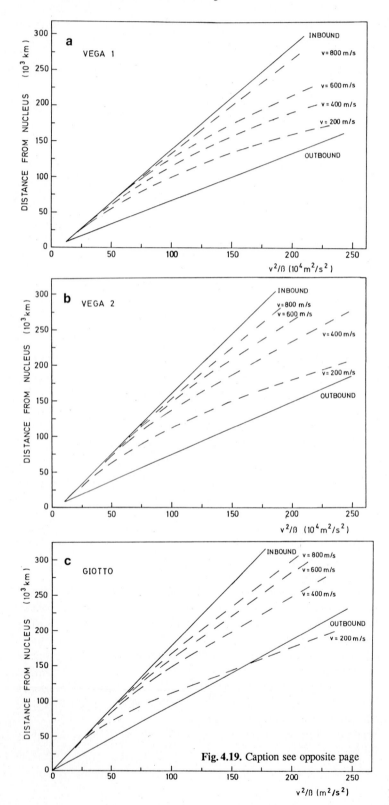

Fig. 4.19. Caption see opposite page

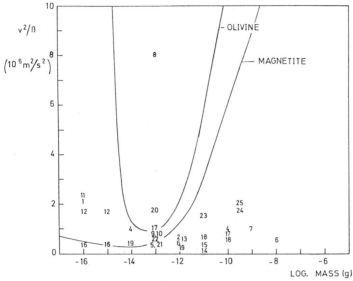

Fig. 4.20. Values of the parameter v^2/β calculated for olivine and magnetite grains as a function of grain mass. The numbers refer to spacecraft measurements shown in Table 4.1. (From Grün et al., 1987a).

the R^{-2} dependence of the impact rate up to a factor of 5 (Mazets et al., 1987; McDonnell et al., 1987; Simpson et al., 1987; Vaisberg et al., 1987b) are caused by spatial fluctuations of the dust production from the nucleus. There is direct evidence that the variations of the spatial dust density diffuse with the distance from the nucleus, i.e., the variation of the dust production from active and inactive areas on the nucleus may be much stronger than what has been observed by the dust detectors at some distance.

As has been discussed by Massonne (1985), during a major part of the approach trajectory of Giotto (from about 20 minutes to one minute before closest approach), only small particles originating from the sunlit hemisphere of the nucleus can reach the spacecraft on their reflected, tailward branch of the trajectory (i.e., particles that have reached the envelope distance and now are pushed back by solar radiation pressure). Only particles coming from the direction of the night side hemisphere of the nucleus can reach the spacecraft on direct trajectories.

Direct evidence (Keller et al., 1986a; see also Chap. 2., Sect. 2.4.8) indicates that dust emission is restricted to the sunlit hemisphere of the nucleus and perhaps a small portion of the surface near the evening terminator (Smith et al., 1986). This observation is supported by the calculated low surface temperatures and the highly reduced sublimation rate on the night side (Weissman and Kieffer, 1981). During that portion of the trajectory, where only reflected particles could reach Giotto, the impact rate should be approximately constant, independent of the distance. Surprisingly,

---

Fig. 4.19. Envelope distances as a function of the parameter v^2/β for the three close flyby Halley missions. The solid lines represent the envelope distances for the simple fountain model. (a) Vega 1, (b) Vega 2, (c) Giotto. (From Grün et al., 1987a).

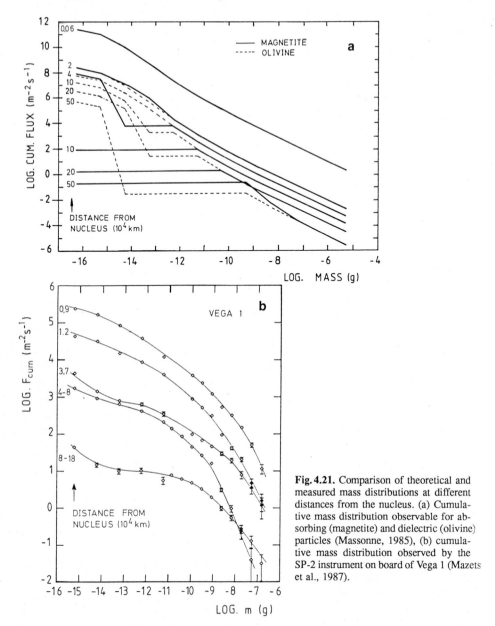

Fig. 4.21. Comparison of theoretical and measured mass distributions at different distances from the nucleus. (a) Cumulative mass distribution observable for absorbing (magnetite) and dielectric (olivine) particles (Massonne, 1985), (b) cumulative mass distribution observed by the SP-2 instrument on board of Vega 1 (Mazets et al., 1987).

even during this time the impact rate was increasing approximately as R^{-2} (see Fig. 4.22; Maas et al., 1989; McDonnell et al., 1987), indicating that many particles arrived from the direction of the night side of the nucleus.

This does not necessarily mean that these particles are emitted from the night side but that they may be carried by lateral winds, driven by the overpressure of gases from active areas on the nucleus (Wallis and Macpherson, 1981; Keller et al., 1987a; see also Chap. 2, Sect. 2.5.5). Because of these lateral winds, the dust

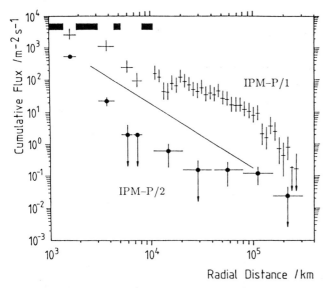

Fig. 4.22. Radial flux profile measured by the DIDSY-IPM-P sensor in two mass channels ($> 10^{-11}$ g, top and $> 10^{-9}$ g, bottom) during the inbound leg of the Giotto trajectory. The profiles are compared to an R^{-2} dependence. (From Maas et al., 1989).

outflow, at least of small grains with masses less than 10^{-10} g, is less anisotropic at distances exceeding several nuclear radii than on the surface.

Within about 100 s of closest approach, the impact rates on both Vega spacecraft varied systematically by less than a factor of 2, consistent with the different flyby distances. Here the time variation of the impact rates is mostly caused by spatial fluctuations of the dust emission along the ground track on the nucleus surface (Fig. 4.23). Strong, short-term variations of up to a factor of 3 are seen in the data (Mazets et al., 1987).

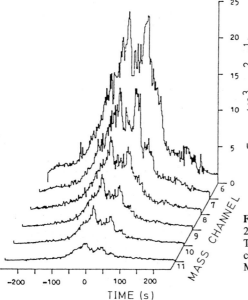

Fig. 4.23. Dust flux profile measured by the SP-2 instrument during closest approach of Vega 2. The different profiles refer to different mass channels from 6×10^{-13} to 10^{-10} g. (From Mazets et al., 1987).

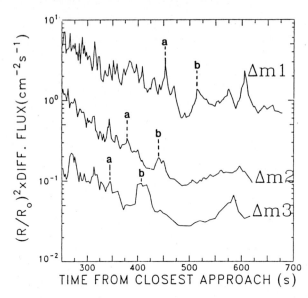

Fig. 4.24. Normalized $(R/R_0)^2$ differential fluxes measured by the DUCMA instrument on the Vega 1 spacecraft. Two jets are identified in three mass intervals. (From Simpson et al., 1987).

Some of the jets show a mass dispersion (Simpson et al., 1987; Vaisberg et al., 1987b) which is caused by the different emission speeds of particles of different masses (Fig. 4.24) in combination with the nucleus rotation. A quantitative analysis of this dispersion, however, awaits the solution of the nuclear spin problem.

Many of the detailed features of the new *in situ* dust measurements in the coma are not yet fully understood. Further modeling and ground-based observations are needed to complete the model of the comet dust coma.

4.6 Dust Tails

4.6.1 General Features

After dust is released from the sunlit hemisphere of the nucleus, entrained by the gas in the general solar direction and then deflected at the coma envelope, it ends up in the comet's dust tail before it is dispersed into interplanetary space. Because of the action of radiation pressure, particles of different sizes and emission times are separated in space and time inside the tail.

In order to understand the formation of dust tails, the concepts of synchrones and syndynes (or syndynames) were introduced by Bredikhin in 1903. Let us first consider a nucleus constantly releasing particles with a certain radiation pressure parameter β and with zero speed relative to the moving nucleus. The particles are pushed back by the radiation pressure and will form a line called "syndyne" (see Fig. 4.25). The syndyne is the locus of dust particles of the same β at a certain observation time t_0, emitted from the nucleus with zero relative velocity at different times $t_0 - t$, where t is the emission time of the particle relative to the time of observation. In a syndyne, the relative emission time, t, of the particles increases with distance from the nucleus. Because the particles experience a central force, the syndyne is located in the plane of the comet orbit. Moreover, syndynes leave the comet head in the antisolar direction. Generally, the initial velocity of the dust particles is not zero but has a certain value v_d. Thus the resulting syndyne tail will have a finite cross section of approximate width $2v_d t$, shown as the shaded portion in Fig. 4.25. In a realistic situation we will have particles of different sizes and, perhaps, different compositions, i.e., particles with different values of β that give rise to different syndynes.

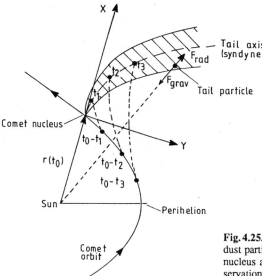

Fig. 4.25. Formation of a dust tail. Trajectories of dust particles are shown that were released from the nucleus at different times t_1, t_2, and t_3 prior to observation time t_o.

Instead of arranging the particles according to their β values, one can also order them according to the release time. This is particularly useful if variations in the dust production rate are considered. Correspondingly, a "synchrone" is defined as the locus of the particles of different β emitted at the same time, i.e., consisting of particles of the same relative emission time t. Synchrones do not leave the comet in the antisolar direction but lag behind it in the opposite sense of the comet orbital motion by an angle dependent on the synchrone's age. Again, particles will actually spread around a given synchrone because their initial velocity is different from zero. Figure 4.26 shows a mesh of syndynes and synchrones, as computed by Beisser and Boehnhardt (1987), superimposed on an image of Comet Halley.

Dust tails must always be outside of the comet orbit because of the repulsive force of the radiation pressure. Nevertheless, there are sometimes situations where the positions of comet, Sun, and Earth are such that part of the comet tail appears to point toward the Sun. The sunward part of the tail is then called an anomalous tail or antitail. A famous case is the antitail of Comet Arend-Roland (1957 III) that was observed several days around April 28, 1957 (see Fig. 4.27). At that time the Earth passed through the orbital plane of the comet. Because the dust in the tail is concentrated in the orbital plane of the comet, the conditions for observing an antitail were favorable. The situation is explained in Fig. 4.28, where synchrones and

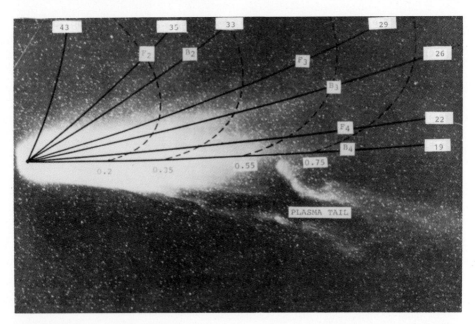

Fig. 4.26. Mesh of synchrones (solid lines, the numbers are the age in days) and syndynes (dashed lines, the number at the bottom are β values) superimposed on an image of comet P/Halley, taken on 10 March, 1986, from the European Southern Observatory. Several faint (F) and bright (B) streamers are indicated. (From Beisser and Boehnhardt, 1987).

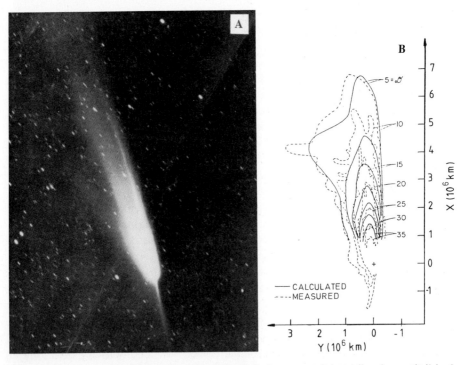

Fig. 4.27. Comet Arend-Roland on 27 April, 1957, displaying a normal dust tail and an antitail in the sunward direction. (a) Photograph taken by H. L. Giclas (1974), (b) measured and computed isophotes. (From Finson and Probstein, 1968b).

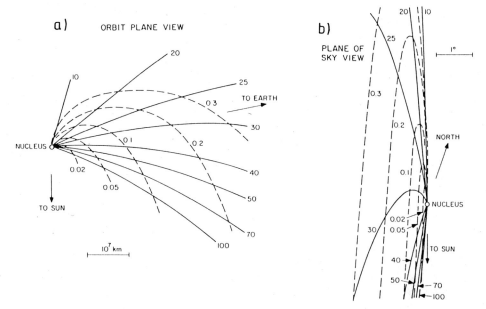

Fig. 4.28. Mesh of synchrones (solid lines with ages in days) and syndynes (dashed lines with β values) for Comet Arend-Roland on 28 April, 1957, (a) in the orbit plane, (b) projected onto the plane of the sky. (From Sekanina, 1976b).

syndynes of Comet Arend-Roland are shown in its orbital plane and in projection on the sky. The diagrams are for a date about three weeks after the comet passed perihelion and three days after the Earth passed through the comet's orbital plane. The "normal" tail, pointing away from the Sun, is composed of small dust particles (large β) that left the comet nucleus recently. On the other hand, the antitail is composed of heavy dust particles (small β) that left the nucleus about two or three months earlier. Because of their small β value, they are still in the vicinity of the nucleus. Smaller particles in the antitail have moved far away from the nucleus so that they are no longer visible. Sekanina (1976b) has estimated diameters of the order of 0.1 to 1 mm for the large particles in the antitail.

When observed edge-on (see Fig. 4.27), the antitail appears as a very thin spike that might suggest very small emission velocities of the heavy particles. However, this is not a necessary condition since particles of the antitail, ejected when the comet was on the opposite side of the Sun, will be approaching the orbital plane of the comet again (Kimura and Liu, 1977). A sharp antitail was also observed from space in Comet Kohoutek (1973 XII), shortly after perihelion (Richter and Keller, 1988). Because of favorable geometric observing conditions, this antitail remained visible for more than one month but became broad and diffuse as the Earth moved away from the comet's orbital plane. For reasons of perspective, antitails are much more likely to be observed after perihelion. P/Halley also displayed an antitail after its perihelion passage (see Brandt and Niedner, 1987).

With the above described "mechanical theory" and under some simplifying assumptions, it is possible to derive properties of the dust size distribution from the observed brightness distribution of cometary dust tails (Finson and Probstein, 1968a,b; Sekanina and Miller, 1973). As we have seen, the relative emission time, t, and the ratio of radiative to gravitational acceleration, β, are the fundamental parameters of the mechanical theory. If the radiation pressure efficiency, η_{pr}, is assumed to be constant, β can be used as a size parameter, and the light received from a dust particle can be expressed as a function of β. Even in the realistic situation of initial velocities of the dust particles different from zero, t and β remain the only parameters of the problem.

The computation of the density distribution of dust particles in the tail (Finson and Probstein, 1968a) involves a double integration of dust particles characterized by different values of β and emission times $t_o - t$. Assuming a dust production rate, Q_d, depending on emission time, $t_o - t$, i.e., $Q_d(t_o - t)$, the number of particles emitted during the time interval from t to $t + dt$ in the size range from $\rho_d a$ to $\rho_d a + d(\rho_d a)$ is given by

$$Q_d(t_o - t)dt\, f(\rho_d a)\, d(\rho_d a) = -C\, Q_d(t_o - t)dt\, f(\beta)\, \beta^{-2} d\beta, \tag{4.22}$$

where $f(\rho_d a)$ is the size distribution function of dust particles and $C = 5.78 \times 10^{-4}\, \eta_{pr}$ is obtained from (4.20). Note that $\rho_d a$ has been used instead of a in (4.22) (which is equivalent, provided ρ_d is constant) to allow introduction of the parameter β by means of (4.22). The scattering cross section of a particle is proportional to $(\rho_d a)^2 \eta_{sc}$, i.e., to $\beta^{-2} \eta_{sc}$, where η_{sc} is the scattering efficiency. The intensity $I(\beta, t)$ of light scattered by the particular set of dust particles given by (4.22) is obtained from

$$I(\beta, t) = C\, Q_d(t_o - t)\, dt\, \beta^{-4}\, \eta_{sc}\, f(\beta)\, d\beta. \tag{4.23}$$

Because of the finite emission speed v_d of the particles, this intensity is distributed over a shell with radius $v_d t$. In order to obtain the surface brightness along a line of sight, the intensity (Eq. 4.23) has to be modified by a factor $2dA/[dA_\perp\, 4\pi\, (v_d t)^2]$, where dA and $dA_\perp$ are the differential surface of this shell and its projection perpendicular to the line of sight, respectively. Integrating the modified (4.23) for particles characterized by different values of β and $(t_o - t)$, one finally obtains the theoretical photometric profile for the dust tail of the comet. The solution of the problem involves three functional parameters: The dust production rate $Q_d(t)$, the size distribution of dust particles $f(\rho_d a)\, d(\rho_d a)$, and the emission velocity of dust particles $v_d(\beta, t; t_o)$ with t_o as the observation time. A trial and error procedure is used to solve the problem where different values of these three functional parameters are tested, and those leading to the best agreement with the observed photometric profile are finally chosen. Figure 4.27b shows measured isophotes of Comet Arend-Roland, together with theoretical isophotes as computed by Finson and Probstein (1968b).

It is important to point out some limitations to the applicability of the Finson and Probstein dynamical model: First, the factors η_{sc} and η_{pr}, both involved in (4.23), are not constant but actually depend on the particle's size and nature. This is an important condition to be considered because there is the possibility that dielectric and absorbing particles are both present. Second, the density ρ_d enters into the calculations as a constant. Again, this condition breaks down if particles of different nature are present. Third, the variable change from β to $\rho_d a$ introduces an ambiguity, since a value of β generally has two possible values of $\rho_d a$ (see Fig. 4.8). Fourth, it is assumed that the dust emission is isotropic. Since we know that dust emission is concentrated on the sunlit hemisphere of the nucleus, some modifications to the results are necessary (Fulle, 1989).

Similar distribution functions of particle sizes are obtained for Comets Arend-Roland and Bennett (1970 II) (Hanner, 1980; Sekanina and Miller, 1973). Finson and Probstein (1968b) derived for the optically important dust particles in Comet Arend-Roland a diameter of 5.6 μm, if a density of $\rho_d = 1\,\mathrm{Mg\ m^{-3}}$ is assumed. Once the dust production rate and the terminal velocities of the particles are found, it is possible to estimate the gas production rate from a model calculation of the dust acceleration by the gas flow. Finson and Probstein (1968b) found that the mass production ratio of dust to gas for Comet Arend-Roland was close to one at the time of observation.

Quite different results for the grain properties were found by Sekanina (1973) when he applied the mechanical theory to the tails of the nearly parabolic Comets Baade (1955 VI) and Haro-Chavira (1956 I), both with perihelion distances near 4 AU. The theoretical synchrones fitted by Sekanina to the observed tails suggest a large dynamical age for the grains. He derived release times ranging between 1500 and 200 d before perihelion for the grains of Comet Baade and between 2000 and 300 d for those of Comet Haro-Chavira, corresponding to heliocentric distances between 5 and 15 AU. The fact that the particles have not moved far away from the comet indicates low acceleration (small β), which corresponds to grain diameters larger than 100 μm. These unusually large grain sizes, as compared to those derived for comets observed closer to the Sun, led Sekanina to the conclusion that such grains would have to be mainly water ice or clathrate hydrates. The large heliocentric distances at which grains were released also suggest that the grains were entrained by the sublimation of another, more volatile substance than water.

4.6.2 Fine Structures

Dust tail inhomogeneities called streamers (see Fig. 4.29) are straight or slightly curved bands converging to the nucleus but following directions that depart perceptibly from the radius vector. These streamers are generally compatible with individual synchrones. In agreement with this interpretation, Sekanina and Farrell (1978) conclude that the streamers displayed by Comet West (1976 VI) can be explained as being produced by particles of different sizes expelled simultaneously from the nucleus (synchrones) that may be identified with outburst events. The presence of streamers in dust tails is therefore an indication of a varying dust production rate of the nucleus. The variation of the dust production rate with the presumed spin period of the nucleus ($\sim 2.2\,\mathrm{d}$) has been observed for P/Halley (Lamy, 1985; Beisser and Boehnhardt, 1987; see Fig. 4.29) after its perihelion passage. For a period of about two months, Comet Halley displayed well separated multiple streamers with centers aligned with synchrones separated by multiples of 2.2 d.

Another type of tail inhomogeneity are striae. They are characterized as a series of parallel, narrow bands displayed at large distances from the nucleus. Unlike streamers, striae do not start from the nucleus, and their extensions in the direction of

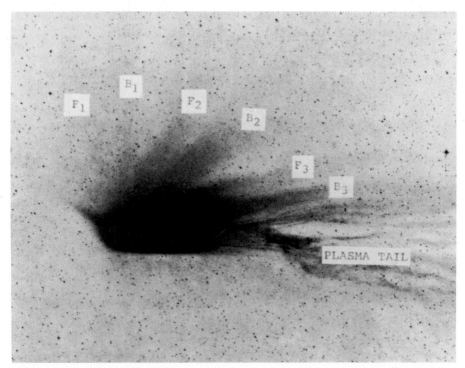

Fig. 4.29. Photograph of Comet P/Halley obtained by K. S. Russell on 22 February, 1986. The field is about $2° \times 2°$; north is up and east to the left. Several nearby straight streamers (see also Fig. 4.26) extending in various directions from the nucleus are superimposed on the extremely broad tail, fanning out from the direction of the plasma tail (to the west) through the north and terminating in a sharply edged antitail that points to the northeast. (From Sekanina and Larson, 1986).

the nucleus usually intersect the comet – Sun line on the sunward side. The formation of striae is still an unsettled problem. Sekanina and Farrell (1980) have suggested that striae are formed when large particles, ejected from the nucleus simultaneously and subjected to the same repulsive force, fragment at the same time. If this is true, striae can be treated as synchrones of the fragments, starting in the fragmentation places rather than in the comet nucleus. The difficulty arises in explaining the simultaneous destruction of dust particles. Lamy and Koutchmy (1979) have suggested that striae are made up of charged dust particles driven by the interplanetary magnetic field carried along with the solar wind. So far, it is not possible to determine whether the striae formation can be solved in a purely mechanical way, whether electromagnetic forces play an essential role or whether a combination of both effects gives rise to the observed phenomenon.

4.6.3 Dust Trails

A quite new type of dust tail has been detected by the Infrared Astronomical Satellite (IRAS, Davies et al., 1984). Inspection of the sky flux maps mainly at wavelengths of 25 and 60 μm revealed narrow trails of dust extending up to 48° of the sky (Sykes

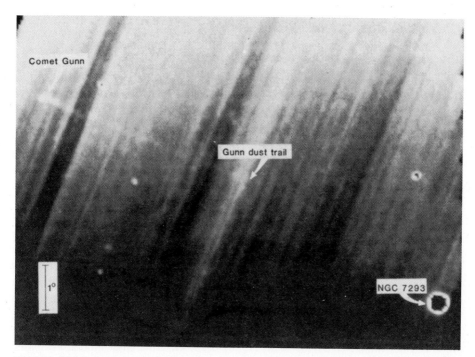

Fig. 4.30. Dust trail along the orbit of Comet Gunn detected by the infrared satellite IRAS at 60 μm wavelength. (From Sykes et al., 1986).

et al., 1986). Several of the brighter trails were found to coincide with the projection on the sky of comet orbits (see Fig. 4.30). On the other hand, at least five of the more than 100 observed "dust trails" are not associated with any known periodic comet (Sykes et al., 1986). The bright dust trails of Comets Tempel 2, Encke, and Gunn exhibited material both in front of and behind the comet's orbital position by several degrees in mean anomaly. The narrowness of the dust trails indicates that large particles, that are influenced very little by radiation pressure, are ejected at low velocities (a few times 1 m/s) from the comet nucleus. Dynamical analysis of the dust trail of Comet Tempel 2 by Eaton et al. (1984) indicates that the emission of dust must have occurred at least 1500 d (about one orbital period) prior to the IRAS observations. Calculations of the time that it would take for material to move ahead of the orbital positions of comets Tempel 2 and Gunn suggest an age for the dust trails of hundreds of years (Sykes et al., 1986). Thus, dust trails would be the product of numerous emission events, superimposed on each other over many tens of orbits. Although no trail could yet be linked to any known meteor stream, dust trails are believed to provide the generic link between meteor streams and comets.

4.7 Mass Distribution

The most readily available information on comet dust stems from ground-based optical observations of the dust coma and tail. A large number of comets have been observed in this way. Since it is light scattered by the dust particles that is observed, these data relate to dust particles with maximum scattering cross sections, i.e., 1 to 100 μm particles. For infrared observations, the situation is somewhat improved with respect to larger particles. In addition, thermal emission depends not on detailed knowledge of the optical properties of the materials involved but only on the gross albedo. However, the mass distribution of cometary dust particles was best determined by the *in situ* detectors on board of the spacecraft. The statistically most significant data are the fluence data because they include all measurements along the trajectory. This averaging of the particle flux along the trajectory smooths out dynamical effects on the mass distribution in the distant parts of the trajectory (Massonne and Grün, 1986) as well as the dispersive effect of nucleus rotation near closest approach (Vaisberg et al., 1986a). Figure 4.31 shows the spatial dust densities at a distance of 1000 km from the nucleus that have been calculated from the measured fluences (Vaisberg et al., 1987b; Mazets et al., 1987; McDonnell et al., 1987), following the procedure described by Lamy et al. (1987). For comparison, the pre-encounter model of Divine et al. (1986) is shown. This latter model was based on ground-based observations. The spacecraft measurements are fairly consistent in slope, but they vary in absolute intensity by a factor of 10. This is caused by the observation times that differ by several days (Vega 1: March 6, Vega 2: March 9, and Giotto: March 14). During this time, the total comet brightness varied accordingly. It demonstrates the strong variability of the dust emission by P/Halley.

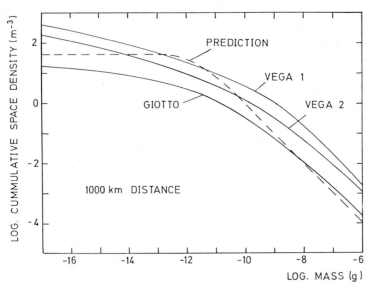

Fig. 4.31. Spatial dust density at the time of the spacecraft encounters given at 1000 km distance from the nucleus. The prediction from Divine et al. (1986) refers to absorbing particles.

Cometary dust grains of all sizes from 10^{-17} g to 10^{-3} g were observed by the *in situ* detectors. The number of particles bigger than the limiting mass m_d is steadily increasing with decreasing m_d. Remote, ground-based observations, on which the pre-encounter models were based, did not allow the detection of particles smaller than about 10^{-13} g. Tiny particles with mass 3×10^{-17} g, or just larger, were detected by the *in situ* detectors in large quantities. These particles are of sizes that are comparable to interstellar dust.

The slope of the cumulative size distribution in the mass range 10^{-9} g to 10^{-5} g is about -0.75 to -1.0. Lamy et al. (1987) obtain a best fit polynomial to the Vega 2 flux curve presented in the form of a cumulative spatial density $n_d(m_d)$ at a reference distance of 10^3 km from the nucleus

$$\log n_d(m_d) = \sum_{i=0}^{3} A_i (\log m_d)^i, \tag{4.24}$$

where n_d is in m^{-3} and m_d is in g. The coefficients A_i are given in Table 4.2. This expression is applicable in the mass range from 10^{-17} to 10^{-5} g. The comparison of the *in situ* measurements with the predicted mass distributions shows some significant differences. Submicrometer-sized particles have not been predicted from optical observations because they did not contribute enough to the scattered light to be observable. The same is true for large particles with masses larger than about 10^{-10} g.

Table 4.2. Polynomial coefficients A_i for the Vega 2 dust distribution

i	A_i
0	-12.04042413
1	-2.063823315
2	-0.1058154891
3	-0.001975492189

Only in the range from 10^{-13} g to 10^{-10} g are predictions and *in situ* measurements in acceptable agreement. This is the range for which particles contribute most to the optically scattered light.

Impacts of very large particles with masses larger than 1 mg were recorded only by the Giotto spacecraft. Because of the very large sensitive area of about 2 m^2 for detecting large particles and because of the close proximity of Giotto to the nucleus, the largest single particles with a mass of about 1 mg were recorded by the DIDSY experiment (McDonnell et al., 1986b). Several seconds before closest approach, Giotto was hit by a very large particle (0.1 to 1 g) that caused a strong nutation of the spacecraft. As a consequence, the data transmission to Earth severely degraded for at least 20 seconds around closest approach and was only fully restored after about 30 minutes. Some data from the scientific experiments were lost during that time. Nevertheless the radio signal, though faint, could be received all the time during the Giotto encounter. Edenhofer et al. (1986) analyzed this signal and found from the Doppler shifted frequency a total velocity decrease of $\Delta v = 0.232$ m/s during the encounter. This velocity decrease occurred in small but distinct steps, indicating either individual hits of large particles or the passage through narrowly confined jets. The largest step of deceleration occurred 7 s prior to closest approach. The simultaneously occurring strong nutation of the spacecraft indicates an off-center hit on the dust shield by a large particle rather than by a multitude of small particles, whose torque on the spacecraft would have canceled out. If one assumes that this particle penetrated the front shield by a large margin and most of the ejecta could only escape sideways between the front and the rear shield, then no momentum enhancement occurs, and one can calculate the mass of the particle

$$m_d = m_{sc} \, \Delta v / v_{sc}. \tag{4.25}$$

The spacecraft mass was $m_{sc} = 573.7$ kg and its velocity relative to the comet $v_{sc} = 68.37$ km/s. Taking a value of $\Delta v = 0.04$ m/s for the event at -7 s to closest approach (Edenhofer et al., 1986), a particle mass of about 0.4 g is obtained. Unfortunately, there is only very limited overlap between the time period of maximum deceleration and direct data from the DIDSY experiment (McDonnell et al., 1986b). The slope of the large particle mass distribution ($m_d > 10^{-5}$ g) at about -0.5 is very flat (McDonnell et al., 1986b). Such a slope for the mass distribution is also required in order to connect the intermediate size particle flux with the flux of big particles that caused the deceleration of the Giotto spacecraft. This flat slope of the cumulative mass distribution implies that most mass is carried by the largest particles. In order to estimate the total dust output from the comet, it is important to know the size of the largest particle emitted from the nucleus. The total amount of the dust output is highly uncertain, because there is no direct experimental evidence for the large mass cutoff (see McDonnell et al., 1987).

A simple theoretical consideration leads to an estimate of the size of the largest particle that can leave the nucleus. By equating the gas drag on a particle of radius a_m with the gravitational attraction on the surface of the nucleus, i.e., zero acceleration and zero initial velocity in (4.11), one obtains the maximum particle size $a_m \simeq 0.1$ m. Inclusion of the centrifugal force from the nuclear spin results in an increase of the limiting size by only about 1% in the case of P/Halley (see Eq. 4.9). However, the uncertainty about the density and the shape of the particles and the variations of the gas production rate introduces an uncertainty of about a factor of 2 in the value of a_m.

Meteor observations show that these sizes are not unreasonable. For example, fireballs in the mass range of 0.1 to 1000 kg were identified by the Prairie Network observations to be members of the meteor showers (Ceplecha and McCrosky, 1976).

4.8 Optical and Infrared Properties

4.8.1 Scattered Light

The visible and near-infrared spectra of comet dust indicate a black body temperature of about 6000 K, corresponding to scattered sunlight (see Fig. 4.32). The color of the scattered light is neutral to somewhat red at wavelengths of 0.36 to 2.2 μm. The lack

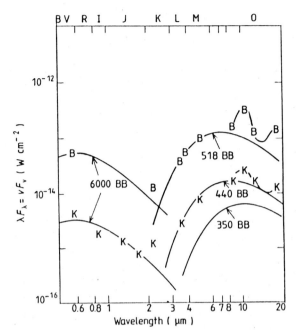

Fig. 4.32. Energy spectra of comets Bennett (1970 II) (4 April, 1970) and Kohoutek (1973 XII) (10 December, 1973) at $r \simeq 0.65$ AU heliocentric distance. Different black body temperatures are fitted to the various portions of the spectra. The value of 350 K corresponds to a fast rotating black body at the same heliocentric distance. (From Ney, 1974).

of observable Rayleigh scattering (blue color, isotropic scattering) implies that the contribution from grains much smaller than these wavelengths is minor. For infrared wavelengths larger than about 3.5 μm, thermal emission from the dust dominates cometary radiation. Here the spectra sometimes indicate temperatures considerably higher than for a fast rotating black body at the same heliocentric distance. This suggests that the dust grains are effectively smaller than the wavelength of maximum thermal emission. The emission features observed around 10 and 18 μm (Ney, 1974) are indicative of silicate as a major component of comet dust. Again, the particles must be of sizes smaller than 10 μm in order to show significant silicate features. Remote observations of electromagnetic radiation from comet dust provide data for the entire ensemble of dust grains within an observing aperture, whereas individual grains may have optical properties differing from grain to grain. Therefore, the information obtained by remote observations pertains to "effective" particle properties that may be realized by a wide range of real particles. The weighting factors are the geometric cross section πa^2 of the particle and the efficiency factors η_{sc}, η_{abs}, and $\eta_{ex} = \eta_{sc} + \eta_{abs}$ for scattering, absorption, and extinction, respectively. These efficiency factors depend on the wavelength and on particle properties such as size, shape, orientation, and refractive index.

The brightness at each point in an image of the scattered light from a comet dust cloud is given by

$$I_\lambda = \frac{F_\odot(\lambda)}{r^2} \int_l \frac{\psi(\lambda, \theta)\, dl}{\Delta^2}, \tag{4.26}$$

where $F_\odot(\lambda)$ is the spectral brightness of the Sun, r and Δ are the heliocentric and geocentric distances of the comet in AU, and $\psi(\lambda, \theta)$ is the volume scattering function. The integral is along the line of sight, l. In most cases for a ground-based observer, Δ is much larger than the extension of the dust cloud that contributes to the brightness. Therefore, Δ is approximately constant over the line of sight. Also the scattering angle θ (Sun – comet – observer) is constant in this case. The volume scattering function is defined as

$$\psi(\lambda, \theta) = \int_{a_1}^{a_2} n(a)\,\pi a^2 \eta_{sc}(a, \lambda)\,p(a, \lambda, \theta)\,da, \tag{4.27}$$

where $p(a, \lambda, \theta)$ is the phase function for scattering and a_1 and a_2 are the limiting sizes of the distribution. Figure 4.33 shows the phase functions for particles of different shapes and roughness in comparison with Mie spheres. If the size distribution is known, e.g., from *in situ* measurements, and the particle shape is assumed, the volume scattering function can be evaluated for different types of materials. The indices of refraction for some materials compatible with the results of the Giotto and Vega particle analyzers (Kissel et al., 1986a,b) are presented in Table 4.3. Four of the materials are basically of the silicate family, ranging from slightly to strongly absorbing. The material denoted "silicate A" is the "astronomical silicate" whose dielectric function has been constructed by Draine (1985) to be consistent with astronomical observations. The material denoted "silicate B" is a hypothetical silicate whose optical constants are in the range determined by Mukai et al. (1986a) to fit the observed polarization. Tholin (Khare et al., 1984) is representative of the class of organic grains. Surprisingly, it has only slightly different optical constants than the "astronomical silicate A," implying that either case may represent a composite grain with a silicate core and a tholin mantle.

The density of the grains ρ_d is a key parameter. Table 4.3 lists the bulk density of materials selected for the present calculation (Lamy et al., 1987). However, it appears plausible that comet grains are porous aggregates and have therefore lower effective densities. Since the elementary grains in the aggregates have a normal bulk density, the aggregate density will be decreasing with increasing particle radius.

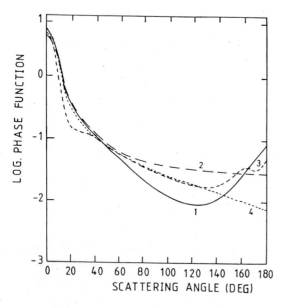

Fig. 4.33. Phase functions $p(a, \lambda, \theta)$ for irregular particle distributions. The phase function is normalized to 1 over the whole solid angle. The particle distributions are characterized by the size parameter $x = 2\pi a/\lambda$, where a is the particle radius and λ is the wavelength. The complex index of refraction is $n + ik$, with $n = 1.55$ and $k \simeq 5 \times 10^{-3}$. (1) Mie spheres with $1.9 < x < 17.8$, (2) cubes with $1.9 < x < 17.8$, (3) convex and concave particles with $5.9 < x < 17.8$, and (4) platelets and flakes with $1 < x < 25$ (Pollack and Cuzzi, 1980).

Table 4.3. Bulk density and complex index of refraction at $\lambda = 0.58$ μm for selected materials

Material	Density [Mg m^{-3}]	Complex index of refraction	
		n	k
Olivine	3.3	1.66	$1.0 \ 10^{-4}$
Chondrite	2.2	1.90	$1.9 \ 10^{-3}$
Silicate A	2.2	1.72	$3.0 \ 10^{-2}$
Silicate B	2.2	1.4	$3.0 \ 10^{-2}$
Tholin	1.45	1.7	$2.0 \ 10^{-2}$
Magnetite	5.2	2.56	$6.0 \ 10^{-1}$
Graphite	2.16	2.7	1.4

The integral in the expression for ψ has been calculated by Lamy et al. (1987) for Mie scattering in the size range 0.01 to 100 μm, taking into account the size distribution as discussed in Sect. 4.7. Figure 4.34 shows ψ as a function of the upper limit a_2 of the integral (Eq. 4.27) for the case of the astronomical silicate A and illustrates how the various size intervals contribute to the total value; for $a < 0.1$ μm, the contribution is negligible. The integral reaches half its final value at $a \simeq 3$ μm if a constant density ($\rho_d = 2.2$ Mg m^{-3}) is assumed and at $a \simeq 8$ μm if the density varies like

$$\rho_d = 2.2 - 1.4 \frac{a}{(a_o + a)}, \qquad (4.28)$$

where $a_o = 1$ μm. In the variable density case, the integration extends to 300 μm since the 100 to 300 μm interval contributes about 5% to the total value.

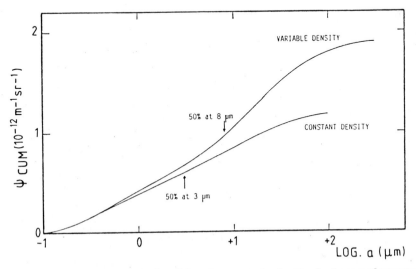

Fig. 4.34. Cumulative representation of the volume scattering function for the case of astronomical silicate (A), Mie scatterers, and the cometary particle size distribution (Lamy et al., 1987).

4.8.2 Polarization

Measurements of the polarization of the scattered light can contribute to the understanding of the nature of dust particles. Defining I_1 and I_2 as the components of the
scattered light having the electric vector perpendicular and parallel to the scattering
plane (Sun – particle – observer), the degree of linear polarization P is

$$P = \frac{(I_1 - I_2)}{I},\qquad(4.29)$$

where $I = I_1 + I_2$ is the total scattered light. As (4.29) shows, P can be positive
or negative depending on the predominance of I_1 or I_2. As expected for scattered
light, linear polarization has been found for all comets observed polarimetrically.

Figure 4.35 presents the volume scattering function ψ at $\theta = 107°$ and the
associated polarization for the selected materials under various conditions. Basically,
ψ ranges from 0.8 to 4.3 10^{-12} m^{-1} sr^{-1} for the Vega 2 flux distribution. Lamy et
al. (1987) chose $\theta = 107°$ because this angle is appropriate for the post encounter
Vega 2 television observations (Abergel and Bertaux, 1987) and the Giotto optical
probe experiment (Levasseur-Regourd et al., 1986). A preliminary calibration of an
image obtained by the Vega 2 camera one day after encounter (Abergel and Bertaux,
1987) yields a volume scattering function $\psi \simeq 2.5 \times 10^{-12}$ m^{-1} sr^{-1} in satisfactory
agreement with the above calculated results.

For the Giotto optical probe experiment (Levasseur-Regourd et al., 1986), a
preliminary calibration gives $\psi \simeq 7.4 \times 10^{-13}$ m^{-1} sr^{-1}. This is a factor of 3 below

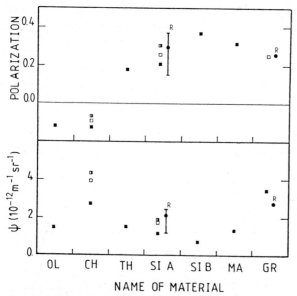

Fig. 4.35. Volume scattering function and polarization at $\theta = 107°$ for selected materials: OL = olivine,
CH = chondrite, TH = tholin, SI A = "astronomical" silicate A, SI B = silicate B, MA = magnetite, GR
= graphite. The squares correspond to Mie spheres of various densities: Filled squares = 2.2 Mg m^{-3},
open squares = 1 Mg m^{-3}, and half filled squares = variable density. The filled circles correspond to
rough grains and are labelled R. (From Lamy et al., 1987).

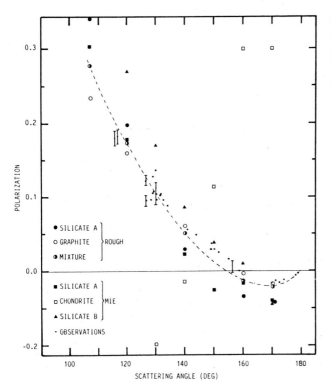

Fig. 4.36. Polarization observed for Comet P/Halley at large scattering angles. For comparison model calculations are shown for several materials. (From Lamy et al., 1987).

that for Vega 2, which is in accordance with the reduced particle flux observed by the Giotto dust experiments (McDonnell et al., 1987). Ground-based observers (e.g., Festou et al., 1986; Cosmovici et al., 1986) also reported that dust activity was minimal at the time of the Giotto encounter.

The polarization P at about 107° (Fig. 4.35) shows large variation from -0.13 to $+0.37$ for different materials. The observations suggest that P lies between 0.20 and 0.24, ruling out several of the selected materials. In fact, one sees that overall, the astronomical silicates, tholin (or optically equivalent materials) and graphite, produce acceptable results.

Figure 4.36 shows the polarization data obtained by several groups (Mukai et al., 1986a; Dollfus et al., 1986; Le Borgne et al., 1987) for scattering angles larger than about 100°. The polarization behavior, characterized by a reversal at an angle of about 160° and a negative branch, is similar to that reported for other comets (Lamy, 1986). The calculated values confirm that olivine and chondrite are not acceptable. The "mixture" is a combination of 50%, by grain number, of astronomical silicate and 50% graphite. From photometric observations with the International Halley Watch filters, Tholen et al. (1986) deduced a neutral color of the dust coma over the wavelength range 0.3 to 0.7 μm. However, these authors have called attention to contamination from the wings of emission bands. High resolution, spectrophotometric observations are far superior to tackle this question, since they allow the selection of clean windows to measure the continuum. Both the IUE spectra (Feldman et al., 1987) and those obtained by Jewitt and Meech (1986) indicate a substantial reddening – 7% per 0.01 μm in the ultraviolet and 10% per 0.1 μm in the visible. This latter result is quite similar to those obtained on Comets P/Stephan-Oterma, P/Neujmin 1, P/Churyumov-Gerasimenko, and P/Arend-Rigaux (Tholen et al., 1986).

4.8.3 Absorption and Thermal Emission

The single scattering albedo A is defined as the ratio of the energy scattered in all directions to the total energy removed from a unidirectional incident beam at a given wavelength, in the form

$$A = \eta_{sc}/\eta_{ex}. \tag{4.30}$$

It is useful to introduce an albedo A_p that depends on scattering angle θ in the form

$$A_p(\theta) = \pi A \eta_{ex} p(a, \lambda, \theta), \tag{4.31}$$

where $p(a, \lambda, \theta)$ is the phase function for scattering (see, e.g., Fig. 4.33), normalized in the form

$$2\pi \int_0^\pi p(a, \lambda, \theta) \sin \theta \, d\theta = 1, \tag{4.32}$$

and A_p is equal to the geometric albedo for $\theta = 180°$. It was first pointed out by O'Dell (1971) that simultaneous observations in the visible and infrared allow computation of the bolometric Bond albedo of dust particles. The Bond albedo is defined in the same way as (4.30). For a more detailed discussion of albedo, see Hanner et al. (1981).

Average values for the albedo A_p for an observed ensemble of cometary dust grains can be derived from simultaneous measurements of scattered thermal radiation. At near infrared wavelengths (e.g., J, H, and K) and scattering angles 120 to 170°, typical values for A_p are 0.02 to 0.05 for short period comets at $r < 2$ AU, comparable to those for the darkest asteroids (Hanner et al., 1985). Measured infrared albedos for Comet Halley of $A_p = 0.032$ at $\lambda = 1.25$ μm and of 0.045 at $\lambda = 1.65$ μm (Tokunaga et al., 1986) compare favorably with a value obtained by Lamy et al. (1987), who found $A_p = 0.036$ for astronomical silicate at a scattering angle of $\theta = 107°$.

The temperature of dust particles is determined by the equilibrium between energy absorbed and emitted. For a spherical grain of radius a and temperature T, one obtains

$$\int_0^\infty \frac{F_\odot(\lambda)}{r^2} \eta_{abs}(\lambda, a) \pi a^2 \, d\lambda = \int_0^\infty \pi B(\lambda, T) \eta_{abs}(\lambda, a) 4\pi a^2 \, d\lambda, \tag{4.33}$$

where $F_\odot(\lambda)$ is the solar spectral brightness which integrates over the whole wavelength range to $F_\odot = 1370$ W m^{-2}, $B(\lambda, T)$ is the Planck function at particle temperature T, and r is in AU. For a black body, $\eta_{abs} = 1$, and (4.33) reduces to

$$T = 280 \, r^{-1/2}, \tag{4.34}$$

where T is in Kelvin and r in AU.

For wavelengths $\lambda \gg 2\pi a$, η_{abs} decreases very quickly, which means that particles cannot radiate efficiently at these wavelengths. Therefore, the grain temperature increases as the grain size decreases below one tenth of the wavelength. Particles of a wide range of sizes are responsible for the thermal emission. Therefore, at shorter wavelengths such as the 5 to 8 μm region, the emission will be dominated by small, warm particles. At longer wavelengths larger, cooler grains will contribute most of the flux. Airborne and ground-based spectrophotometry from 5 to 30 μm of Comet Halley indicated at least two differently sized components of dust contributing to the thermal emission (Bregman et al., 1987; Herter et al., 1987). To date, no model calculations have been made to check whether the recorded size distribution alone can give rise to such an emission.

Radio continuum emission in the cm wavelength region of the electromagnetic spectrum has been successfully detected from several comets: Kohoutek (1973 XII), West (1976 VI), IRAS-Araki-Alcock (1983 VII), and P/Halley (Falchi et al., 1987). Strong variability in flux density was observed on time scales of a day, indicating that the continuum may be a transient phenomenon. Sporadic outbursts and the formation of an icy grain halo surrounding the nucleus (Gibson and Hobbs, 1981) may explain some observations.

4.9 Dust Composition

4.9.1 Introduction

It is a popular assumption that comets are formed within the same dusty gas nebula as the larger bodies of the Solar System, like the planets, except at a different place. However, because of their small size, which implies the absence of large-scale geologic differentiation and mixing processes, one may expect that the isotopic, chemical, and molecular characteristics of the presolar material are best preserved in comets. Thus experiments to study the composition of dust and gas in the coma were central in the missions to Comet Halley in 1986. Here an overview is given of what has been found so far about the dust composition.

4.9.2 Cometary Dust Composition Before the Halley Encounters

Before the *in situ* analyses of comet dust were possible in 1986, its composition was inferred from meteor spectra and laboratory analysis of interplanetary dust particles thought to be related to comets (Rahe, 1981). Millman (1977) presented a comparison of available data with carbonaceous chondrites of petrographic type 1 and with the Earth's crust (Fig. 4.37). He concluded that comet dust more closely resembles the composition of the average Solar System mixture (C1) than terrestrial composition. Yavnel (1977), using the same data from the Draconides, pointed out that the compositional variations between different carbonaceous groups and Draconide dust follow systematic trends. The degree of volatile – refractory element fractionation in

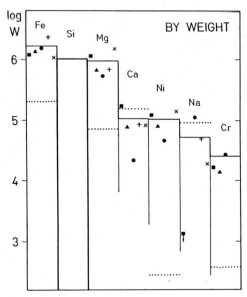

Fig. 4.37. The relative abundances, by weight, of seven elements commonly found in cometary meteoroids. Values are normalized to Si = 10^6. Since Si is not available for meteor spectra, there the normalization was to the mean of Fe and Mg. Solid line: Solar abundance; dotted line: Earth crust; squares: micrometeorites; triangles: microcrater residue; circles: E-region ions; + signs: 12 meteor spectra; × signs: 4 meteor spectra. Figure is taken from Millman (1977), where also the sources of the data are given.

comets was less than even in the most unfractionated chondrites (Fig. 4.38). These important findings are corroborated by the 1986 results from Comet P/Halley, as will be seen below.

4.9.3 The PIA and PUMA Experiments

To facilitate the *in situ* chemical and isotopic analysis of comet dust, which had a velocity of about 70 to 80 km/s relative to the spacecaft, a unique and new type of instrument had to be designed. Based on previous experience with dust detectors in the Helios mission (Grün et al., 1980) an impact-ionization time-of-flight mass analyzer was developed. On the European spacecraft Giotto, this experiment was

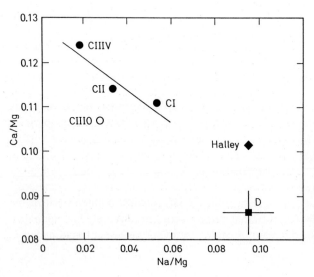

Fig. 4.38. Ca/Mg versus Na/Mg in Draconides (D) and various types of carbonaceous chondrites. Figure is taken from Yavnel (1977). Newly added is the average composition of Comet P/Halley dust from Table 4.6.

called PIA (Particulate Impact Analyser) and on the Soviet spacecraft VEGA 1 and 2, it was called PUMA 1 and 2, respectively. A detailed description of the instrument has been given by Kissel (1986) and, together with first results, by Kissel et al. (1986a,b). Ions originate when a cometary dust particle hits a metal target of the instrument with high velocity. They are then mass-separated in the reflector-type time-of-flight tube and detected by an electron multiplier. The reflector voltage of PIA is set such that ions with an initial energy of up to 150 eV are reflected. In the PUMA instruments, the reflector voltage is switched every 30 s from that setting (the wide energy window that produces the "short" spectra) to one where ions with initial energies of only up to about 50 eV are reflected (the narrow energy window that produces the "long" spectra). The long-mode setting provides a better mass resolution and higher transmission for heavy elements (mass to charge ratio $M/q >$ 20 amu/e) relative to that for the lighter elements (see Table 4.5). Each particle results in a mass spectrum that reflects its chemical, isotopic, and to some extent also the molecular composition as well as its size and density (Kissel and Krueger, 1987a).

Because of the limited data transfer rate from the spacecraft to the Earth, the mass spectra were transmitted in differently compressed modes, with mode 0 being comparable to a recorder-like mode (about four data points per mass unit), while in the higher modes 1 through 3, gradually more information had to be sacrificed (for details see Kissel, 1986). The most complete analyses (Kissel et al., 1986a,b; Jessberger et al., 1986; Sagdeev et al., 1986e; Šolc et al., 1986; Jessberger and Kissel, 1987; Jessberger et al., 1987; Brownlee et al., 1987; Mukhin et al., 1987; Kissel and Krueger, 1987a; Krueger and Kissel, 1987; Jessberger et al., 1988a,b) have been performed on PUMA 1 mode 0 data, on which many of the results presented here are based. The analysis of the higher mode spectra and also of the size and density of the grains, which requires elaborate and extensive computer work, is still in progress. However, some preliminary results have been published (Clark et al., 1986; Langevin et al., 1987a; Šolc et al., 1987a).

4.9.4 Principal Particle Types

Immediately after the Halley encounters in March, 1986, when the data were first inspected, it was noted that some spectra are dominated by ions from the elements H, C, N, and O, while in other spectra these ions were virtually absent, and instead the rock-forming elements Mg, Si, and Fe were the most abundant ions (Kissel et al., 1986a,b). This led to the designation "CHON" particles and "silicate" particles, respectively. Here we use the term CHON in a rather qualitative fashion, merely indicating that the intensity of the H, C, N, and O ions in some spectra is higher than the ions from the rock-forming elements. Clark et al. (1986) use a more stringent definition of the term CHON. Figure 4.39 shows that the Mg/C and Si/C ion ratios range over more than four orders of magnitude, which demonstrates the highly variable proportions of the CHON and silicate components. On the other hand, the correlation of Mg/C with Si/C (Fig. 4.39) is evidence for a rather constant Mg/Si ion ratio in the silicate component. The same conclusion has been reached by Langevin et al. (1987a).

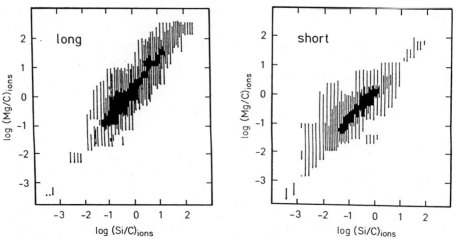

Fig. 4.39. Mg/C versus Si/C from PUMA 1 mode 3 spectra. The darkness reflects abundance of occurrence. Note that the variations range over 5 orders of magnitude. (From Šolc et al., 1987a).

In an analysis of 2204 mode 1 to 3 PUMA 1 spectra, Langevin et al. (1987a) found one-third each of CHON, silicate, and mixed (silicate and CHON) particles. For 2784 spectra from PIA, these authors arrive at 20% silicate, 40% CHON, and 40% mixed particles, which is in good agreement with Clark et al. (1986), who report a ratio of CHON to mixed particle of 0.8. According to Šolc et al. (1987a), who analyzed 1200 highly compressed spectra of PUMA 1, about 70% of the grains are mixed while the abundances of the end-member particle types are 10% silicates and 6% CHON. The remaining 14% of the spectra are difficult to interpret because major element peaks are not recognizable. A detailed study of 79 mode 0 spectra involving cluster analysis (Jessberger et al., 1988a) revealed that 72% of the particles are CHON-dominated with a mean (C+O) to (Mg+Si+Fe) ratio of about 5, while 28% of the particles are dominated by Mg, Si, and Fe with a (C+O) to (Mg+Si+Fe) ratio of about 0.1. The different values given by the various authors are mostly due to the different definitions of CHON, mixed, and silicate particles and should not be taken as contradictions. Common to all studies is the clear evidence that the cometary dust is a mixture of CHON-rich and silicate-rich material present in highly variable proportions. It would be interesting to see if there is a systematic trend in the CHON to silicate ratio with distance from the comet. Clark et al. (1986) and Langevin et al. (1987a) have attempted to study this question. They both reveal possibly significant compositional variations in time, which is equivalent to distance from the nucleus. Clark et al. (1986) report that during the Giotto flyby the abundance of CHON particles gradually decreased towards closest approach to the nucleus and increased slightly afterwards (see Fig. 4.40). This trend agrees with the finding of Langevin et al. that on the average the proportion of CHON is 50% lower in the time period beginning 50 s before closest approach than during the time before. In addition, Clark et al. (1987) noted that during small time intervals of about 2 s, clusters of grains of similar compositional type were found. The clusters comprise up to six grains. Clark et al. estimate that an original particle of about 1 mg would have to explode into

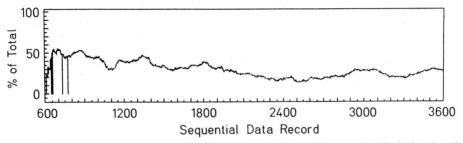

Fig. 4.40. Moving averages of the ratio CHON/(CHON + mixed) particle types plotted as a function of sequential data record. Each average is 256 records long and the graph shows that the organic CHON type material is dominating the silicate material further away from the comet. Closest approach spacecraft to P/Halley is about at the 2400th record. (From Clark et al., 1986).

about 10^{12} daughter particles in order to record a cluster. Because no such effective disruption mechanism is known, they argue that these clusters probably reflect dust jets ejected from the nucleus rather than fragmentation of larger grains.

4.9.5 Sizes and Densities of Grains

It is possible to estimate the size of the impinging grains from the intensity of the target ions (Ag) produced upon impact and the density from the ratio of target to projectile ions (Kissel et al., 1986a). No full analysis of the data is available yet. Šolc et al. (1987a), presenting preliminary data, state that the masses of silicate particles are in the range 10^{-12} to 10^{-14} g, while those of the CHON grains typically are less than 10^{-14} g. The densities of CHON particles range from 0.1 to 0.3 Mg m^{-3} and those of silicate grains from 0.5 to 3 Mg m^{-3}. It should be stressed that the quoted values may be subject to revision.

4.9.6 The Organic Component

In the early stages of data analysis it was argued that the joint occurrence of the CHON elements with silicates strongly points to the presence of an organic component in the dust particles and that this component is refractory relative to the frozen gases in the nucleus (Jessberger et al., 1986). A refractory organic component has been predicted by Greenberg (1982). The available information on the molecular composition of this component has been inferred by Kissel and Krueger (1987a) from PUMA 1 mode 0 spectra. They subtracted from the measured spectra singly charged ions of the elements that were, whenever possible, identified by their isotopic composition. They then summed the remaining spectra. This sum-spectrum was the basis of their study and was analyzed by means of a coincidence diagram of molecular mass lines. Especially the lower mass lines ($M/q < 50$) were accessible to a consistent interpretation. The results of this exercise are given in Table 4.4. Accordingly, the organic component consists of highly unsaturated hydrocarbon polymers. Based on the mass spectra and the energy distribution of the ions, the authors argue that the particles are composed of a somewhat fluffy mineral core

Table 4.4. Types of organic molecules as inferred from PUMA 1 mass spectra (From Kissel and Krueger, 1987a)

C−H− Compounds

(Only high-molecular probable due to volatility; hints only to unsaturated)

HC≡C(CH₂)₂CH₃   Pentyne
HC≡C (CH₂)₃CH₃   Hexyne
H₂C=CH−CH=CH₂   Butadiene
H₂C=CH−CH₂−CH=CH₂   Pentadiene

Cyclopentene, cyclopentadiene

Cyclohexene, cyclohexadiene

Benzene, toluene

C−N−H− Compounds

(Mostly of high extensity; also higher homologues possible)

H−C≡N   Hydrocyanic acid
H₃C−C≡N   Ethanenitrile (acetonitrile)
H₃C−CH₂≡N   Propanenitrile
H₂C=N−H   Iminomethane
H₃C−CH=NH   Iminoethane ⎫
H₂C=CH−NH₂   Aminoethene ⎬ (tautomeric)
H₂C=CH−CH=NH   Iminopropene ⎭

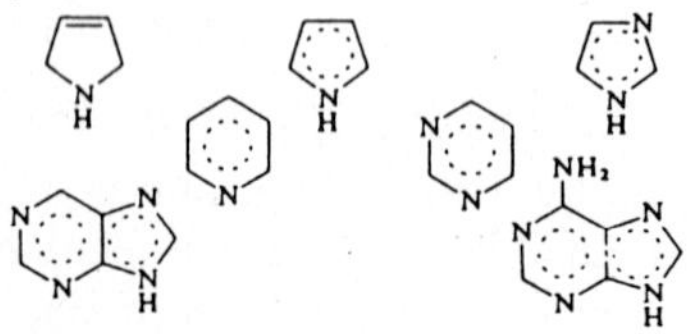

Pyrroline, pyrrole, imidazole

Pyridine, pyrimidine

(and derivatives)

Purine, adenine

C−O−H− Compounds

(Only very few hints to existence)

HC=OH   Methanal (formaldehyde)
H₃C−C=OH   Ethanal (acetaldehyde)
HCOOH,H₃C−COOH   Methanoic (formic) and ethanoic (acetic) acid (?)

C−N−O−H− Compounds

(Amino-, Imino-, Nitrile of -ole, -ale, -keto- only probable with higher C-numbers of -anes, -enes, and -ines or cyclic aromates)

N≡C−OH O=C=NH   (Iso-) cyanic acid (?)
N≡C−CH₂−OH   Methanolnitrile ⎫
HN=CH−CH=O   Methanalimine ⎬ (tautomeric)

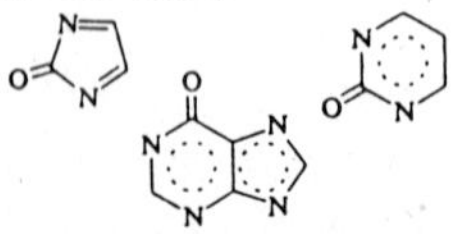

Oxyimidazole, oxypyrimidine

Xanthine

Structure isomers are additionally possible. Several types may form tautomers, mesomers and conformational isomers. Thus the molecules given here serve only as examples of the class of substances possibly present in the organic component of the dust. We are not sure yet if oxygen-containing species are present.

with a density of 1 to 2 Mg m⁻³, which is imbedded in an even more fluffy mantle of refractory organic material with a density of 0.3 to 1 Mg m⁻³.

In the mass spectra of the CHON-rich particles, the intensities of the elements with M/q = 1, 12, 14, and 16 (corresponding to the major isotopes of H, C, N, and O) are generally higher than 100 times the intensities of possible molecular species, because upon impact most of the projectile molecules decompose into their atomic constituents. Thus, it is of interest to analyze the variation of the elemental constituents. Figure 4.41 presents in a diagram O/C versus H/C from a subset of PUMA 1 mode 0 data (Jessberger et al., 1988a). In studying that diagram, it is to

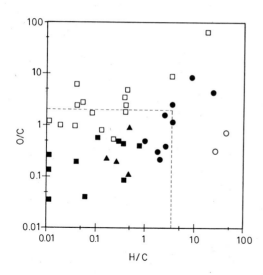

Fig. 4.41. O/C versus H/C in CHON-rich grains from PUMA 1. The area underneath the dotted lines represents possible stoichiometric compositions of H, C, and O. Higher O/C ratios may be due to O from silicates, excessive H may come from the Ag target. (From Jessberger et al., 1988a).

be noted that H may be released not only from the projectile but also in unknown amounts from the Ag target, and O may not only stem from the CHON component but also from the silicate material. Most of the high O to C and low H to C ratios in Fig. 4.41 are accompanied by high Si and Mg contents. It is obvious that the range of CHON-element variation is much larger than encountered in interstellar molecules (Irvine et al., 1985). Organic species inferred to be present in interplanetary dust particles, such as polycyclic aromatic hydrocarbons (Allamandola et al., 1987), or in Halley dust, such as polyoxymethylenes (Huebner, 1987), populate rather small areas in such a diagram. When fully interpreted, the CHON-element variations will set boundary conditions for the composition of Halley's refractory organic component.

4.9.7 The Isotopes

The experiments PUMA 1 and 2 and PIA, being M/q-analyzers, principally allow the determination of isotope ratios. The isotopic composition of the elements may yield information on the origin, source, and subsequent history of comet dust. Deviations from the normal isotopic composition generally are caused by a relatively small number of atoms and therefore are on the order of per mill. Deviations are large only for less frequent elements, such as the rare gases, that are irrelevant in the context of the PIA and PUMA experiments. Exceptions encountered in meteorites and interplanetary dust particles are enrichments of D (Zinner et al., 1983), ^{13}C (Yang and Epstein, 1983), and ^{26}Mg (Lee et al., 1976). The magnitude of the reported isotope anomalies and the relatively high abundances of the three elements motivated a search for isotopic anomalies in the PIA and PUMA mass spectra. Deuterium is only accessible in mass spectra of negative ions and thus not with the experiments on hand.

The Solar System $^{12}C/^{13}C$ ratio is 89, while lower values, down to 4, are observed in the interstellar medium (Audouze, 1977) and in certain components of primitive meteorites and interplanetary dust particles (McKeegan et al., 1985; Yang and Epstein, 1983; Zinner and Epstein, 1987) and are interpreted to reflect the presolar

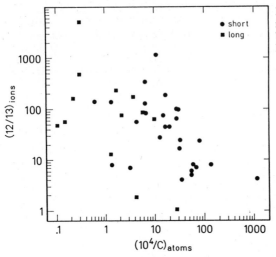

Fig. 4.42. Measured intensity ratio of mass 12/13 ions vs the reciprocal number of C-atoms determined for each grain. Note the lack of an anticorrelation.

origin of these components. Spectroscopic observations of cometary carbon indicate an isotopic composition compatible with the Solar System value, although the uncertainties are rather large (Rahe and Vanýsek, 1978). The *in situ* measured intensity ratios of mass 12 to 13 are shown in Fig. 4.42, with the reciprocal C-abundances on the abscissa. This ratio ranges from about 1 to 5000, a span never encountered before. It is, however, questionable whether measured mass ratios below 89 correspond to $^{12}C/^{13}C$ since at $M/q = 13$ the molecular ion $^{12}CH^+$ may interfere with $^{13}C^+$. A possible test, searching for an anticorrelation of mass 12 to 13 with the abundance of atomic hydrogen, proved to be inconclusive (Jessberger et al., 1987). Noise or background peaks could also decrease the mass 12 to 13 ratio at low C intensities, but Fig. 4.42 demonstrates that the ratio varies independently from the C abundance. The upper right corner of the graph is devoid of data points, since there the number of mass 13 counts would be less than 1. Higher than normal mass 12 to 13 ratios could be caused by $^{24}Mg^{++}$, but this is excluded by the lack of correlation in Fig. 4.43. Although speculations on the wide variety of possible interpretations of the low and the high mass 12 to 13 ratios and thus of the interstellar origin and

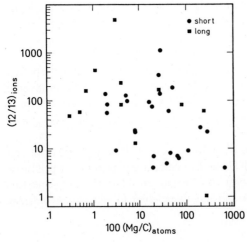

Fig. 4.43. Measured intensity ratio of mass 12/13 vs the Mg/C ratio. If an excessive (> 100) 12/13 mass ratio were caused by $^{24}Mg^{++}$, a correlation would be expected; no correlation is evident in the data.

subsequent history of comet Halley appear intriguing, we feel that they are premature and should wait until high-resolution mass spectrometric analysis of cometary material, either *in situ* or in the laboratory, will be possible.

The Mg isotopes are especially interesting. In certain Al-rich components of meteorites ^{26}Mg excesses of several percent have been detected. They are correlated with the Al/Mg ratio and are interpreted to indicate the presence of life ^{26}Al in the early Solar System (Lee et al., 1976). ^{26}Al, with a half-life of 0.72 My, could have been the most powerful heat source in primitive bodies. With the accuracy of the PIA and PUMA 1 and 2 instruments no such small deviations from the normal Mg isotopic composition can be detected. In addition, no Al-rich, Mg-poor grains were encountered in P/Halley's dust (Jessberger et al., 1988b). The upper limit given by Jessberger and Kissel (1987), which excludes a ^{26}Mg excess of greater than a factor of two, does not rule out nor does it support significant heating of the nucleus of Comet Halley. The measured intensity ratios of masses 24/25/26, however, do not correspond to the normal ^{24}Mg/^{25}Mg/^{26}Mg ratios. The most obvious deviation from the normal Mg isotopic composition was found for the ratio of masses 24 to 25 and is related to varying additions of ^{12}C$_2^+$. Therefore, it can be expected that the mass 24 to 25 ratio increases with the amount of carbon relative to magnesium. But, as shown in Fig. 4.44, there is no correlation. This then may indicate that ^{12}C$_2^+$ is not formed during the impact of the projectile but C$_2$ already existed in the dust grain. It also might be pointed out that Fig. 4.44 may serve to demonstrate the ability of the instruments to measure the isotopic composition of Mg with the given restrictions of possible molecular interference and only 4 data points per amu. Excluding the two extreme data points, the geometric mean value for mass ratio 24 to 25 is 7.3, close to the normal ^{24}Mg/^{25}Mg value of 7.77.

In summary, isotopic information has been obtained by PUMA 1 for the elements C and Mg. It is, however, difficult to distinguish isotopic peculiarities from molecular interferences. For C the few high ^{12}C/^{13}C ratios, up to 5000, which are not easily explained as artifacts, need independent confirmation before drawing far reaching conclusions about their origin.

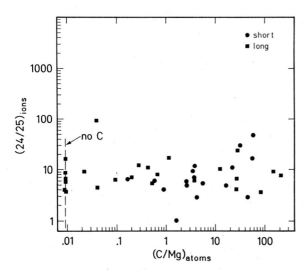

Fig. 4.44. Measured intensity ratio of mass 24/25 ions vs C/Mg. Although an excessive (> 8) mass ratio of 24/25 is probably caused by ^{12}C$_2^+$ interference, there is no obvious correlation of the mass ratio 24/25 with the relative C content.

4.9.8 The Average Chemical Composition

To obtain the average chemical composition, the ion intensities measured in all spectra are added for each element. Table 4.5 gives the results obtained in two independent studies. Langevin et al. (1987a) analyzed 2784 compressed spectra for PIA, 2204 for PUMA 1, and 627 for Puma 2. The intensity measured at the mass of the major isotope of an element is taken as the ion number of that element. In the results presented here, which are the basis of the study of Jessberger et al. (1988a), 79 PUMA 1 mode 0 spectra are individually analyzed, and the ion numbers are obtained by additionally taking into account the isotopic composition of the elements. For example, for Mg with ^{24}Mg: ^{25}Mg: ^{26}Mg = 8:1:1.1, only that portion of the mass 24 intensity is used that corresponds to about 8 times the mass 25 intensity (Jessberger et al., 1988a,b). Inspection of Table 4.5 shows that (a) the CHON ions are more abundant in short spectra than in long spectra, (b) the relative abundances of the ions from the rock-forming elements are the same in both types of spectra, and (c) with the exception of H, the data sets from Langevin et al. (1987a) and the present work agree for most elements within less than a factor of two. In comparing the ion numbers obtained from the three experiments, one should take into account that PIA produced only short spectra. A very significant difference between the three experiments, which has to be taken into account but is not yet fully explored, is the fact that they were equipped with different targets – PIA with a silver-doped platinum foil, PUMA 1 with a corrugated (60° sawtooth) silver plate, and PUMA 2 with a flat, solid silver plate.

In order to convert ion abundances to element abundances, ion yields have been introduced that were derived from calibration measurements with lower projectile speeds by Kissel and Krueger (1987b) and Krueger and Kissel (1987). The calibration

Table 4.5. Mean ion abundances normalized to Mg = 100 in the PIA and Puma 1 and 2 mass spectra of P/Halley's dust particles

Experiment	H	C	N	O	Mg	Si	S	Ca	Fe	Ref.
PIA	240	140	7.1	195	100	48	7.6	2.4	9.5	a
PUMA 1 short	1100	440	–	260	100	55	17	9.5	18	a
PUMA 1 long	160	40	–	28	100	40	10	7.2	11	a
short/long	6.9	11	–	9.3	1	1.4	1.7	1.3	1.6	
PUMA 1 short	8200	710	27	440	100	79	19	3.8	20	b
PUMA 1 long	280	85	1	54	100	49	15	6.2	20	b
short/long	29	8.4	27	8.2	1	1.6	1.3	0.6	1.0	
PUMA 2 short	170	120	5.3	200	100	67	19	8.7	33	a
PUMA 2 long	49	37	2.3	57	100	57	25	9.1	34	a
short/long	3.5	3.3	2.3	3.5	1	1.2	0.8	0.9	0.9	

[a]Langevin et al., 1987a
[b]These data are the basis for the elemental abundances in Table 4.6

Table 4.6. Average atomic abundances of 17 elements in dust grains of P/Halley, in carbonaceous (CI) chondrites, and the solar abundances of the CHON elements, all normalized to Mg

Element	Halley dust	CI[a]	Halley/CI	Solar/CI[a]
H[b]	2025. ± 385.	492	4.1	5143
C	815. ± 165.	70.5	11.6	16.0
N	42. ± 14.	5.6	7.5	41.2
O	890. ± 110.	712	1.3	2.63
Na	10. ± 6.	5.3	.9	–
Mg	100.	100	1.00	1.00
Al	6.8 ± 1.7	7.9	0.9	–
Si	185. ± 19.	93	2.0	–
S	72. ± 23.	47.9	1.5	–
K	0.2 ± 0.1	0.35	0.5	–
Ca	6.3 ± 1.9	5.68	1.2	–
Ti	0.4 ± 0.2	0.22	1.9	–
Cr	0.9 ± 0.2	1.25	.7	–
Mn	0.5 ± 0.2	0.88	0.6	–
Fe	52. ± 9.	83.7	0.6	–
Co	0.3 ± 0.2	0.21	1.2	–
Ni	4.1 ± 2.1	4.59	0.9	–

[a] From Anders and Ebihara (1982)
[b] From short spectra only

problem, in the context of the average elemental composition derived from PUMA 1 mode 0 spectra, has been discussed by Jessberger et al. (1988a). The resulting abundances of 17 elements in Halley dust are given in Table 4.6. The data on comet dust are based on a total of 79 high-quality mass spectra from PUMA 1. For the derivation and the meaning of the quoted uncertainties, see Jessberger et al. (1988a).

4.9.9 Comparison to Carbonaceous Chondrites and the Sun

The abundances of the elements heavier than and including Na in carbonaceous chondrites of type 1 (CI chondrites) and in the solar photosphere, normalized to the Si abundances, are very similar over the enormous spread of about seven orders of magnitude (Wood, 1979). This is the reason why the cosmic abundances of these elements are derived from CI chondrites (Anders and Ebihara, 1982). The abundances of the CHON elements, in the solar photosphere and, by inference, in the whole Sun are derived spectroscopically (Anders and Ebihara, 1982). Relative to their abundance in the Sun, the CHON elements are depleted in CI chondrites. This means that even though CI chondrites contain the least altered early Solar System material available for laboratory analyses, they did not retain the full inventory of the volatile elements.

In Table 4.6 the average atomic abundances (normalized to Mg = 100) of the elements found in P/Halley dust can be compared to those in CI chondrites and, for the CHON elements, to the solar values. The relative abundances of all elements heavier than oxygen in the comet are, on the average, within a factor of two the same as in CI chondrites, while the CHON elements, notably C and N, are highly enriched. Viewed differently, the CHON elements are less depleted in comet dust than in CI chondrites. The CHON abundance ratios in the dust approach the solar values. This is the reason for our earlier statement (Jessberger et al., 1988a,b) that P/Halley dust is, on the average, less metamorphosed, i.e., more primitive than CI chondrites, and to a greater extent reflects the original composition of the material in the Solar System. Concerning the other elements, the deviations of cometary dust from CI chondrites, especially the apparent Fe deficiency and Si overabundance, have been discussed by Jessberger et al. (1988a). It was argued that the deviations are more likely due to the insufficiently well known ion yields in the instruments rather than to reality.

4.9.10 Variability of Mg, Si, and Fe

The distributions of the main rock-forming elements Mg, Si, and Fe serve as distinctive features of extraterrestrial rocks. Figures 4.45a and b show the distributions of the ratios of Fe/(Fe + Mg) in two interplanetary dust particles (Bradley, 1988). They were determined by point count analysis of micro-sections for which the analyzed volumes are only about $10 \times 10 \times 80$ nm, even smaller than typical individual dust grains of Comet Halley. The distribution of Fe/(Fe + Mg) in a layer lattice silicate

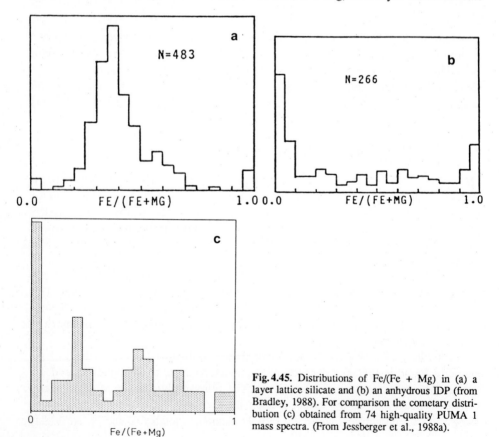

Fig. 4.45. Distributions of Fe/(Fe + Mg) in (a) a layer lattice silicate and (b) an anhydrous IDP (from Bradley, 1988). For comparison the cometary distribution (c) obtained from 74 high-quality PUMA 1 mass spectra. (From Jessberger et al., 1988a).

interplanetary dust particle (IDP) (Fig. 4.45a) centers strongly around 0.5 and is very similar to the distribution in a carbonaceous chondrite matrix (Brownlee et al., 1987; Bradley, 1988). The distribution of Fe/(Fe + Mg) in an anhydrous IDP (Fig. 4.45b) has a strong maximum at low values and then is smeared out over the whole scale. Figure 4.45c shows the distribution of Fe/(Fe + Mg) obtained from 74 high quality PUMA 1 spectra (Jessberger et al., 1988a). It resembles more closely that of the anhydrous IDP than that of the layer lattice IDP. The local maxima at about 0.3 and 0.5 in the P/Halley dust particle distribution may point to an admixture of layer silicates to the dominant anhydrous grain population, in accord with the results from infrared studies (McKeegan et al., 1987). However, grains rich in OH and poor in C – a signature of layer silicates – have not yet been identified in the spectra. The Fe/(Fe + Mg) distribution similar to that found in P/Halley and the anhydrous IDPs is very rare in meteoritic matrix material (Rietmeijer, 1988) and clearly constitutes a major difference of bulk meteoritic and cometary compositions, as had also been noted by Brownlee et al. (1987).

The ternary system Mg-Si-Fe can be used to classify different extraterrestrial compositions. In Fig. 4.46 the signatures of a layer silicate IDP (Fig. 4.46a), of an

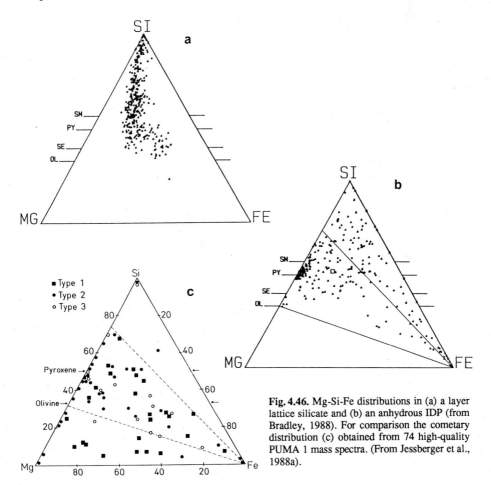

Fig. 4.46. Mg-Si-Fe distributions in (a) a layer lattice silicate and (b) an anhydrous IDP (from Bradley, 1988). For comparison the cometary distribution (c) obtained from 74 high-quality PUMA 1 mass spectra. (From Jessberger et al., 1988a).

anhydrous IDP (Fig. 4.46b) and of P/Halley dust (Jessberger et al., 1988a) are compared. Again, as with the Fe/(Fe + Mg) distributions, cometary dust differs greatly from the layer lattice silicate IDP and matches better with the anhydrous IDP. The similarity is most obvious in the rather dense population of the near-pyroxene field and the extension to Fe-rich compositions. There is, however, a notable difference between the comet and the anhydrous IDP; the latter is practically devoid of data below the olivine-Fe line, while in the former quite a number of Si-poor and Mg-rich grains occur. The distribution indicates a lack of both Fe-rich pyroxenes and Fe-rich olivines. The grains with compositions close to Fe = 0, Mg = Si = 0.5 may indicate the presence of primary Mg-silicates. This is what one would expect for primitive Solar System material, i.e., silicates, that never had been exposed to high temperatures.

4.10 Dust Production Rate and the Dust to Gas Ratio

The 1986 Halley missions allowed, for the first time, collection of comprehensive and direct information on the gas and dust emissions from a comet. Mass spectrometric measurements of the spatial density of the parent molecules were obtained along the Giotto trajectory. Krankowsky et al. (1986) derived a spatial H_2O density of 4.7×10^{13} molecules/m^3, with an uncertainty of $\pm 40\%$ at 1000 km from the nucleus and a neutral gas expansion velocity of 0.9 ± 0.2 km/s. The water vapor accounts for more than 80% of the gases escaping from the comet. Assuming isotropic emission of the gas (at least at large distances), the total gas production rate amounts to 6.7×10^{29} molecules/s, or about 20 Mg/s. The uncertainty of $\pm 50\%$ estimated by Krankowsky et al. (1986) lies in the extrapolation from the measurement of the flux along one line to the flux over a whole surface around the nucleus.

The determination of the total dust production rate is more complicated. First, the dust emission is highly anisotropic as evidenced by all *in situ* dust experiments and camera observations (Keller et al., 1986a; Sagdeev et al., 1986a). McDonnell et al. (1987) assume that a total active area of 50 km^2 emits dust at a constant rate. Another uncertainty arises from the emission speed and its variation with particle size, for which only theoretical estimates are available. Because of the variable particle speed, the "ages" of particles observed at some distance from the nucleus vary with the same magnitude. The "age" of a particle is the time interval between its emission from the nucleus and the time that it is recorded. Figure 4.47 shows some examples of the age of dust particles encountered by the Giotto spacecraft.

Particles of different masses recorded at the same time have been emitted from the nucleus at different times and from different locations. Therefore, the mass distribution observed, even its integral over the trajectory, will be modulated by this effect. From astronomical observations (Festou et al., 1986; Schleicher et al., 1986), as well as from spacecraft measurements, it is known that the gas and dust production of Comet Halley varied on time scales of a day by a factor of 3 to 5. As a consequence, large particles, e.g., the ones detected by Giotto (McDonnell et al., 1987), belong to a different emission epoch than the small particles. It has

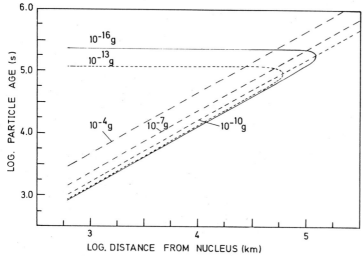

Fig. 4.47. Age of dust particles encountered by Giotto at different distances. (From Grün et al., 1987a).

been proposed that the excess in the flux of large particles with mass more than 1 μg, observed by the DIDSY experiment, is caused by an elevated production rate about a day before the encounter. A time variability of the observed dust flux cannot be distinguished from a spatial variability across the surface of the comet. This time and spatial variability of the dust production sets limits to the accuracy of the estimates of the total dust production from the comet. The ground track of the spacecraft trajectories (i.e., the position on the nucleus from which a particle of given size can reach the spacecraft position, taking the travel time of the particle into account) covers only a very small portion of the nuclear surface. In addition, the dust production rate from each surface element is dependent on its own solar illumination condition. For the night side of the nucleus, it can be safely assumed that it does not significantly contribute to the total dust production of the comet. From three dimensional modeling of the jets seen by Vega 2, Smith et al. (1986) show that the major dust emission occurs from a linear region on the nucleus, including the subsolar point. Similar analysis of the jet activity during the flybys of Vega 1 and Giotto may show the variability of the emission pattern.

The question of the total dust production from Comet P/Halley is even more complicated; there are indications that the particles carrying most of the mass have not been detected. Because of the flat slope of the mass distribution, most of the mass is carried by larger particles. According to McDonnell et al. (1987), the slope of the cumulative mass distribution is -0.5 for masses larger than $10\,\mu$g. Therefore, the total mass carried by the dust particles depends critically on the largest existing particle.

The largest particle recorded by the DIDSY experiment had a mass of 1.1 mg (McDonnell et al., 1986b). Assuming that the deceleration encountered by the Giotto spacecraft was caused by impacts of a few large particles, their mass was about 1 g. If the slope of the mass distribution stays the same (-0.54) to larger masses, the integration up to 1 g yields a total mass flux of 2.9 Mg/s (McDonnell et al., 1987).

Integration of the mass distribution up to 1 kg yields a total mass flux of 21 Mg/s. The corresponding dust to gas ratios are 1:7 and 1:1, respectively. Extrapolation of the mass distribution to the largest mass that can be lifted from the nucleus (see Chap. 2, Sect. 2.5.4) leads to much larger dust to gas ratios (Crifo, 1987). Therefore, we conclude that the dust to gas mass ratios quoted above have to be increased by a factor, possibly as high as 3.

Because of all these uncertainties, it is appropriate to explore other means to obtain independent evidence for the dust to gas mass ratio. This involves the chemical composition of the gas and the dust. It is commonly assumed that comets were formed within the solar system, far from the Sun. They remained cold throughout their entire history of about 4.5 Gy. They contain essentially unaltered material from an outer region of the solar nebula. To obtain information on its bulk gas and dust composition is a major goal, since even the most unaltered meteorites, the CI chondrites, have a history of metamorphisms changing their "original" composition (Wasson, 1985).

The bulk cometary abundances of the elements with $Z > 8$ that are not present in the gas, are already given in Table 4.6; therefore, only the relative abundances of the elements H, C, N, and O need to be discussed here. For this purpose, Geiss (1988) assumed a dust to gas mass ratio of 2:1 and the composition of the dust and the gas are as measured by instruments on the spacecraft. The bulk composition obtained is listed in Table 4.7. Jessberger et al. (1988b) followed a different argument and estimated, at the same time, the gas to dust ratio and the bulk composition. From Table 4.6 it is evident that the ratio of C to Mg in the dust is almost, but not quite, as high as the solar ratio. Assuming that the bulk C to Mg ratio of P/Halley is indeed solar, which is the only assumption made, they ascribe the C deficit in the dust to the gas. With the known gas (Krankowsky and Eberhardt, 1990) and dust composition (Table 4.6), they derive the bulk composition of P/Halley as given in Table 4.7. This composition can be compared to that derived by Geiss (1988) and the solar and CI compositions. Considering the different approaches, the agreement with the values obtained by Geiss is excellent. Compared to solar composition, nitrogen is underabundant by a factor of 3 and hydrogen is deficient by more than a factor of 700. However, the other element ratios are very similar to solar values and differ significantly from the CI values. This finding corroborates again our previously made contention that comets are, at most, slightly altered relics from the early Solar

Table 4.7. Relative atomic element abundances of the material (gas and dust) released by Halley's comet. The results of a study by Geiss (1988) renormalized to Mg with the solar Mg/Si-ratio are given for comparison. Also the abundances in the primordial Solar System and in CI-chondrites are listed (Anders and Ebihara, 1982).

	This work	Geiss (1988)	Solar System	CI
H/Mg	31	39	25200	4.9
C/Mg	11.3	12	11.3	0.71
N/Mg	0.7	0.4–0.8	2.3	0.06
O/Mg	15	22.3	18.5	7.1
N/C	0.06	0.03–0.06	0.2	0.08
O/C	1.3	1.8	1.6	10.0

System. The dust to gas mass ratio, resulting from the above argument, is 1:1 with an uncertainty of a factor of two. This value is well within the much wider range of values derived from direct measurements.

4.11 Summary

What is seen of a comet with the unaided eye is mostly dust. Although comets have been studied from the ground for centuries, only very little knowledge about the dust could be gathered because the process of light scattering produces little information about the dust. Only particles that are the most effective light scatterers, with a radius of 1 to 10 μm in radius, leave an imprint. From this very limited information the amount of dust released had been estimated and a dust to gas mass ratio of about 0.2 had been derived for Comet P/Halley (Divine et al., 1986).

However, there is another source of information on the particles released from comets. The close relationship between some meteor streams and comets was recognized long ago, because of their similarity of orbits. Studies of stream meteors indicate the release from cometary nuclei of large, cm- to dm-sized, particles of low density that are not found in our meteorite collections, not even in the IDP collections.

Only recently have new astronomical techniques been developed that improved our knowledge on cometary dust. Computer processing of digitized photographs or CCD images reveal details in the structure of the inner dust coma that were not recognized before. Jet-like dust features emitted from isolated areas on the nucleus have been analyzed to yield information on the rotational state of the nucleus and on the distribution of dust emitting surfaces, as well as on the size of the emitted particles themselves.

The development of infrared astronomy gave new insights in the dust emission from comets. Spectral features at 10 μm wavelength allowed the identification of silicates in comet dust. Dust trails along the orbits of some comets, observable between 25 and 60 μm wavelength have been interpreted as large particles that were emitted several perihelion passages ago.

The fleet of spacecraft approaching Comet Halley in the spring of 1986 for the first time allowed us a close look on a comet nucleus and to collect unprecedented information. Dust particles from 100 nm to several mm in size have been analyzed directly. The smallest of these suggest the connection between comet dust and interstellar dust that is known to be of similar size. The biggest particles detected confirm the supposition that particles are released from comets that are much bigger than those that dominate the scattered light. The sizes of the particles that can be lifted from the nucleus are probably only limited by the gas drag that has to compete against the gravitational attraction from the comet itself. Luckily, none of these several decimeter big particles was encountered by the spacecraft during their flyby of Comet Halley.

The structure of the dust coma proved even more complex than it had been assumed previously. The dust emission from the nucleus is very heterogeneous and time variable; it could not have been disentangled by the images taken close to the nucleus. More complications arose from the intricate rotational state of the nucleus. The variety of different materials that are comprised in comet dust add even more complexity to its dynamics and hence to the structure of the dust coma. Although the foundation for the determination of the dust to gas mass ratio has been much improved, its present value of 1 for Comet Halley includes still major uncertainties.

The molecular, elemental, and isotopic composition of P/Halley dust was analyzed with the impact-ionization time-of-flight mass spectrometer instruments PIA on Giotto and PUMA 1 and 2 on the two Vega spacecraft. Two principal particle types were recognized: Those rich in the elements H, C, N, and O are called CHON particles and the others, rich in Mg, Si, and Fe, are called silicates. Most particles are variable mixtures of the extremes. The CHON to silicate ratio was noted to change considerably during the flyby, probably reflecting dust jets ejected from different locations on P/Halley's nucleus.

Both, the mass and density of CHON particles, estimated from the amount of target and projectile ions measured in individual spectra, tend to be less than those of silicate grains. The CHON component consists of refractory organic material that is probably highly unsaturated hydrocarbon polymer. Apparent deviations from normal values of the isotopic composition of elements most probably are due to molecular interference with the possible exception of a few grains with high $^{12}C/^{13}C$ ratios.

The average abundances of the rock-forming elements are, within a factor of two, the same as in the CI chondrites and the Sun, while the light, volatile elements are enriched relative to the CI chondrites and approach solar abundances. Assuming a solar ratio of C/Mg for the whole comet, i.e., gas and dust, and using the measured gas and dust compositions, we estimate a dust to gas mass ratio of 1. The overall composition of the comet is almost solar, but with a nitrogen deficit of a factor of three. The observed abundance variations of the major rock-forming elements Mg, Si, and Fe prove that the solid dust differs from carbonaceous chondrite matrix, but is similar to anhydrous interplanetary dust particles. Many grains appear to contain primary Mg silicates that have never been exposed to high temperatures and that testify to the pristine nature of comet dust.

Still much remains to be learned about comet particles. However, new techniques are currently being developed and will be available in the not too distant future. Astronomical observations with the Hubble Space Telescope will enable us to observe the structure of the dust coma almost comparable to observations from a spacecraft. The availability of infrared space observatories and powerful ground-based microwave and radar telescopes will boost our knowledge about average properties of large cometary grains.

NASA's Comet Rendezvous Asteroid Flyby (CRAF) mission to a short-period comet will improve our knowledge about the spatial and time variability of the dust emission from the nucleus. Through collection of cometary particles in Earth orbit, it is hoped to bring the first nonvolatile cometary material to laboratories for analysis. The ultimate method for analysis of cometary material, however, is a Comet Nucleus Sample Return (CNSR) mission like the Rosetta mission that is currently under study by ESA and NASA. Such a mission will bring back to Earth unaltered nucleus material, hopefully within the next 20 years. Only such a mission will finally allow to settle major questions about the relationship of cometary material with materials from interstellar and galactic regions.

5. The Plasma

Wing-H. Ip and W. Ian Axford

5.1 Introduction

The study of comet plasma physics has long been a fascinating field of astronomical research and it has become more so after the flyby observations of Comet P/Giacobini-Zinner by the International Cometary Explorer (ICE) spacecraft in 1985 and the flotilla of space probes to Comet P/Halley in 1986. This point may be appreciated just by noting the dramatic surge in publications in space physics concerned with comet plasma. Furthermore, while the solar wind interaction with comets and the novel physical processes involved therein should be investigated in their own rights, comet comae also provide a natural laboratory for the simulations of many intriguing phenomena in astrophysical environments. Some of the most sophisticated magnetohydrodynamic (MHD) numerical simulations have indeed been carried out to study the morphologies and dynamics of comet plasmae.

Before comet plasma physics matured as an important discipline of space physics, contributions were made by several pioneering researchers. In a historical context, the modern study of comet–solar wind interaction began with the statistical investigation of the pointing direction of cometary CO^+ and N_2^+ ion tails by Biermann (1951). From the aberration angle of about $3°$ relative to the radial direction from the Sun, Biermann deduced that there must exist a solar corpuscular radiation in order to sweep away the cometary ions. The radial velocity of these solar charged particles (i.e., the solar wind) was deduced to be on the order of a few 100 km/s. However, to facilitate the momentum coupling between the solar plasma and the cometary ions by Coulomb collisions, a very high plasma number density, far exceeding the limit set by the coronal white light measurements, had to be invoked. To overcome this difficulty, Alfvén (1957) advocated the presence of a magnetic field in the interplanetary space so that the comet ionosphere would drape the field lines into a magnetic tail. In this classical picture, shown in Fig. 5.1, the cometary ions would be channeled into the antisunward direction by the magnetic field. The general properties of the magnetic field configuration have been confirmed by later numerical simulations as well as measurements at Comets Giacobini-Zinner and Halley.

Several excellent reviews have been written just prior to or immediately after the spacecraft flybys at Comets Giacobini-Zinner and Halley. These include several chapters in the book *Comets* (Wilkening, 1982) and a monograph by Mendis et al. (1985). McComas et al. (1987) produced a comprehensive review of the plasma measurements at Comet Giacobini-Zinner by the ICE spacecraft. Galeev's (1987) concise summary of the major results from the Vega and Giotto spacecraft at Comet Halley is also very useful. The reader is referred to them for a more detailed coverage of the many interesting results.

DYNAMICS OF PLASMA TAILS

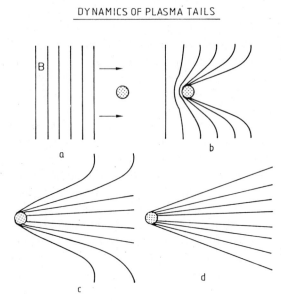

Fig. 5.1a–d. The field draping model of Alfvén (1957) in which the IMF field lines are supposed to be "caught up" by the comet ionosphere. The sequence from (a–d) denotes the gradual draping of the interplanetary magnetic field, **B**, into a magnetic tail.

Two years after the main event of the P/Halley encounters, the data analysis activities and theoretical efforts have stepped up further. It is thus impossible for us to describe every item in detail. Instead, we concentrate on topics that are most interesting from our point of view. They are: Large-scale processes of the solar wind–comet interaction, ion composition, ionospheric processes, plasma wave turbulence, and particle acceleration.

It should also be mentioned that the information on the ion composition described in Sect. 5.3 will be mainly used as a tracer for plasma dynamics. For more comprehensive discussions on the composition measurements and their implications, the reader should consult the reviews by Balsiger (1990), Krankowsky and Eberhardt (1990), and Ip (1989).

In this chapter neutral and ion species have different flow parameters. Therefore variables referring to neutral species are designated with a subscript n and ions with a subscript i.

5.2 Large-Scale Processes
of the Solar Wind–Comet Interaction

Prior to the advent of large-scale model computations, a topic of major interest dealt with the comet bow shock formation. For example, Axford (1964) pointed out that with the ionization of the expanding coma gas, a comet should act as an obstacle to the supersonic solar wind flow. As a result, a bow shock was expected to form a few 10^4 to 10^5 km upstream of the comet despite the small size (about 1 to

10 km) of the nucleus. This issue was taken up by Biermann et al. (1967) in a one-dimensional calculation. In analogy with the solar wind interaction with the Earth's magnetosphere, these workers had found that the Mach number at the bow shock should be $\mathcal{M} \simeq 10$. This result was later disputed by Wallis (1971) who proposed that the comet bow shock could be weak with $\mathcal{M} \simeq 2$ or even nonexistent. The basic idea was that because of the assimilation of new cometary ions with thermal velocity initially comparable to the solar wind speed, the upstream solar wind plasma would be heated and slowed down by the continuous mass loading by the cometary ions.

The behavior of the two-component fluid (solar wind protons and cometary ions of CO, CO_2 and the water-group molecules) along the comet–Sun axis, X, can be described approximately by the one-dimensional equations (Wallis and Ong, 1975; Galeev et al., 1985)

$$\frac{d(\rho_i v_i)}{dX} = \tilde{q}, \tag{5.1}$$

$$\frac{d}{dX}\left(\rho_i v_i^2 + p_\perp + \frac{B^2}{2\mu_o}\right) = 0, \tag{5.2}$$

$$\frac{d}{dX}[\rho_i v_i f(v_i, \mu)] = \tilde{q}\,\delta\left(\mu - \frac{m_i v_i^2}{2B}\right), \tag{5.3}$$

and

$$\tilde{q} = \frac{Q m_i}{4\pi v_n \tau_i X^2}. \tag{5.4}$$

Here v_i is the mass-loaded solar wind plasma flow speed, ρ_i the mass density of the ions of mass m_i (the cometary ions are assumed to co-move with the solar wind plasma and $\rho_{i\infty}$ and $v_{i\infty}$ are values at infinity), $p_\perp$ the thermal pressure perpendicular to the magnetic field B, Q the comet gas production rate, v_n the expansion speed of the neutral coma gas, τ_i the ionization time scale, μ the magnetic moment of the cometary ions at creation, and $f(v_i, \mu)$ the distribution function of μ at flow speed v_i. The permeability constant $\mu_o = 4\pi\,10^{-7}\,\mathrm{T\,m\,A^{-1}}$. It is assumed that the magnetic moment is invariant and the magnetic field is perpendicular to the flow direction. Finally, to estimate the strength of the magnetic field, we may use the relationship between the magnetic field and the flow velocity that holds for the conserved component of the subsonic flow, i.e., the solar wind protons, for the case of axisymmetric flow along the central axis (Lees, 1964; Schmidt and Wegmann, 1982)

$$\frac{B^2 v_i}{n_{sw}} = \text{const.} \tag{5.5}$$

Galeev et al. (1985) discussed the fact that the additional effect of the magnetic pressure gradient, i.e., $d(B^2/2\mu_o)/dX$ would accelerate the mass-loaded comet plasma in the subsonic region towards the comet center. This point will become more apparent in the following discussion.

In the supersonic flow, the effect of the magnetic field is small. Hence with $B = 0$ and γ the ratio of specific heats, (5.1) and (5.2) yield

$$v_i^2 \left[\frac{\gamma+1}{2(\gamma-1)} \rho_i v_i \right] - \left(\frac{\gamma}{\gamma-1} \right) \rho_{i\infty} v_{i\infty}^2 v_i + \frac{\rho_{i\infty} v_{i\infty}^3}{2} = 0, \tag{5.6}$$

with solutions

$$v_{i\pm} = v_{i\infty} \left(\frac{\gamma}{\gamma+1} \right) \left(\frac{\rho_{i\infty} v_{i\infty}}{\rho_i v_i} \right) \left[1 \pm \sqrt{1 - \frac{(\gamma^2-1)\rho_i v_i}{\gamma^2 \rho_{i\infty} v_{i\infty}}} \right], \tag{5.7}$$

where

$$\rho_i v_i = \int \tilde{q} \, dX. \tag{5.8}$$

Thus in the fluid approximation, a shock must form in the cometary accretion flow before the condition

$$\rho_i v_i \geq [\gamma^2/(\gamma^2-1)] \rho_{i\infty} v_{i\infty}, \tag{5.9}$$

is met (Biermann et al., 1967; Wallis, 1971). The spacecraft observations at P/Halley have found very clear signatures for the bow shock formation with $\mathcal{M} \simeq 2$ (Gringauz et al., 1986a; Galeev et al., 1986; Mukai et al., 1986b; Johnstone et al., 1986; Coates et al., 1987). The ICE measurements at P/Giacobini-Zinner are not as definite (Bame et al., 1986); however, as discussed below, this might be a consequence of the basic characteristics of a comet bow shock involving a mixture of heavy cometary ions.

The full analytical solutions for the combined fluid, i.e., solar wind plasma plus comet ions, with $\mathcal{M} = 2$ and $B = 0$ are illustrated in Fig. 5.2. Several important

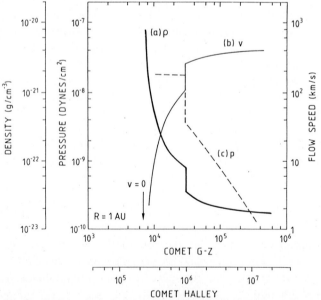

Fig. 5.2. Radial variations of (a) the number density, (b) the axial flow velocity, and (c) the thermal pressure, from a one-dimensional analytical model of the cometary accretion flow. Values are scaled to Comets Giacobini-Zinner and Halley at a heliocentric distance of 1 AU.

features can be recognized: Most notable is the rapid increase of the thermal pressure, p, at the bow shock; the thermal pressure has increased about 30 times while the flow speed has decreased by only about 25%. At the bow shock crossing, there is a further step-wise increase of the thermal pressure and thus, in the subsonic region, the plasma is essentially incompressible. To be noted also is the fact that the flow speed continues to decrease rapidly up to a cometocentric distance of 10^4 km for P/Giacobini-Zinner and 2×10^5 km for P/Halley. At these locations v_i approaches zero. In other words, there is a singularity in the comet plasma flow which limits the possibility of acceleration of cometary ions by momentum transfer from the solar wind plasma alone. Addition of the magnetic field (i.e., $B \neq 0$) and the inclusion of cooling effects resulting from charge exchange between the hot pickup ions and the neutral gas in the coma permit the plasma flow to continue beyond this point (Galeev et al., 1985). The analytical work suggests that the global comet–solar wind interaction, as sketched in Fig. 5.3, can be broadly divided into two regions:

(a) The strong momentum coupling region in the outer coma where comet ions are assimilated into the solar wind flow as a result of pickup and wave–particle interaction and

(b) the $J \times B$ acceleration region where the solar wind plasma will decouple from the bulk flow of the cometary ions created locally – as the wave turbulence therein is insufficient to enforce pickup of the new ions. There is a relative drift between the solar wind plasma and cometary ions in the direction parallel to

(a)

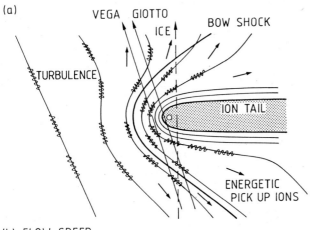

(b) FLOW SPEED

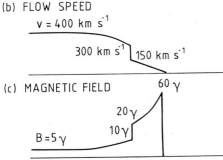

Fig. 5.3a–c. A sketch of the general characteristics of comet–solar wind interaction. (a) The general plasma environment of a cometary coma in which high levels of magnetic field and plasma turbulences are generated by the ionization of the new cometary ions. (b) The variation of the solar wind velocity as a function of axial distance from the nucleus. (c) A sketch of the gradual increase of the magnetic field strength and the formation of a magnetic field-free cavity in the inner coma where the magnetic drops to zero.

B. The comet plasma could be accelerated tailward via the $J \times B$ force acting against ion–neutral friction.

In the latter region, where the magnetic field direction is parallel to the relative motion of the two plasma components, a plasma instability (firehose instability) could occur so that the solar wind plasma is retarded by the slowly moving comet ions. Galeev et al. (1988) proposed that the formation of the cometopause observed by the Vega spacecraft could be related to this process.

The function $f(v_i, \mu)$ described in (5.3) directs attention to another important aspect of the comet–solar wind interaction, namely, the generation of energetic ions. As will be seen in Sect. 5.3, if the interplanetary magnetic field (IMF) is nearly perpendicular to the solar wind flow, newly produced ions have a ring-like velocity distribution in the solar wind frame with their guiding centers moving at the solar wind velocity v_{sw}. On the other hand, if the IMF is nearly parallel to v_{sw}, the new-borne ions initially remain stationary with respect to the comet and thus form a cold beam drifting at velocity $-v_{sw}$ in the solar wind frame. Both velocity distributions are unstable to the growth of waves that eventually pitch-angle scatter the comet ions into shell-like velocity distributions. In spacecraft measurements, these cometary ions appear as an intense flux of energetic particles with a strong anisotropy, as was first seen by ICE at P/Giacobini-Zinner. Figure 5.4 illustrates the detection of cometary ions with $M/q = 13 \pm 1$ amu/e, the water-group, and those with 28 ± 3 amu/e by the electrostatic analyzer on the Suisei probe at P/Halley (Mukai et al., 1986b). Similar results were obtained by the Vega and Giotto spacecraft; in particular the Implanted

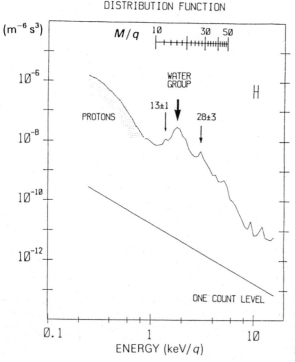

Fig. 5.4. Example of the energy (M/q) distribution function from the Suisei spacecraft at about 2×10^5 km from P/Halley. Peaks corresponding to the new-borne cometary ion groups ($M/q = 13 \pm 1$, water-group, and $M/q = 28 \pm 3$) are prominent (Mukai et al., 1986b).

Ion Sensor (IIS) on Giotto was designed to analyze this ion population with high mass resolution (Johnstone et al., 1986).

Except for details concerning the two-fluid nature of the comet plasma flow, the formation of the cometopause, and the relative drift of the solar wind plasma and cometary ions, MHD simulations have been quite successful in providing a global view of the whole interaction process. This assessment can be best seen in the work of Wegmann et al. (1987), which describes the global properties of the plasma flow, magnetic field structures, plasma densities, and ion and electron temperatures using a three-dimensional numerical code (see Fig. 5.5).

Attempts have been made to simulate the small-scale structures, i.e., ion rays and the time-dependent behavior of plasma tails often observed from ground-based observations. The linear ion streamers emanating from the central region of a coma, as shown in Fig. 5.6, have attracted much attention in theoretical work, invoking mostly the presence of magnetic field changes in the solar wind (see Schmidt and Wegmann, 1982; Schmidt-Voigt, 1987). Following an earlier suggestion by Ness and Donn (1966), Schmidt and Wegmann (1982) proposed that a directional change in the interplanetary magnetic field (IMF) by 90° could bring about the formation of a system of narrow streamers. They tested this hypothesis by fitting two three-dimensional, stationary solutions of different field directions along an isochronic surface. Schmidt-Voigt (1987) pointed out that in his time-dependent calculations, a 90° twist of the magnetic field direction would result in the formation of a condensation propagating down the tail; however, no symmetrical ion rays resulted.

Numerical results from various groups also differ on the effect of a sharp magnetic field reversal by 180°. Niedner and Brandt (1978) proposed that the production of ion tail disconnection events (DE) [or tail condensations in Jockers's (1985) terminology] results from the interception of a comet ionosphere with a sector boundary of the IMF at which the magnetic field direction reverses by 180°. The basic physical ingredient is that the IMF following the sector boundary reconnects at the front side of the ionosphere, hence generating ion streamer structures as indicated in Fig. 5.7a. Much work has been done in this topic using time-dependent models. While significant density enhancements in the neutral sheet separating the opposite magnetic polarizations had been suggested, Schmidt-Voigt (1987) found that the density enhancement effect, if any, is actually quite small. On the other hand, in the two-dimensional simulations by Ogino et al. (1986), a reversal of the direction of the IMF created a clear reconnection process in the subsolar region of the comet ionosphere, as well as large-scale disturbances in the central tail.

The contradictory results described above, are, to a certain extent, the result of the different methods used in the numerical simulations. This suggests that the generation of plasma structures in ion tails could be a complicated process not necessarily conforming to such simple scenarios. In fact, soon after the publication of the Niedner-Brandt model for the ion tail disruption event, several alternative proposals were made. These include the interaction effect of the ionosphere with the compression regions of solar wind streams, where the solar wind particle flux and ram pressure increase abruptly, probably leading to a Rayleigh-Taylor instability (Ip and Mendis, 1978; Ip and Axford, 1982; Jockers, 1981, 1985). The possibility of magnetic field reconnection in the tailward side has been noted as well (Ip and

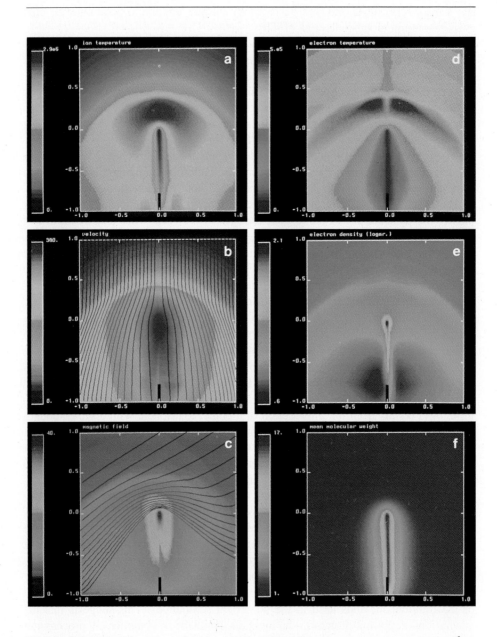

Fig. 5.5a–f. Numerical simulations of the bow shock structure (the color contour maps extend to 10^6 km on each side of the nucleus): (a) ion temperature (K); (b) velocity (km/s) with streamlines; (c) magnitude of the magnetic field (nT) with field lines separated by time intervals of 600 s; (d) electron temperature (K); (e) electron number density (cm^{-3}); (f) mean molecular weight of the ions (including solar wind protons). Results numerically improved and temperatures corrected. (Wegmann et al., 1987).

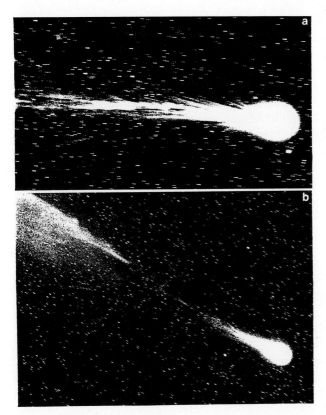

Fig. 5.6a, b. Photographic observations of Comet Halley on Jan. 9–10, 1986: **(a)** ion tail with formation of a system of symmetric ion rays; **(b)** formation of large plasma condensation in the ion tail (Photograph courtesy Dr. Kurt Birkle, Max-Planck-Institut für Astronomie, Heidelberg, FRG).

(a) FRONT-SIDE RECONNECTION

(b) TAILWARD RECONNECTION

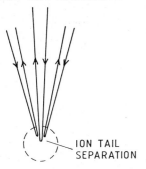

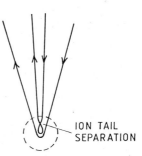

Fig. 5.7a, b. The Niedner-Brandt model **(a)** of reconnection on the frontside comet ionosphere as the comet intercepts a reversal of magnetic field polarity associated with an IMF sector boundary. Reconnection at the tailward side as shown in **(b)** might also be possible.

Axford, 1982; Ip, 1985; Russell et al., 1986; see also Fig. 5.7b). The spacecraft flyby observations in coordination with the extensive ground-based network for observing the large-scale ion tail phenomena of P/Halley thus provided a good opportunity to clarify this issue (Jockers, 1981; Brandt et al., 1982).

Using the data from the magnetometer experiment on Vega and observational material from several ground observatories, Niedner and Schwingenschuh (1987) and Schwingenschuh et al. (1987) have systematically studied the ion tail activity at the time of the Vega encounters (March 7 to 8, 1986). Their conclusion is that the onset of a DE observed on March 8 to 10 was related to a reversal of the IMF as detected by the Vega 1 and 2 spacecraft. On the other hand, Saito et al. (1986) have compared the plasma data obtained by the Sakigake spacecraft with ground-based observations and found that no distinct DE can be seen in P/Halley's ion tail during 11 to 14 March, even though Sakigake crossed the neutral sheet of the IMF at least four times. The global dynamics of plasma tails thus remains a field with a number of important questions not yet answered.

Finally we note that Celnik and Schmidt-Kaler (1987) have reported that time-varying plasma structures were seen in P/Halley's ion tail, having a periodicity of 50.5 ± 0.6 h. As this is very close to the 53 h periodicity determined for coma activity (Kaneda et al., 1986b), it is possible that ion production and plasma dynamics in the ion tail could be modulated by nucleus sublimation processes in addition to the solar wind conditions.

Of all the comet missions, only ICE passed through the onset of the ion tail region of a comet (at a distance of 7800 km from the nucleus). The ICE plasma observations are therefore of particular importance in the present context. The major findings may be summarized as follows (see also McComas et al., 1987):

(a) The field draping model is basically confirmed except for the detection of the formation of a magnetosheath at the ion tail boundary (Smith et al., 1986; see Fig. 5.8). The magnetic field strength in the lobes of the ion tail is on the order of 60 nT. This relatively high field may be explained in terms of pressure balance at the tail boundary where the total external pressure of the cometary ions was as large as the solar wind ram pressure (Siscoe et al., 1986).

(b) A thin plasma sheet with a total thickness of about 2000 km and a width of about 1.6×10^4 km was found at the center of the ion tail (Bame et al., 1986; Meyer-

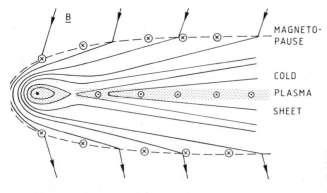

Fig. 5.8. The general nature of the magnetic field configuration as observed by the ICE spacecraft at P/Giacobini-Zinner. The kinky structure of the magnetic field lines at the "magnetopause" is generated by a current layer required to couple the slow moving comet plasma with the external solar wind.

Vernet et al., 1986; Smith et al., 1986; Daly et al., 1986). A peak electron density of $n_e = 6.5 \ 10^8 \text{ m}^{-3}$ and an average electron temperature of $T_e = 1.3 \times 10^4$ K were determined by the plasma wave instrument (Meyer-Vernet et al., 1986). As pointed out by Bame et al., 1986, a better fit is obtained with a two-temperature approximation for the cometary electrons with $T_e = 2.6 \times 10^4$ K and 1.1×10^5 K.

(c) The plasma flow velocity gradually decreased to zero at the ion tail center (Baker et al., 1986; Richardson et al., 1986). A significant amount of electron heating was seen between the ion tail and the bow shock (Baker et al., 1986; see Fig. 5.9).

(d) The degree of magnetic field turbulence and plasma wave activity was seen to correlate with the production of cometary ions in the solar wind. Outside the bow shock, the anisotropy of the energetic heavy ions was found to be very high (Sanderson et al., 1986). The ion flux became more isotropic inside the bow shock presumably because of stronger pitch-angle scattering and lower flow speeds.

The "dayside" encounter of P/Halley by several spacecraft produced similar observations to those described in (c) and (d). As for the formation of the magnetosheath, first detected at P/Giacobini-Zinner, it is worth noting that during the inbound passage of the Giotto probe a sharp magnetic field discontinuity was observed at about 1.3×10^5 km from the nucleus (Neubauer et al., 1986). The magnetic field strength increase by about 20 nT has led to the terminology "magnetic field pile-up region." This structure was not seen outbound nor was it observed by the Vega magnetometer experiments.

A unique feature, first observed by the Vega plasma instruments, is the so-called cometopause at $R \simeq 10^5$ km from the nucleus where the solar wind proton intensity drops suddenly while the comet ion density increases rapidly with a radial dependence of $1/R^2$ (Gringauz et al., 1986a,b). The transition has a thickness of about 10^4 km. The boundary separating the fast-moving solar wind plasma from the cold, slow plasma flow of cometary origin was much more diffuse during the Giotto encounter (Balsiger et al., 1986). The physical nature of the cometopause will be discussed in more detail in Sect. 5.4.

One other new feature discovered in Comet Halley involves the relative drift motion between the solar wind plasma and the cometary ions. As shown by Formisano et al. (1990), the newly-created cometary ions take on a slower velocity in the coma beginning at a cometocentric distance $R \simeq 2 \ 10^5$ km. This behavior probably indicates that in the inner coma the solar wind plasma is not able to accelerate the new ions to full speed immediately as the mass loading rate increases (e.g., Wallis and Johnstone, 1982). The details depend on the momentum coupling via wave–particle interaction as well as magnetohydrodynamics in a global scale. The Giotto results thus demonstrated that future numerical simulations of the comet–solar wind interaction process must take this effect into account.

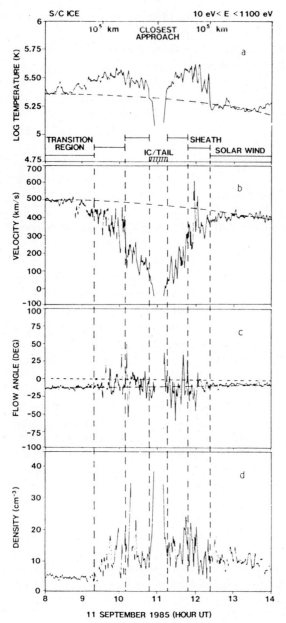

Fig. 5.9a–d. Electron behavior obtained from the solar wind plasma experiment on ICE showing (**a**) the electron temperature, (**b**) the plasma velocity, (**c**) the flow angle and (**d**) the electron number density (Baker et al., 1986).

5.3 Ion Composition

A major result of the spacecraft observations made at P/Halley concerns the *in situ* measurements of the neutral and ion composition of the coma. Whereas the Vega spacecraft were well-equipped with remote-sensing instruments such as the infrared spectrometer experiment (IKS) for detection of molecular emissions, the Giotto probe

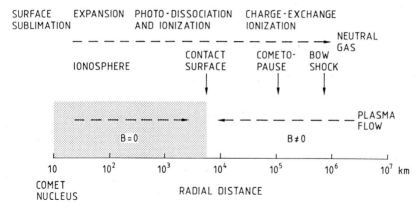

Fig. 5.10. A schematic view of the physical processes associated with the expansion of the cometary neutral coma and the ionospheric and solar wind plasma flows.

carried a number of neutral and ion mass spectrometers enabling *in situ* observations of the gas composition and plasma dynamics of the coma. In the following, a brief summary of the many new results from the mass spectrometer observations will be given with most of the detailed descriptions referring to the original papers and the reviews by Balsiger (1990) and Krankowsky and Eberhardt (1990).

As an introduction to our discussion, Fig. 5.10 shows the general processes associated with the expansion of the neutral coma gas followed by photodissociation, photoionization, and then charge exchange with the solar wind ions at different regions in the coma. The ion mass spectrometer instruments on Giotto therefore would first see the ions of mostly atomic composition, i.e., H^+, O^+ and C^+, at large distances from the comet nucleus (see also Sect. 6.1). As the spacecraft moves closer to the comet, molecular ions more representative of the original gas phase composition of the neutral gas would be seen. Figure 5.11 depicts the evolution of the water-group ions, O^+, OH^+, H_2O^+ and H_3O^+, across the bow shock at 19:30 UT into the sheath region of the comet plasma flow, as registered by the implanted ion sensor (IIS) instrument (Johnstone et al., 1986; Wilken et al., 1987). To be noted is the sudden enhancement of the count rates of the water-group ions (G1) following the bow shock. The broadening and shift of the solar wind proton profiles are also indicative of the thermalization and deceleration effects to be expected at a shock.

The proton population in the vicinity of the bow shock contains already a significant fraction of H^+ ions from ionization of the coma gas. Their presence can be recognized by the phase space density distributions; the Ion Mass Spectrometer (IMS) experiment on Giotto, for example, detected the shell-like structure of H^+ ions of cometary origin at about 8×10^6 km upstream of the bow shock (Neugebauer et al., 1987). As illustrated in Fig. 5.12, the original velocity distribution of the new pickup ions should be in the form of a ring with velocity component $v_\perp = v_{sw} \sin \alpha$ perpendicular to the interplanetary magnetic field B. There is also an initial drift along B with a velocity of $v_\parallel = v_{sw} \cos \alpha$ relative to the solar wind plasma. Because such a ring-beam distribution is unstable (Wu and Davidson, 1972), plasma instabilities with various time scales will first isotropize the H^+ ions such that the ring

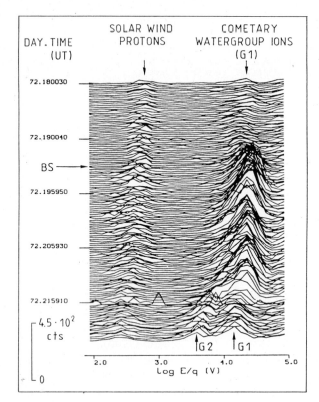

DAY. TIME
(UT)

SOLAR WIND
PROTONS

COMETARY
WATERGROUP IONS
(G1)

72.180030

72.190040

BS

72.195950

72.205930

72.215910

4.5 · 10²
cts

0

G2   G1

2.0   3.0   4.0   5.0
Log E/q (V)

Fig. 5.11. Stacked profiles of counting rate versus energy of solar wind and cometary protons and cometary water group ions (G1) (summed over all angles) in P/Halley. Passage of the comet bow shock (BS) is clearly marked by the intensification and broadening of the peaks of the distribution. A second water group ion population (G2) appears at 2130 UT (Wilken et al., 1987).

configuration transforms into a shell-like structure, i.e., the pickup shell in Fig. 5.12; subsequently the drift velocity between the cometary ions and the solar wind plasma flow parallel to B may be reduced. Indeed, the IMS observations of the gradual filling of the shell in velocity space, as shown in Fig. 5.13, confirm this prediction and the data are consequently very important for the understanding of the initial stages of the comet–solar wind interaction.

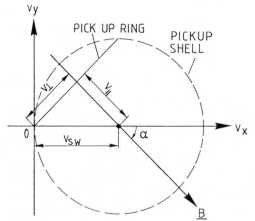

Fig. 5.12. The two initial velocity components, $v_\perp$ and $v_{\parallel}$, of the cometary pickup ions. Pitch-angle scattering transfors the ring distribution into a thin spherical shell.

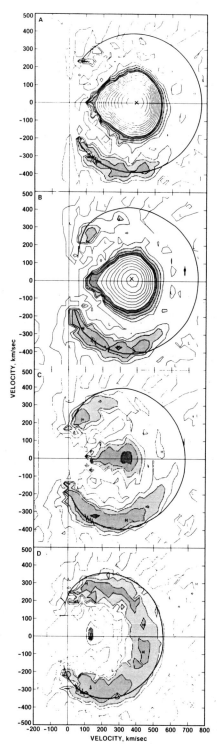

Fig. 5.13a–d. Phase space density distributions of cometary hydrogen ions upstream of the bow shock of Comet Halley as observed by the ion mass spectrometer experiment on Giotto. (a–d) cover the time period 0803–0908 UT, March 13, 1986, when the Giotto spacecraft was a distance of about 8×10^6 km from the nucleus (Neugebauer et al., 1987).

As a result of mass loading of the plasma, the flow speed is reduced as the cometocentric distance decreases. This behavior can be recognized in Fig. 5.11 where the profiles of the count rates of both protons and the water-group ions tilt toward lower values of energy per charge, E/q, with increasing time. The appearance of a second population of water-group ions (G2) becomes noticeable at 72.21:59 UT corresponding to a distance of about 5×10^5 km from the nucleus. The bifurcation of the water-group ion velocity distribution is a reflection of continuous loading of the plasma flow with new ions. Both branches of the water-group ions were seen to terminate at 72.23:50 UT (8.6×10^4 km inbound) and 73.00:29 (7.4 10^4 km outbound); see Johnstone et al. (1986). This transition may be understood in terms of charge exchange loss, in the dense comet coma, of the ions picked up upstream. We return to this point later.

At cometocentric distances $R < 2 \times 10^5$ km, the plasma instruments on Giotto designed to detect medium- and low-energy ions began to measure heavy cometary ions in increasing fluxes. Figure 5.14 shows the radial profiles of several main species of comet ions obtained by the IMS experiment (Balsiger et al., 1986). At the same time, the Neutral Mass Spectrometer (NMS) experiment with operation in both ion and neutral modes were also accumulating firsthand information on the comet ion composition (Krankowsky et al., 1986). Some of the main features of the results obtained from the IMS and NMS experiment may be summarized as follows: As the Giotto probe approached the comet nucleus to radial distances $R < 2\ 10^5$ km, the number density of the water-group ions, H_2O^+, H_3O^+, and O^+, was seen to increase rapidly, with the H_3O^+ ions displaying the steepest gradient. The atomic ions, C^+ (at $M/q = 12$ amu/e) and S^+ (at $M/q = 32$ amu/e), were observed to be surprisingly abundant (Balsiger et al., 1986; Balsiger, 1990) and their radial gradients appear to

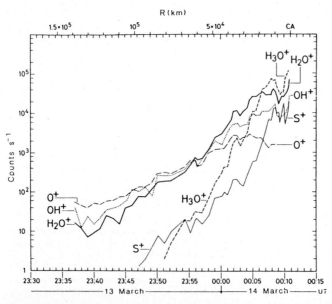

Fig. 5.14. Time profiles of HIS counting rates for masses 16, 17, 18 and 19 amu and the 32 to 34 amu mass group as measured by the IMS experiment on Giotto. Time is ground-station receive time in UT; R is the distance of Giotto from the nucleus (Balsiger et al., 1986).

be more gradual than those of the water-group ions. One possibility is that the C and S atoms might have been emitted from distributed sources such as the CHON dust grains in the coma (see Kissel et al., 1986a); however, the total mass contained in these grains appears to be too small to account for the observed production of C^+ ions. A more likely source of carbon atoms is photodissociation of CH, CH_2, CO, and CO_2 molecules.

A significant fraction of the CO molecules was found to be coming from a zone with radius 5×10^3 to 10^4 km surrounding the comet nucleus (Eberhardt et al., 1987; see also Chaps. 3 and 4). While the total production rate of CO from Comet Halley at Giotto encounter was determined to be about 15% of the water production rate, only 5 to 7% came directly from the nucleus. There is as yet no definite conclusion as to whether this extra component of CO molecules was injected into the coma by fragmentation of parent molecules, by surface sublimation from organic grains, or simply originated from a CO-rich jet (Krankowsky and Eberhardt, 1990). On the basis of the infrared spectrometer observations made by the IKS experiment on Vega (Combes et al., 1988), it has been determined that the production rate of H_2CO should be about 4% of that of water and hence the photodissociation of H_2CO molecules, with a time scale of about $3\ 10^3$ s at 0.9 AU (Levine, 1985b), might contribute to the distributed source of CO. The rest could result from fragmentation of formaldehyde polymers (POM) as postulated by Huebner (1987).

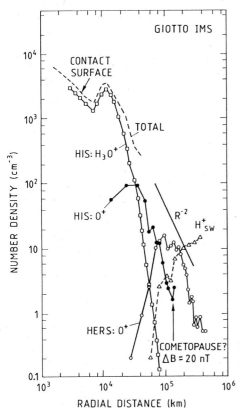

Fig. 5.15. A summary of the number density profiles of several water-group ions as observed by the two sensors of the IMS experiment on Giotto: Solar wind protons and hot O^+ from Balsiger (1990) and Shelley et al. (1987); cold O^+ and H_3O^+ from Balsiger et al. (1986).

At a distance of about 10^5 km, energetic comet ions created in the solar wind flow at large distances upstream begin to disappear and be replaced by cold ions produced locally. Figure 5.15 summarizes the behavior of both the hot and cold ions as detected by the IMS experiment. Several features are of interest: First, at the cometopause where the magnetic field strength jumped by 20 nT (Neubauer et al., 1986), the number density of solar wind protons was observed to drop by a factor of 2 (Balsiger, 1990) and a similar reduction was detected in the He^{++} flux (Fuselier et al., 1988). This feature might be related to the magnetic field pile-up region as a tangential discontinuity or propagating rotational discontinuity in the solar wind (Neubauer, 1987). That is, the reduction in the number density of solar wind particles need not be caused by charge exchange loss as has been proposed previously (Gringauz et al., 1986a,b; Gombosi, 1987a). In any event, near the cometopause, there is a rapid increase of the cold O^+ ions detected by the High-Intensity Spectrometer (HIS) sensor of the IMS experiment (Balsiger et al., 1987) together with the appearance of a maximum in the number density of the hot O^+ ions measured by the High Energy Range Spectrometer (HERS) sensor of the same experiment. The H_3O^+ ions, which are produced from ion–neutral reactions between the water molecules and other

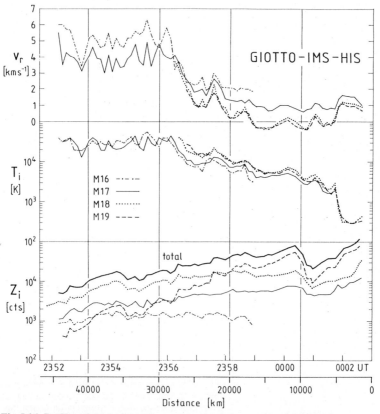

Fig. 5.16. Profiles of the ion velocity components (v_r) opposite to the Giotto probe ram direction in the rest frame of the nucleus, temperature (T_i) and the relative count rates (Z_i) for ion masses 16, 17, 18 and 19 amu (Schwenn et al., 1987).

ion species, become more and more abundant inside the "cometopause" until they dominate the ion composition in the inner coma at a few 10^4 km. It should be noted that the solar wind protons and the hot oxygen ions disappeared completely at $R \simeq$ 6 to 8×10^4 km, essentially as a result of charge exchange losses (Gombosi, 1987a; Ip, 1989).

Once inside a cometocentric distance of a few 10^4 km, the plasma flow becomes nearly stagnant with the force balance determined by the piled up magnetic field and ion–neutral friction (Ip and Axford, 1982, 1987a; Cravens, 1986). The ion temperature also is very low with $kT_i \leq 1$ eV (see Fig. 5.16). Several mass spectrometers for low energy particles observed increasing fluxes of comet ions. The IMS, NMS and the Positive Ion Cluster Composition Analyzer (PICCA) found groups of several populations, i.e., the water-group with peak count rates at 18 to 19 amu, the "CO-group" peaking at 31 to 32 amu and the "CO_2-group" peaking at 44 to 45 amu (see Fig. 5.17).

The electrostatic analyzer experiment on Vega covered the ion mass range up to $E/q = 3$ keV/e (Gringauz et al., 1986a). Near closest approach, the cometary ions were almost stagnant with respect to the comet nucleus. The ion temperature

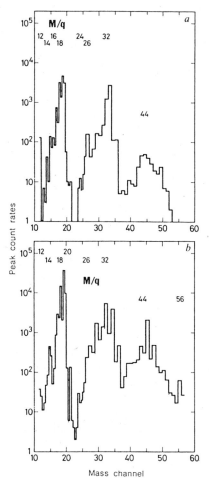

Fig. 5.17a, b. Peak count rates of comet ions measured in the H-mode of the IMS experiment in one spin of Giotto: (a) 14 March, 00:09:32 UT, $R = 6000$ km (outside the contact surface); (b) 14 March, 00:10:42 UT, $R = 1500$ km (inside the contact surface) (Balsiger et al., 1986).

was also very low. As a result, the measurements by the Vega plasma analyzer could be directly used to infer the ionospheric composition of the comet in a very comprehensive manner, i.e., for mass to charge ratios from 1 to 85 amu/e. Besides mass peaks at about 18, 28, and 44 amu/e, broad count rate maxima were found and labeled approximately as 56, 70 and 85 amu/e. The Vega observations were thus the first to reveal the presence of ions more massive than CO_2^+. This unexpected result was subsequently confirmed by the PICCA experiment on Giotto with mass coverage extending to 213 amu/e (Korth et al., 1986). At high masses the spectra showed a sequence of mass peaks with alternating differences of 14 and 16 amu/e at 45, 61, 75, 91, and 105 amu/e that led Huebner (1987) to construct a cracking sequence of the formaldehyde oligomer $(H_2CO)_n$, or polyoxymethylene (POM). More recently Mitchell et al. (1989) suggested that ion groups with units of CH_2, NH, O, and H connected by single bonds to some complex hydrocarbon molecules could also provide an explanation of the spectral signatures seen by the PICCA instrument.

Because of ion chemistry in the comet ionosphere, ions do not necessarily have the same relative abundances as the neutral species. For example, in the inner iono-sphere with $R < 1000$ km, NH_4^+ could be a major species and H_2CN^+ could easily dominate CO^+ at mass 28 amu. Figure 5.18 depicts the chemical network by which the densities of NH_4^+, CH_3^+, and CH_5^+ ions might build up in the comet ionosphere. The important point is that NH_3 and CH_4, though only minor species, react very effi-ciently with H_2O^+ and H_3O^+ ions and become protonated in the collision-dominated region of the ionosphere. At the same time, neither NH_4^+ nor CH_3^+ react with H_2O such that they can be accumulated, with electron dissociative recombination as the only loss mechanism. In principle, the count rates at masses 19 amu (H_3O^+) and 18 amu (H_2O^+ and NH_4^+) could provide information on the ratios of the NH_4^+ ions to the H_3O^+ ions and hence the relative abundance of ammonia. Allen et al. (1987), using a set of preliminary data from the IMS experiment, have compared theoretical results to the observed ratios of the count rates in these two mass channels and esti-mate the relative abundance of NH_3 to be about 1 to 2% of that of H_2O. A similar treatment using the mass 15 amu channel suggests that the production rate of CH_4 should be about 1% of that of H_2O. Wegmann et al. (1987) obtained very similar results.

The exact value of the ammonia abundance is still debated at the present time. Using different physical parameters, i.e., ionospheric electron temperature profile, photoionization rates, and reaction rates, Marconi and Mendis (1988) showed that under certain circumstances, the experimental data could be consistent with complete absence of NH_3 in the coma of Comet Halley. In support of this, Wyckoff et al. (1988) have deduced from high-spectral resolution ground-based measurements of the NH_2 emission that the production rate of ammonia should not be larger than 0.3% of $Q(H_2O)$ while Magee-Sauer et al. (1989) reported a value between 0.1 and 0.4 %.

Wyckoff and Lindholm (1989) reported a production rate of $7 \times 10^{-3} Q(H_2O)$ for the CH radical, or its parent molecule, by assuming a photodissociation time scale of 5000 s for CH. However, using the photodissociation lifetime of $\tau_d = 200$ s (Singh and Dalgarno, 1987; see also Levine, 1985b), one obtains $Q(CH) \simeq 0.18$ $Q(H_2O)$. The emission of the C-H stretch bond at 3.2 to 3.5 μm was detected by the IR spectrometer experiment (IKS) on Vega. It leads to two estimates for the

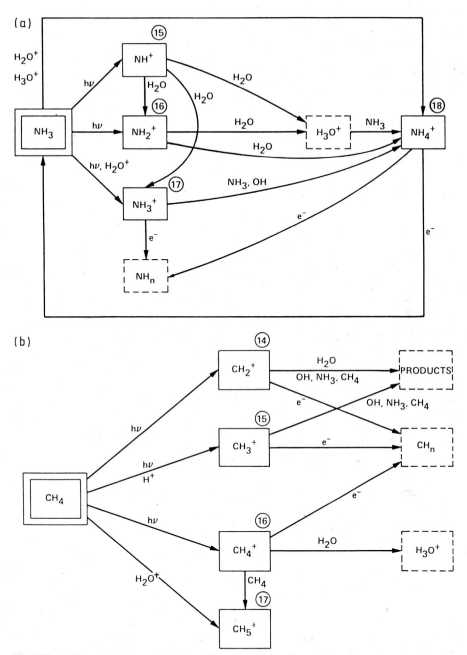

Fig. 5.18a, b. The chemical networks describing the ion chemistry in the comet coma leading to the formation of (a) NH_4^+ ions and (b) CH_5^+ ions (Allen et al., 1987).

production rate of CH (Combes et al., 1988): If the $3.2\,\mu$m emission is caused by thermal emission from small grains or polycyclic aromatic hydrocarbons (PAHs) heated to 500 K (Knacke et al., 1986), the production rate would be $Q(\text{CH}) \simeq 3$ $10^{-3}Q(\text{H}_2\text{O})$; on the other hand, if infrared fluorescence in the gas phase from saturated hydrocarbons is the major mechanism of the CH emission, $Q(\text{CH}) \simeq 0.2$ $Q(\text{H}_2\text{O})$.

5.4 Ionospheric Processes

The radial expansion of cometary ionospheric plasma has some similarities to the interaction of the solar wind with the interstellar medium in which the solar wind flow is supposed to be terminated by the formation of a heliopause (Axford, 1972). In this context, Wallis and Dryer (1976) were the first to propose that a similar structure could exist in the interface between the ionospheric flow and the mass-loaded solar wind plasma. In other words, a contact discontinuity[1] should be formed between a pair of shocks (the inner and the outer shock) such that the supersonic flow of the ionospheric plasma could be diverted into the lateral direction near the boundary. In a subsequent study, Houpis and Mendis (1980) proposed an analytical model for a hypersonic ionospheric outflow. In their model, the distance between the inner shock and the tangential discontinuity on the flanks should be relatively thick ($\Delta R \simeq 1000$ to 2000 km). Such a structure, however, was not detected in the plasma observations of Giotto; instead, the interface separating the radial ionospheric outflow from the external plasma flow was found to be very thin with $R < 100$ km (Balsiger et al., 1986; Neubauer et al., 1986). The best signature of the contact discontinuity comes from magnetometer measurements in which the magnetic field strength was observed to drop from a value of about 20 nT to zero over a distance of 25 km (Neubauer, 1988). The concept of a magnetic field-free cavity was not a new idea in studies of solar wind–neutral atmosphere interaction (see Ip and Axford, 1982; Luhmann, 1986); but the main issue, starting with an interesting stability analysis by Ershkovich and Mendis (1983), has to do with the stability of the interface. These authors argued that the ion–neutral frictional force can be extremely destabilizing; hence a fully magnetized ionosphere was predicted. The Giotto observations of a magnetic field-free cavity thus brought riddles as well as many new results in this connection.

Even though a complete picture of the physical processes involved in the formation of the ionospheric boundary layer is still to be investigated, several major features can be identified. In this section, we consider (1) the electrodynamics of the ionospheric plasma, (2) the energetics of the ionospheric plasma and (3) the MHD instability of the surface of the magnetic field-free cavity.

---

[1] The contact surface divides plasma from two different sources: The mass-loaded solar wind on the outside and the comet ions on the inside.

5.4.1 Ion Dynamics

Because of the lack of any intrinsic magnetic field, the expanding neutral gas resulting from sublimation of the nucleus is directly exposed to the interplanetary medium in all regions except inside a small cavity whose size is determined by a balance between the Lorentz, $J \times B$, force and the ion–neutral drag. Assuming there is a region of almost stagnant cold plasma surrounding the nucleus in which the magnetic field magnitude is amplified such that the corresponding magnetic pressure is approximately equal to the supersonic solar wind ram pressure, then, neglecting curvature effects,

$$\frac{B_s^2}{2\mu_0} = n_{sw}m_{sw}v_{sw}^2. \tag{5.10}$$

We obtain for the field of the stagnant cold plasma $B_s \simeq 60\,\mathrm{nT}$, if $n_{sw} \simeq 5 \; 10^6\,\mathrm{m}^{-3}$ and $v_{sw} \simeq 400\,\mathrm{km/s}$. At the point where the magnetic field has its maximum, i.e., $dB/dr = 0$, the radially outward streaming neutral gas exchanges appreciable frictional momentum with the plasma, which has to be balanced by the curvature force

$$\frac{B_s^2}{\mu_0 R_s} \simeq k_{in}n_i m_i n_n v_n, \tag{5.11}$$

where k_{in} is the ion–neutral collision rate, n_n the neutral number density, n_i the ion number density, v_n the neutral speed, and R_s the radius of the stagnant ionosphere. If the radius of curvature is taken to be similar to the radial distance, and we assume a spherically-symmetric coma with radial flow speed v_n, then $n_n = Q/(4\pi v_n R_s^2)$, and we obtain as an approximation to the size of the stagnant ionosphere

$$R_s \simeq \frac{\mu_0 k_{in} Q m_i n_i}{4\pi B_s^2}. \tag{5.12}$$

This relationship was derived when it was realized that the thermal pressure of the ionospheric plasma might not be sufficient to stand off the solar wind ram pressure (Ip and Axford, 1982). Thus, for Comet Halley with a gas production rate $Q \simeq 10^{30}\mathrm{H_2O}\,\mathrm{s}^{-1}$ and with $n_i \simeq 10^9$ to $3 \times 10^9\,\mathrm{m}^{-3}$, $m_i \simeq 18\;m_u \simeq 3 \times 10^{-26}\,\mathrm{kg}$, $B_s \simeq 50\,\mathrm{nT}$, and $k_{in} \simeq 2 \times 10^{-15}\,\mathrm{m^3/s}$, the radius of the stagnant region at the subsolar point would be $R_s \simeq 2400$ to $7300\,\mathrm{km}$. In general, the collision rate should also include the photoionization rate (e.g., Haerendel, 1987), but this is only a small correction in the present situation.

The Giotto magnetometer experiment (Neubauer et al., 1986) showed the presence of a well-defined magnetic cavity with the field-free region separated from the external magnetized plasma by a relatively sharp boundary (see Fig. 5.19). This discovery prompted Cravens (1986) and Ip and Axford (1987a) to reexamine the neutral drag process by incorporating the magnetic pressure gradient term in (5.11) so that the structure of the boundary could be determined

$$\frac{1}{\mu_0}B\frac{dB}{dR} + \frac{1}{\mu_0}\frac{B^2}{R} = k_{in}n_i m_i n_n v_n. \tag{5.13}$$

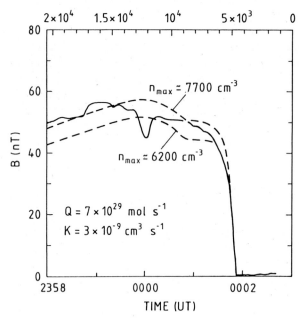

Fig. 5.19. Magnetic field profile obtained by considering the force balance between the magnetic field curvature force and the ion–neutral friction using the measured ion density profile obtained by the IMS experiment on Giotto. The peak ion number density at about 1.3×10^4 km are assumed to be $n_{max} = 6.2 \times 10^9 \, \text{m}^{-3}$ and $7.7 \times 10^9 \, \text{m}^{-3}$.

The plasma can be taken to be stationary as a first approximation since it is "frozen" into the field and it cannot move radially if the cavity boundary is assumed to be fixed. If it is further assumed that the ionospheric plasma is in photochemical steady state such that the ion density is determined by balancing photoionization with electron dissociative recombination, then the ion number density can be expressed as

$$n_i \simeq \left(\frac{\beta Q}{4\pi \alpha v_n} \right)^{1/2} \frac{1}{R}, \tag{5.14}$$

where α is the electron dissociative recombination coefficient and β the photoionization rate (both assumed to be constant). In this case (5.13) has a very simple solution for the magnetic field profile

$$B(R/R_{max}) = B_{max} \frac{[1 + 2\ln(R/R_{max})]^{1/2}}{R/R_{max}}, \tag{5.15}$$

where R_{max}, proportional to $Q^{3/4}$, is the radial distance at which the piled up magnetic field reaches its maximum value B_{max}. The essential feature of a rather rapid but not discontinuous decrease of the magnetic field by about 40 to 60 nT is reproduced, as shown in Fig. 5.19.

Physically, the "soft" action of the ion–neutral friction creates a broad region of width $\Delta R \simeq 1000$ km in which the required force balance is achieved. The formation of the ionopause of Venus, either identified by the plasma signature or by the magnetic field signature (see Luhmann, 1986), is somewhat similar since neutral gas-drag also plays a role. In contrast, the magnetopause boundary of the Earth's magnetosphere has a thickness of only a few proton gyroradii and is determined by pressure balance.

By generalizing the above equations to a two-dimensional configuration, Wu (1987) has been able to produce a model of a teardrop shaped ionospheric cavity. By considering the force balance normal to the boundary and neglecting curvature effects (cf. Cravens, 1986), the appropriate equation reads

$$|\boldsymbol{J}_d \times \boldsymbol{B}_d| = k_{in} n_i m_i n_n v_n \cos\phi,$$ (5.16)

with

$$\boldsymbol{J}_d = |\frac{1}{\mu_o}\nabla \times \boldsymbol{B}_d| = \frac{1}{\mu_o \cos\phi}\frac{dB_d}{dR_d},$$ (5.17)

where ϕ is the angle of incidence of the neutral wind at the boundary. Combining (5.14), (5.16), and (5.17) we have

$$B_d\frac{dB_d}{dR_d} = A^2\frac{\cos^2\phi}{R_d^3},$$ (5.18)

where $A^2 = \mu_o (\beta/\alpha)^{1/2}(Q/4\pi v_n)^{3/2} k_{in} m_i v_n$. The magnetic field profile is thus given as

$$B(R) = \left(B_\infty^2 - \frac{A^2\cos^2\phi}{R^2}\right)^{1/2},$$ (5.19)

where B_∞ is the magnetic field magnitude at large distances from the nucleus. The radius of the boundary surface at angle ϕ, where $B = 0$, is expressed as

$$R_d = \frac{A\cos\phi}{B_\infty}.$$ (5.20)

The next step is to approximate the value of B_∞ as

$$B_\infty(\psi) = B_\infty(0) + B_\infty(\pi/2)(1 - \cos\psi),$$ (5.21)

where $\psi = \Theta + \phi$ is the angle between the normal of the contact surface and the comet–Sun line. Wu's result is reproduced in Fig. 5.20 where the sizes of the field-free cavities of Comets Halley and Giacobini-Zinner are compared. This shows that the ICE spacecraft may have been far outside the cavity at its closest approach of about 8000 km on the tail side. Note that both R_{max} in (5.15) and R_d in (5.20) are proportional to $Q^{3/4}$. A teardrop shaped cavity boundary has also been found in the MHD simulations of Schmidt et al. (1988).

Neubauer (1988) showed that, by analyzing the magnetometer data at the finest time resolution, the magnetic field magnitude jumped by 20 nT within a short distance of about 25 km from the field-free region (see Fig. 5.21). Such a sharp increase in B cannot be explained by ion–neutral drag since this can contribute no more than a few percent of the magnetic pressure change in such a short distance. There might be an additional pressure gradient caused by a hot electron component inside the cavity; for $n_e \simeq 1.5 \times 10^9$ m^{-3}, an average electron temperature of $T_e \simeq 1$ eV would be needed. However, this explanation encounters several potential problems

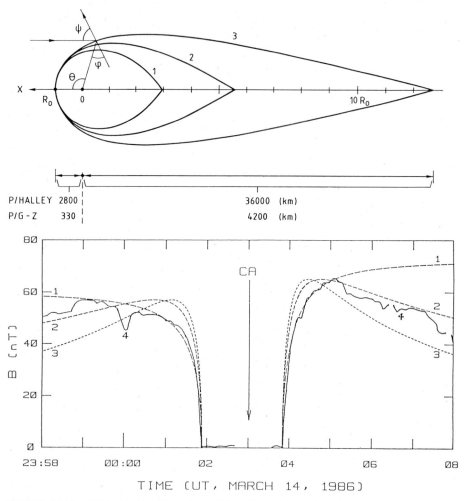

Fig. 5.20. (a) Two-dimensional configurations of the diamagnetic cavity according to Wu (1987). (b) Magnetic field measurements made by the magnetometer experiment on Giotto showing the inner pile-up region inbound and outbound and the magnetic cavity region. Curve (1) is obtained by taking into account the pressure gradient effect in the force balance, (2) and (3) are obtained by including the curvature force term but with different ion number density profiles. The observational results by the magnetometer experiment on Giotto is denoted by curve (4). The experimental curves are from Neubauer et al. (1986) and the theoretical curves from Wu (1987).

(Neubauer, 1988). If photoelectrons are the only source of heating, one would expect the electron temperature to increase steadily with distance from the nucleus and the sense of the pressure gradient required is therefore not easily achieved. In fact, photochemical calculations (Körösmezey et al., 1986), using improved values of the rates of electron cooling from H_2O collisions (Cravens, 1986), have shown that the ionospheric electron temperature should be relatively low with $T_e < 300\,\mathrm{K}$ within the cavity. The $1/R$ dependence of the ion density profile (Balsiger et al., 1986;

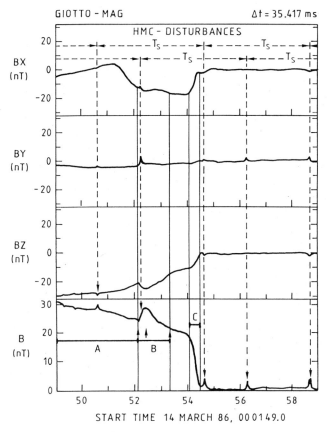

Fig. 5.21. High-time resolution measurements of the vector magnetic field components B_x, B_y, B_z, in Halley-centered solar ecliptic (HSE) coordinates and of the field magnitude B, covering the inbound ionopause boundary layers and the ionopause (C) (Neubauer, 1988). Feature A denotes the region where there is a smooth decrease of the magnetic field before interruption by a bulge (B). The abrupt decrease of magnetic filed at the contact surface is marked (C).

Krankowsky et al., 1986) as measured by Giotto also suggests that photochemical equilibrium with no sudden electron temperature variation was maintained up to a radial distance of about 10^4 km. An additional electron heating mechanism internal to the cavity must therefore be invoked; however, there is no obvious candidate, except perhaps for magnetic field reconnection which still is very much an open issue. One must wait for further analyses of the ion data so that the plasma flow and density might be more accurately known.

In addition to the electron pressure, the ion flow pattern could also contribute to the formation of the sharp rise of the magnetic field strength. In a simple picture, a comet ion should execute a gyration, reversing its radial streaming at the cavity boundary (see Fig. 5.22). The change in momentum flux has to be balanced by a corresponding change in the magnetic field pressure

$$\frac{(\Delta B)^2}{2\mu_0} \simeq c_r n_i m_i v_i^2. \tag{5.22}$$

Taking the reflection coefficient $c_r = 2$ for specular reflection by 180°, $v \simeq 1$ km/s, and $m_i \simeq 18 \cdot 1.6 \ 10^{-27}$ kg (for H_2O^+), we find that a plasma density of $n_i \simeq 10^9 \, \mathrm{m}^{-3}$ yields $\Delta B \simeq 12$ nT and $n_i \simeq 2 \times 10^9 \, \mathrm{m}^{-3}$ yields $\Delta B \simeq 17$ nT. Thus ion reflection

(a)

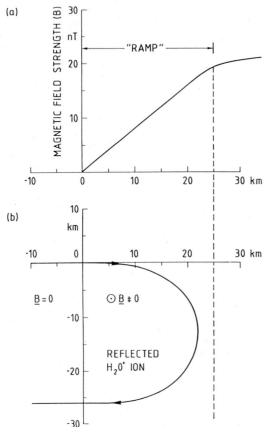

Fig. 5.22a, b. The idealized trajectory of a water ion with a velocity of 1 km/s normal to the ionospheric magnetic boundary. The electrostatic field is assumed to be zero. The model field profile is given in (a) and the trajectory of a water ion with initial motion parallel to the magnetic field is shown in (b).

could contribute up to 50% of the force balance. In fact, it is more appropriate to take $c_r \simeq 1$ so that these values of ΔB are upper limits.

Figure 5.22 illustrates the penetration of an H_2O^+ ion in the "ramp" with a constant magnetic field gradient. The value of ΔR_d is comparable to the observed thickness of 25 km if the ion speed is about 1 km/s. The penetration depth of the ions depends to some extent on the electric field established by the charge separation effect caused by the difference in the proton and electron gyroradii (Neugebauer et al., 1974). However, in the present situation, the effects of continual photoionization and recombination guarantee that the plasma is electrically neutral so that no electric field should exist.

The transit time for a comet ion in the thin layer is on the order of $\tau_t \simeq 60\,s$. In comparison, the electron dissociative recombination time of H_2O^+ ions with $T_e \simeq 300\,K$ would be $\tau_r \simeq (n_e\, 2 \times 10^{-13})^{-1} \simeq 2500\,s$ if $n_e \simeq 2 \times 10^9\,m^{-3}$. However, according to Allen et al. (1987) the electron dissociative recombination coefficient for H_3O^+ ions has a lower limit given as $\alpha > 2 \times 10^{-12}\,m^3/s$; hence, if the ionospheric composition in the inner coma is dominated by the protonated water ions (H_3O^+), as seems to be the case, the appropriate value of τ_r would be smaller than 250 s. There is therefore an interesting possibility that most of the comet ions would be effectively lost by recombination close to or within the ramp.

Another parameter of interest is the ion–neutral reaction time scale which is $\tau_{in} \simeq (k_{in}n_n)^{-1} \simeq 100\,\mathrm{s}$ if $n_n \simeq 5 \times 10^{12}\,\mathrm{m}^{-3}$ and $k_{in} \simeq 2\ 10^{-15}\,\mathrm{m}^3/\mathrm{s}$. This indicates that during the traversal of the magnetized region about 60% of the comet ions could undergo a collision with a neutral gas molecule. The situation is obviously complicated and deserves a more careful investigation.

Cravens (1990) has made a detailed, one-dimensional calculation on a fine grid in which he examines the three-dimensional fluid behavior of a supersonic ionospheric flow in spherically symmetric expansion in the vicinity of the cavity boundary. A simplified model is given below to illustrate the basic physical ingredients (see Fig. 5.23). The abrupt stop of the supersonic flow at the edge of the magnetized region leads to the formation of a shock with compression ratio $r_c = 4/(1 + 3/\mathcal{M}^2)$. For large Mach number $\mathcal{M}$, $r_c \simeq 3$ or 4; hence a thin, high-density shell can be formed. The thickness of this compressed fluid shell may be approximated by equating the sum of the particle flux into the shell plus the photoionization source strength to the electron recombination loss rate

$$n_i v_i + (n_n/\tau_i) \cdot \Delta R_d \simeq \alpha(r_c n_i)^2 \Delta R_d, \tag{5.23}$$

where n_i and v_i are the ion number density and flow speed just before the shock. The thickness of the "spherical boundary layer" can be expressed as

$$\Delta R_d = \frac{n_i v_i}{\alpha r_c^2 n_i^2 - n_n/\tau_i} \simeq \frac{v_i}{\alpha r_c^2 n_i}. \tag{5.24}$$

For $r_c \simeq 3$ and $n_i \simeq 2 \times 10^9\,\mathrm{m}^{-3}$, $\Delta R_d \simeq 28\,\mathrm{km}$ if $\alpha \simeq 2 \times 10^{-12}\,\mathrm{m}^3/\mathrm{s}$ (i.e., ion composition dominantly H_3O^+) or $\Delta R_d \simeq 280\,\mathrm{km}$ if $\alpha \simeq 2 \times 10^{-13}\,\mathrm{m}^3/\mathrm{s}$ (i.e., H_2O^+ dominant in the ionosphere). The first case therefore leads to the formation of a very narrow shell of compressed plasma at a point where, according to Cravens's

SPHERICAL MODEL OF THE
COMETARY DIAMAGNETIC CAVITY

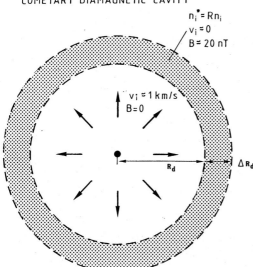

Fig. 5.23. A schematic representation of the stopping of the ionospheric outflow at the boundary of the magnetic cavity.

calculations, there may also be a very steep magnetic field gradient. As with the simple kinetic description given above, this originates from the requirement that the dynamic pressure of the ionospheric supersonic flow be balanced by the magnetic pressure (see Fig. 5.22). It is nevertheless still difficult to achieve the necessary pressure. The remainder of the force balance would have to be produced by ion–neutral friction in a much broader region as described by (5.13). Additional physics and at least a two-dimensional, axisymmetric calculation are needed to make further progress.

Note that in (5.24), ΔR_d is proportional to r_c^{-2}; therefore, for weakly supersonic ionospheric flow, the boundary layer could be as wide as 250 km even for an H_3O^+ dominated flow. For a plasma with temperatures $T_i \simeq T_e \simeq 300$ K, and $v_i \simeq 1$ km/s, $r_c \simeq 1.7$; hence a fluid model would predict a typical thickness of $\Delta R_d \simeq 120$ km. A density spike with a density enhancement of a factor of about 3.5 and a width of about 47 km was indeed detected by the IMS experiment on Giotto in the vicinity of the contact discontinuity (B. Goldstein et al., 1989). At the same time, a curious burst of energetic ions was observed near the same location (Johnstone et al., 1986; R. Goldstein et al., 1987).

The current density in the narrow ramp region can be shown to be far below the critical value required to generate an ion–acoustic instability and therefore we should not expect to find plasma instabilities such as the modified two-stream instability at this location. However, since the exact physical nature of the magnetic ramp is still not clear, numerical simulation experiment addressing the microscopic effects and kinetic processes might reveal other interesting phenomena.

5.4.2 Thermal Structure of the Ionosphere

Before the spacecraft measurements at a comet were made, a conventional view held that a field-aligned plasma flow away from the central coma should result along the interplanetary magnetic field lines draped around the ionosphere. Such a picture of "tooth-paste squeezing" (Ioffe, 1968) is based on the assumption that the pressure gradient of the piled up magnetic field should be the driving force dictating the field-aligned motion. However, as discussed in the work of Wegmann et al. (1987), charge exchange may become dominant in the inner coma for $R \leq 10^4$ km and the incoming plasma could be cooled very efficiently (Wallis and Ong, 1975). As a consequence, there would be a thermal pressure gradient acting to push the ionospheric plasma inward along the magnetic flux tubes. The resultant motion of the comet plasma in the plane of the magnetic field is shown in Fig. 5.24. The individual magnetic flux tubes, on the other hand, would flow out of this plane with the stream lines diverging from the central coma (see Fig. 5.24 and Wegmann et al., 1987).

In the MHD simulations by Wegmann et al. (1987), ion heating, cooling, and momentum exchange from ion–neutral collisions and photoionization have all been included. The numerical results, in agreement with the Giotto observations and analytical work, demonstrate the formation of a magnetic cavity outside of which, in the region $R \simeq 10^4$ to 2×10^4 km, the ion temperature $T_i \simeq 10^3$ to 10^4 K and $v_i \simeq$ 3 to 6 km/s (Fig. 5.25). In equilibrium, the ion temperature should be related to the neutral temperature by the following expression (Banks and Kockarts, 1973)

(a)

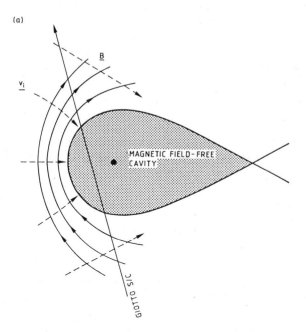

Fig. 5.24. The flow pattern of ionospheric plasma in the vicinity of the magnetic cavity. Because of the strong cooling resulting from charge exchange the plasma flow tends to converge towards the central axis with the magnetic field lines draped around the magnetic cavity boundary.

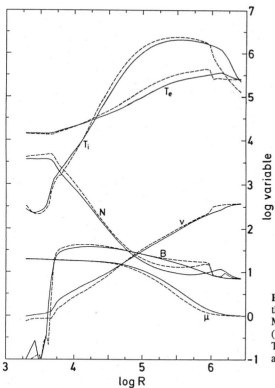

Fig. 5.25. Plasma flow parameters along the Giotto trajectory from the three-dimensional MHD calculations of Wegmann et al. (1987): Inbound (*solid*), outbound (*dashed*). The abscissa is the distance from closest approach in km.

$$T_i \simeq T_n + \frac{m_i}{3k}(v_i - v_n)^2. \tag{5.25}$$

In agreement with Wegmann et al. (1987), Haerendel (1987) and Cravens (1987) found that a relative flow speed between ions and neutrals of 1 to 3 km/s is needed to maintain the observed ion temperature outside the cavity. One result to be noted in the theoretical simulations of Wegmann et al. (1987) and Schmidt et al. (1988) is that the electron temperature remains on the order of 1 eV just inside of the magnetic cavity. In addition there is a sharp decrease of the ion density from about $5 \times 10^9 \, \mathrm{m}^{-3}$ to $1.6 \times 10^9 \, \mathrm{m}^{-3}$ at the boundary ($R \simeq 4700 \, \mathrm{km}$). The resultant change in the electron thermal pressure corresponds to an increase of the magnetic field strength of about 40 nT. This means that the formation of the magnetic cavity results largely from a balance between the magnetic pressure outside and the electron thermal pressure inside.

However, there are several difficulties with such a scenario for the following reasons. Model calculations of the electron temperature profile in a comet ionosphere (Mendis et al., 1985; Ip, 1986; Cravens and Körsmezey, 1986; Körösmezey et al., 1986; Wegmann et al., 1987; Marconi and Mendis, 1988; Schmidt et al., 1988) show that the value of T_e should have a value of a few 10 K within 100 km of the comet nucleus because water molecules are very good coolants (Shimizu, 1976); in the region where electron–molecule collisions are very frequent, the electrons should have the same temperature as the expanding neutral gas. As collisions become less frequent, the electron temperature rises as a result of the excess energy of new photoelectrons; there may be a gradual increase to a temperature of several 100 K out to a radial distance of 4000 to 5000 km (Fig. 5.26). The exact form of the

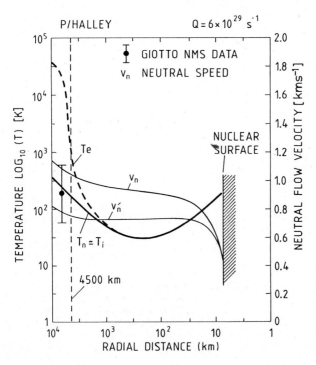

Fig. 5.26. A model for the gas velocity and the neutral gas and electron temperatures in the inner coma of Comet Halley (Ip, 1986). Two neutral flow speeds are given: Coma expansion without IR radiative cooling by water molecules (v_n) and with IR radiative cooling (v'_n).

electron temperature profiles depends on a number of factors: The cooling function of the H_2O molecules, the thermal conductivity of the electron gas, and the shape of the magnetic cavity. The essential result that the ionospheric electron temperature is kept at a value $T_e < 1000\,K$ should remain valid. Moreover, the measurements by the IMS and NMS experiments on Giotto (Balsiger et al., 1986; Krankowsky et al., 1986) did not detect any large-scale jump in the ion density (or count rate) at the contact surface, as illustrated in Fig. 5.27.

If an electron temperature profile as shown in Fig. 5.26 is used in a photochemical calculation, the ionospheric ion density in the vicinity of the contact surface is found to be on the order of $10^9\,m^{-3}$ with the nominal value for the photoionization rate of water vapor. Higher ion densities would be possible only if larger ionization rates were adopted (see Marconi and Mendis, 1988).

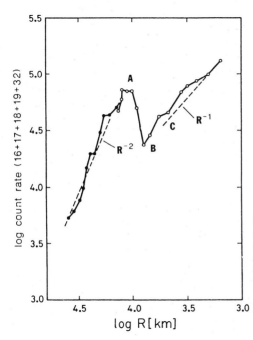

Fig. 5.27. The radial profile of the sum of the ion counting rates for masses 16 to 19 and 32 amu, showing that inside the magnetic cavity (C at $R = 4600\,km$) the total counting rate tends to follow a 1/R dependence and in the outer part ($R > 16000\,km$) a $1/R^2$ dependence (Balsiger et al., 1986). The peak in the ion count rate is at A and the minimum is at B.

Another feature not reproduced in the model calculations reported by Wegmann et al. (1987) concerns the decrease of plasma density observed at about $10^4\,km$ (see Fig. 5.27). When this was first detected by the IMS and NMS experiments, several possible explanations were given:

(a) A sudden enhancement of electron temperature at $R > 10^4\,km$ leads to a reduction in the electron recombination loss (Ip et al., 1987);

(b) a contribution from electron impact ionization at $R > 10^4\,km$ if the electron temperature is large enough, e.g., as a consequence of the critical velocity ionization effect (Formisano et al., 1982; Galeev and Lipatov, 1984; Haerendel, 1986);

(c) a rapid loss of plasma along the draped magnetic field lines at $R < 10^4\,km$ in the region of decreased density (Ioffe, 1968);

(d) a time-dependent effect reflecting a change in the solar wind conditions or the outgassing rate of the comet.

As discussed by Ip et al. (1987), the Giotto measurements of the electron population with energy larger than 10 eV showed that the electron impact ionization rate along the inbound trajectory was typically on the order of 10^{-7} s^{-1} or less. This then eliminates the possibility (b) since the photoionization rate is far more significant. The IMS observations of very low plasma velocity (< 1 km/s) in the ram direction between $5\ 10^3$ to 10^4 km suggest that the effectiveness of the plasma transport loss mechanism (c) is also limited. As for the possible effects of time-dependence (d), it should be noted that a bump in the plasma density at distance similar to that observed from Giotto was apparently detected by Vega 1 at its close approach to Comet Halley (Vaisberg et al., 1987a). In addition, a ground-based observation of the H_2O^+ ion emission at 619.8 nm, 13 h after the Giotto encounter showed that the H_2O^+ ion density distribution appeared to have a cutoff at $R \simeq 1.6 \times 10^4$ km (Ip et al., 1988). Thus the ion density maximum at around 10^4 km, depicted in Fig. 5.27, appears to have been a quasi-stationary feature in the coma of Comet Halley during the period of the Vega and Giotto encounters.

From these considerations it appears that item (a) might be the correct explanation: There is a sharp change in the electron temperature from $T_e \simeq 10^4$ K to 300 K

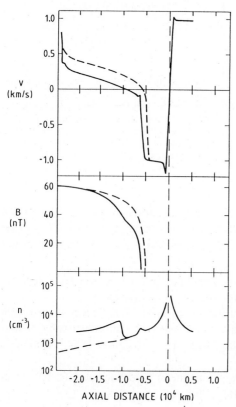

Fig. 5.28. Plasma velocity, magnetic field, and plasma density from a one-dimensional steady state calculation assuming a constant electron dissociative recombination coefficient (*a*) $\alpha = 3 \times 10^{-13}$ m^3/s for $R < 10^4$ km and $\alpha = 0$ for $R > 10^4$ km (*solid curves*) and (*b*) $\alpha = 3 \times 10^{-13}$ m^3/s everywhere (*dashed curves*) (Baumgärtel and Sauer, 1987).

at $R \simeq 10^4$ km. Unfortunately, the electron instrument on Giotto (Rème et al., 1986) was not designed to measure electrons with $kT_e < 10\,\mathrm{eV}$; thus there is no direct experimental support for this conjecture.

In a one-dimensional simulation of comet plasma flow, Baumgärtel and Sauer (1987) investigated the effects of electron recombination on the plasma density upstream of the magnetic cavity. The calculations include the effects of spherically symmetric expansion of the neutral gas, ion–neutral friction, and the piled up magnetic field. Hence, both the magnetic cavity and the plasma accumulation layer can be simulated in a consistent manner (see Fig. 5.28). Assuming that the electron dissociative recombination coefficient should go to zero for $R > 10^4$ km ($\alpha \simeq 3 \times 10^{-13}\,\mathrm{m^3/s}$ for $R < 10^4$ km), the numerical results show the formation of a density discontinuity near this location.

Even if there were a sudden decrease in T_e near 10^4 km, what might be the physical mechanisms leading to this feature? One possibility, as illustrated in Fig. 5.29, is that the electron temperature could depend on the configuration of the magnetic field, whether part of it is, or is not, connected to the interplanetary magnetic field. In a magnetically closed region, the electron temperature is largely determined by a balance between photoelectron heating and collisional cooling by the neutral coma. In contrast, there should be an extra heat input via thermal conduction along field lines from the external interplanetary medium, in the case of an open region. An investigation of the electron temperature distribution along a flux tube in an open field geometry indicates that, together with photoelectron heating, this is sufficient to maintain T_e on the order of a few 10^3 K.

5.4.3 Stability of the Magnetic Cavity

As mentioned before, the simplest model for the cometary magnetic cavity assumes that its boundary is a tangential discontinuity separating plasma flowing freely from

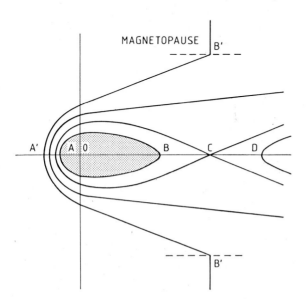

Fig. 5.29. A schematic view of the cometary magnetic field topology which is divided into the open region (i.e., connected to the IMF) and the closed region. The electrons in the open region are thermally coupled to the external solar wind via field-aligned heat flux. The markers in the figure, $B' - B'$ denote the foot points of a magnetic field line connected to the magnetopause, while C stands for the magnetic neutral point in the tail.

the direction of the nucleus from the approximately stationary plasma held by the magnetic field permeating the coma (Wallis and Dryer, 1976). The configuration is similar to that of the solar wind interacting with the interstellar plasma and magnetic field (e.g., Axford, 1972) and is subject to various magnetohydrodynamic instabilities, notably the Kelvin-Helmholtz, Rayleigh-Taylor, and fluting instabilities (Ershkovich et al., 1986, 1989; Ershkovich and Flammer, 1988). If the boundary is treated as a simple discontinuity between two uniform regions, ion–neutral friction is destabilizing.

In fact, as shown by the observations, the magnetic cavity in the case of Comet Halley does not obey this description since the magnetic field transition is rather broad except for a small "discontinuity" on the inner side (Fig. 5.21). Furthermore, the region within the boundary is completely free of magnetic field which would not be the case if the boundary itself were unstable, since this would permit some leakage of the field into the cavity. However, the description of the cavity formation on the basis of neutral gas friction, which seems to be a good approximation in the case of P/Halley, suggests that there could be a quite different type of instability from those mentioned above: Namely an overturning instability, similar to that which occurs in an unstable atmosphere, resulting from the delicate balance between frictional and magnetic field stresses, including curvature, and allowing for the effects of photoionization and recombination (Ershkovich et al., 1989). Not surprising, the analysis of this instability, based on model magnetic field profiles (Cravens, 1986; Ip and Axford, 1987a), has certain formal similarities to atmospheric problems. In general there is instability if the effects of photoionization and recombination are neglected, but stability (except very close to the cavity boundary) if they are included. The growth rate of the remaining instability is sufficiently low so that, with any reasonable time for convection along the cavity boundary, a significant increase would not be expected to occur.

5.5 Plasma Wave Turbulence

The first indication that the plasma environment near a comet could be extremely turbulent came from the ICE encounter with Comet Giacobini-Zinner. As shown in Fig. 5.30, very strong wave activity was seen even at a distance of about 2×10^6 km from the nucleus together with energetic particle fluxes (Scarf et al., 1986; Ipavich et al., 1986; Hynds et al., 1986). In addition to the high level of plasma wave turbulence, large-amplitude magnetic field variations were observed to be very strong with $\Delta B/B \simeq 1$ (Smith et al., 1986). The component of the magnetic field fluctuations along the average magnetic field direction was found to have a maximum in the wave power spectral densities which peaked at a period of 75 to 135 s. As discussed later, the ultra low frequency (ULF) waves observed at Comet Halley had somewhat different characteristics (see Glassmeier et al., 1987). These 100 s waves have been suggested to correspond to ion cyclotron waves for water-group ions ionized and picked up in the solar wind (Tsurutani and Smith, 1986).

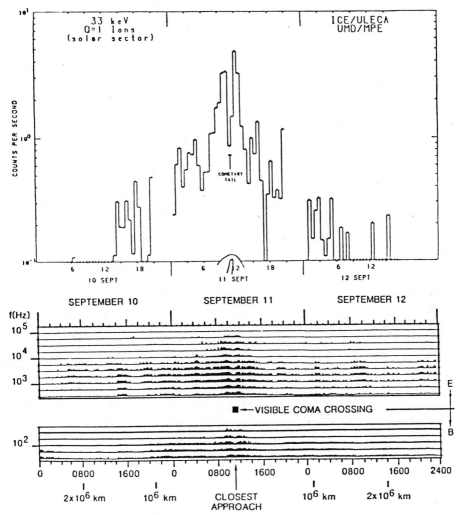

Fig. 5.30. A composite view of the plasma wave turbulence and the energetic particle flux in the vicinity of Comet Giacobini-Zinner as observed by the ICE spacecraft (Scarf et al., 1986; Ipavich et al., 1986).

Similar ULF fluctuations were also observed in the solar wind electron number density and the solar wind flow velocity (Gosling et al., 1986; Tsurutani et al., 1987). Figure 5.31 shows the correlation between such magnetic field variations and the solar wind plasma parameters which suggests that the waves are fast-mode magnetosonic waves. Note that the magnitude of the fluctuations increases as the bow shock was crossed (at $\simeq$ 09:30 UT) at a radial distance of about 2×10^5 km. Furthermore, at large distances from the comet, $R > 3 \times 10^5$ km, the 100 s waves were left-hand polarized; the polarization changed from left-hand elliptical to linear in regions near and inside the bow shock at $R < 2 \times 10^5$ km. Superposed on the long-period linearly polarized waves, left-hand polarized waves of shorter periods of 1 to 3 s were detected. Finally, the magnetic field turbulence is strongest at about

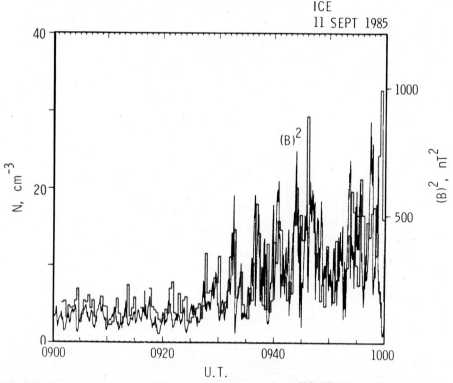

Fig. 5.31. A comparison of the variations in the square of the magnetic field magnitude (*in black*) and the electron plasma density (*in red*) as observed at Comet Giacobini-Zinner. For the interval illustrated the two parameters are strongly correlated, indicating that the fluctuations are fast mode MHD waves (Tsurutani et al., 1987).

09:00 to 10:30 UT and about 11:40 to 12:20 UT. The reduction in wave activity in the central region of the coma presumably is associated with the decrease in the plasma flow velocity that may become too low for the thresholds of the ion cyclotron or firehose instabilities to be reached.

The observations of strong hydromagnetic turbulence at P/Giacobini-Zinner by the ICE spacecraft have resulted in an avalanche of theoretical work addressing this issue. Both analytical treatments and numerical simulations have been applied to interpret the wave generation mechanisms and wave forms. With more results from the ICE, Vega, Sakigake, and Giotto probes becoming available, theoretical studies have also become increasingly sophisticated so that it is difficult to attempt a comprehensive coverage here. In the following, we shall summarize first some of the main points of the analytical considerations and follow with highlights from numerical simulations.

5.5.1 Ultra Low Frequency Waves

As described first by Wu and Davidson (1972) and Wu and Hartle (1974), the basic cause of the plasma turbulence is associated with free energy produced by pickup of cometary ions. Initially, the velocity of these ions relative to the solar wind plasma flow has two components, $v_\parallel$ and $v_\perp$, as shown in Fig. 5.12. In the quasi-perpendicular case with $\alpha \simeq \pi/2$, the comet ions are accelerated by the $v_{\rm sw} \times B$ electric field into cycloidal trajectories. In the solar wind frame, the new ions gyrate around the magnetic field lines with large pitch angles. Under the assumption that the initial distribution function of the new ions can be presented as $\delta(v_\parallel - v_{\parallel\rm o})\,\delta(v - v_{\rm o})$, the dispersion equation has its two most unstable roots at

$$\omega \simeq kv_{\parallel\rm o} - \Omega_i \simeq \pm kV_{\rm A}, \tag{5.26}$$

representing left and right polarized Alfvén (L and R) waves ($\omega \ll \Omega_i$) with a growth rate given by (Wu and Davidson, 1972; Lee and Ip, 1987; Lee, 1989)

$$\tilde{\gamma} \sim |\Omega_i| \left[\frac{V_{\rm A}}{|v_{\parallel\rm o}|} \frac{\frac{1}{2}\tilde{n}_i m_i v_\perp^2}{B_{\rm o}^2/(2\mu_{\rm o})} \right]^{1/3}. \tag{5.27}$$

In the above equations, k is the wave number, Ω_i the gyration angular velocity of ion species i, $V_{\rm A}$ the Alfvén speed, $v_{\parallel\rm o}$ the plasma velocity component along the magnetic field direction with unit vector $e_\parallel$, and $\tilde{n}_i$ the number density of the new cometary ions. In the spacecraft frame with $v_{\rm sw} \cdot e_\parallel > V_{\rm A}$, both R- and L-mode waves are left-hand polarized since the wave polarization in this frame is dominated by the Doppler shift. The propagation direction of these waves should also be sunward along the average, spiral magnetic field direction as observed in the 100 s waves.

If the average magnetic field is nearly parallel to the solar wind flow direction, the new comet ions are at first stationary with respect to the comet. However, the resultant cold beam of heavy ions having speed about $v_{\rm sw}$ relative to the solar wind leads to the nonresonant firehose instability (Gary et al., 1984; Sagdeev et al., 1986f; Brinca and Tsurutani, 1987),

$$p_\parallel > p_\perp + \frac{B^2}{2\mu_{\rm o}}, \tag{5.28}$$

where $p_\parallel$ is the thermal plasma pressure component parallel to the magnetic field and $p_\perp$ the component in the perpendicular direction.

This implies that the firehose instability, which excites long wavelength waves with $k < \Omega_i/v_{\rm sw}$, would become possible when the cometary ion pressure exceeds the sum of the magnetic field and solar wind thermal pressure. For example, in the case of Comet Halley, the firehose instability might occur if $v_{\rm sw}$ is nearly parallel to B beginning at a few 10^6 km upstream (Sagdeev et al., 1986f).

Galeev et al. (1988) have invoked the firehose instability as a possible mechanism for causing the formation of the cometopause, discovered from Vega 2 observations (Gringauz et al., 1986a,b). The nonresonant firehose instability generally has a lower growth rate than the resonant ion cyclotron instabilities. Furthermore, a large beam velocity, or a high concentration of beam ions, is needed for this instability. It therefore has been concluded that the nonresonant effect is generally not important except in some localized regions. In contrast to the resonant ion cyclotron waves, the nonresonant R-mode propagates antiparallel to the ion beam such that it appears as a right-hand mode in the spacecraft frame.

Another type of unstable wave to be expected from the cometary ion pickup process is the whistler mode (Wu and Davidson, 1972; Goldstein and Wong, 1987; Lee and Ip, 1987; Lee, 1989). The cyclotron relation of whistler waves ($\omega \gg \Omega_i$) is simply $\omega_r \simeq k v_z$. If the pickup ion distribution is a narrow ring, the approximation for the wave frequency in the spacecraft frame is $\omega_{sc} = -\Omega_i \ll \omega$. This property has led Goldstein and Wong (1987) to suggest that the 1 to 7 s wave packets seen at P/Giacobini-Zinner have their origin in such water-group ion whistler waves. Another explanation is that the 1 to 7 s wave packets were excited by the hydrogen pickup ion ring with $\omega_{sc} = -\Omega_p$. Thorne and Tsurutani (1987) have pointed out that for $v_{sw} \cos\theta / V_A \simeq 3$ to 5, whistler mode waves with $\omega/\Omega_i \simeq 10$ to 30 and the fast MHD mode with $\omega/\Omega_p \simeq 0.01$ to 0.02 are possible; since the wave packets are detected only at the steepened edges of compressive magnetosonic waves with periods of about 100 s, they could be right-hand whistler waves as proposed by Goldstein and Wong (1987).

The main consequence for resonating ions of the various plasma instabilities is to rapidly isotropize of the ring–beam distribution into a spherical shell. The growth rate is usually on the order of 0.1 Ω_i to Ω_i. The diffusion of the new ions in the phase space distribution is illustrated in Fig. 5.32 where it is shown that the particle scattering to smaller pitch angles is caused by the normal Doppler cyclotron resonance

$$\omega - k v_{sw} = +\Omega_i, \qquad\qquad (5.29)$$

with the L-mode waves; and the pitch-angle scattering to larger angles is caused by the anomalous Doppler resonance

$$\omega - k v_{sw} = -\Omega_i, \qquad\qquad (5.30)$$

yielding R-mode waves (Sadgeev et al., 1986). The L-mode waves are subject to strong attenuation (Winske et al., 1985a; Thorne and Tsurutani, 1987), whereas the R-mode waves propagating toward the Sun remain as the only growing resonant ion cyclotron waves. Because the phase velocity $V_{ph} = \omega/k < v_{sw}$, these R-mode waves generated upstream of the comet are carried back to the spacecraft by the solar wind and observed as being left-hand polarized because of the Doppler effect. According

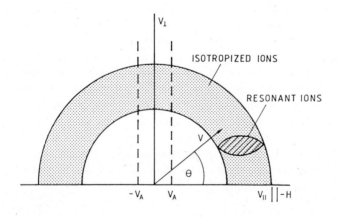

Fig. 5.32. The diffusion of the pickup comet ions in the velocity phase space as a result of wave–particle scattering; V_A is the Alfvén velocity and $v_\perp$ and $v_\parallel$ are, respectively, the velocity components perpendicular and parallel to the magnetic field (cf. Sagdeev et al., 1986f).

to this scenario, Tsurutani et al. (1987) have proposed that the continuous growth of the R-mode waves over a distance of about 10^6 km could allow the wave energy to increase by a factor of 30. The steepening of the large-amplitude magnetosonic waves close to the bow shock is one consequence of wave amplification.

The hydromagnetic turbulence measured at P/Halley by the Vega and Giotto spacecraft differs in several respects from that observed at P/Giacobini-Zinner (Riedler et al., 1986; Neubauer et al., 1986; Glassmeier et al., 1987,1988). First, the amplitude of the magnetic field fluctuations was smaller ($\Delta B/B \sim 0.2$) at P/Halley as compared with the large-amplitude variations ($\Delta B/B \sim 1$) at P/Giacobini-Zinner. Second, neither the steepening of the 100 s waves nor the packets of 3 s waves, that played such a prominent role in the plasma turbulence at P/Giacobini-Zinner, were observed at P/Halley. On the other hand, a pronounced peak in the wave power spectrum at about 140 s was seen, consistent with a pump wave from the water-group ion cyclotron resonant waves detected first at P/Giacobini-Zinner (Tsurutani and Smith, 1986; see also Fig. 5.33).

Careful analyses of the magnetometer and solar wind plasma measurements obtained in a quasi-parallel region upstream of the comet bow shock by Glassmeier et al. (1988) have shown that the 140 s waves, exhibiting a good correlation between magnetic field variations and the solar wind velocity, are characteristic of Alfvén-type turbulence and that they propagate towards the Sun in the plasma frame. In a quasi-perpendicular region upstream of the bow shock, a shoulder was detected in the power spectrum at around 30 s (see Fig. 5.34). This spectral component could be caused by a left-hand ion cyclotron (Sharma and Patel, 1986) or an Alfvén–ion cyclotron anisotropy instability (Gary et al., 1984). If this wave peak is to be inter-

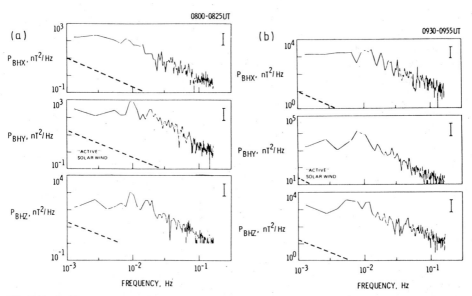

Fig. 5.33a, b. The wave power spectral density along (BHX) and orthogonal to (BHY and BHZ) the average field direction as observed by the magnetometer experiment on ICE at Comet Giacobini-Zinner (Tsurutani and Smith, 1986): (a) within a time interval of about one hour upstream of the bow shock; and (b) downstream of the bow shock.

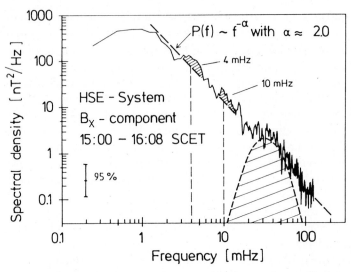

Fig. 5.34. The magnetic field power spectrum of B_X in a quasi-perpendicular region as observed by Giotto at Comet Halley. The hatched area marks the spectral range where the spectrum exhibits a shoulder type variation assumed to indicate additional wave growth (Glassmeier et al., 1988).

preted in terms of an ion cyclotron instability, a period of 30 s implies beam ions with $M/q = 2$, i.e., He^{2+} or H_2^+; however, broadening of the ring distribution of water-group ions could also lead to unstable waves in this frequency range.

The magnetic field turbulence spectrum follows a power law with a spectral index of about 2. That this value is significantly different from the value of 5/3 expected from a fully developed Kolmogroff spectrum could mean that the turbulent cascade has not progressed far enough (Glassmeier et al., 1988).

As the Giotto spacecraft entered the magnetosheath region, the ULF fluctuations increased to a higher level in comparison with the variations in the upstream region. This could be related to the presence of the comet bow shock, where still anisotropic comet ions picked up at large distances would be a source of extra free energy. In addition, as shown by numerical simulations (Galeev and Lipatov, 1984; Omidi and Winske, 1987; Omidi et al., 1986), the generation of wave turbulence could be directly coupled to the formation of the shock itself as a result of wave amplification associated with compression of the medium. The detailed processes of the pump wave generation and subsequent turbulence cascading in the vicinity of the bow shock and in the magnetosheath have not yet been clarified. Whereas in the upstream region, the magnetic field fluctuations are dominated by variations in the transverse component, the magnetosheath fluctuations contain a large compressional component. These compressional magnetosonic waves should make transit time damping acceleration an interesting possibility in addition to first- and second-order Fermi acceleration (see Sect. 5.6).

Using a quasi-linear analysis in which the anisotropy of the pitch angle distribution of the new ions is proportional to the ion production rate and inversely proportional to the intensity of the resonant waves, Galeev et al. (1986) have estimated the intensity of the excited waves in the following way

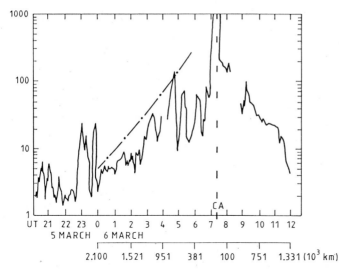

Fig. 5.35. The energy of magnetic field fluctuations $\int (B_{X\nu}^2 + B_{Y\nu}^2 + B_{Z\nu}^2) d\nu$ in the frequency range 10^{-3} Hz $< \nu < 10^{-2}$ Hz for various distances from Comet Halley (Galeev et al., 1986).

$$\int \frac{|B_k|^2}{\mu_o} \frac{dk}{2\pi} \simeq \frac{V_A(\rho v_i - \rho_\infty v_{i\infty})}{2}, \tag{5.31}$$

where $|B_k|^2/(2\mu_o)$ is the spectral energy density of the resonant waves, ρ_∞ and $v_{i\infty}$ the density and flow speed of the solar wind at infinity and ρ, and v_i the corresponding local values. This result was compared to the Vega 2 measurements of the magnetic field variations with frequency range between 10^{-2} and 10^{-3} Hz. As shown in Fig. 5.35, the agreement in the radial behavior of both the theoretical and the experimental values of $(\Delta B/B)^2$ is quite good. In contrast, such a radial dependence of the magnetic field turbulence was not observed by the Giotto magnetic field experiment (Glassmeier et al., 1987).

More recently, Gary et al. (1988) have investigated wave–particle interactions of two pickup ion species, i.e., H^+ and O^+ ions. The time evolution of new comet ions in velocity space was studied by means of one-dimensional electromagnetic hybrid simulations of homogeneous plasmas. It was found that: (a) With weak injection rates, at large cometocentric distances the R-mode H^+ ion cyclotron waves have larger growth rates, higher frequencies, and shorter wavelengths in comparison with the O^+ R-mode waves; (b) initially, pitch-angle scattering promotes the formation of thin shell-like velocity distributions of the pickup cometary H^+ ions in a time scale shorter than that for oxygen ion scattering; and (c) later, when sufficient energy has been accumulated by the O^+ ions to permit fast growth of the oxygen ion cyclotron resonant instability, the corresponding wave energy $(\Delta B/B)^2$ increases in an exponential fashion.

As the O^+ ions are scattered into a shell-like distribution, the pickup protons are found to be subject to stochastic acceleration. Gary et al. (1988) have developed a model in which the protons gain energy through cyclotron resonances with a number of waves with increasingly longer wavelength. Hence, with $v_\parallel \simeq \Omega_p/k$ and the k value determined by the oxygen ion cyclotron resonant instability, protons with gyration angular velocity Ω_p might be accelerated to $(m_O + m_p)^2$ times their injection energy. Thus, an H^+ ion picked up in the fast moving solar wind flow of about 300 to 600 km/s might easily be accelerated to an energy of 250 keV to 1 MeV. On the other hand, stochastic acceleration of the oxygen ions would not be as efficient. The results from this study are in qualitative agreement with the model of stochastic acceleration of cometary ions proposed by Ip and Axford (1986) and others.

From such one-dimensional computer simulations of the R-mode ion cyclotron resonant instability, an empirical relation could be found for the peak energy density of the magnetic field fluctuations (Gary et al., 1988)

$$\frac{|\Delta B|^2}{B^2} \simeq 10 \left(\frac{m_b}{m_c} \right)^2 \left(\frac{v_b}{V_A} \right)^2 \Omega_p \tau_b, \tag{5.32}$$

where m_b is the mass of the beam ions, m_c that of the background or core ions, v_b the relative drift velocity between beam ions and core ions, and τ_b the injection time scale of new ions. Gary et al. (1988) have compared the theoretical values as predicted by (5.32) with the experimental results obtained at P/Giacobini-Zinner and found good agreement, whereas there is a factor of 10 difference between the theoretical predictions and the magnetic field data obtained by Giotto at P/Halley.

5.5.2 High-Frequency Waves

As for high-frequency waves, one somewhat surprising discovery of cometary kilometric radiation (CKR) was made by the Sakigake spacecraft as it flew by P/Halley (Oya et al., 1986). The 10 m tip-to-tip dipole antenna of the plasma wave experiment picked up discrete radio emissions in the frequency range of 30 to 195 kHz. These emissions occurring at the local plasma frequency may be the result of conversion of the electrostatic plasma waves to electromagnetic waves in the turbulent plasma environment of Comet Halley. The generation of the CKR thus may be similar to the Type II solar radio bursts from coronal shock waves (Benz and Thejappa, 1988). In fact, Oya et al. (1986) identified the comet bow shock as the source region of the CKR. It was proposed that motion of the comet bow shock could excite Alfvén and Langmuir turbulences that eventually lead to the CKR emissions. Lakhina and Buti (1988), on the other hand, invoke nonlinear interaction of Alfvén solitons and Langmuir waves as the main mechanism.

The analog between the comet bow shock and interplanetary shocks can also be found in plasma wave measurements by the ICE spacecraft at P/Giacobini-Zinner (Scarf et al., 1986). Kennel et al. (1986) compared the electric field power spectra obtained in the shock interaction regions of P/Giacobini-Zinner and those by ISEE 3 at quasi-parallel interplanetary and Earth's bow shocks and found that they bear many similarities. The electric field spectra upstream of the bow shock of P/Giacobini-Zinner had two peaks, one near 1.78 Hz and the other near the electron plasma frequency at 17.8 to 31.6 kHz (see Fig. 5.36). The waves near the electron plasma frequency are related to upstream electron heat flux (Fuselier et al., 1986). The

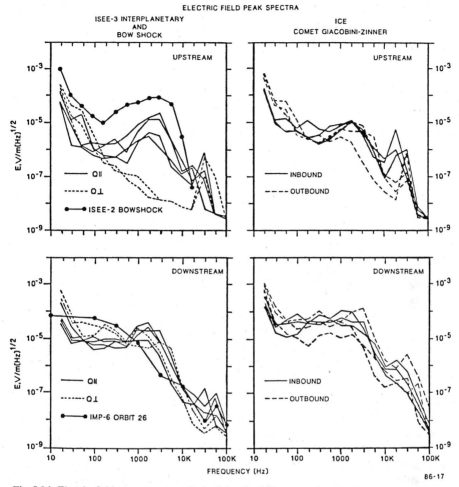

Fig. 5.36. Electric field power spectra obtained by the ICE spacecraft at P/Giacobini-Zinner. The top row compares interplanetary and bow shock (*left*) with cometary (*right*) peak E-field upstream spectra. The bottom row compares downstream spectra in the same format. The ISEE 3 spectra are the peaks registered in the 16 s time interval; the ICE spectra are derived from the peaks in a 5 minute interval. The upstream inbound (*solid line*) peak ICE spectra were taken from the intervals 0916–0921, 0905–0910, and 0855–0900 UT; the upstream outbound (*dashed line*), from the time intervals 1237–1242, 1247–1252, and 1257–1302 UT. The downstream inbound (solid line) ICE spectra were taken from 0931–0936, 0941–0946, and 0951–0956 UT; the downstream outbound (*dashed line*) spectra, from 1217–1222, 1207–1212, and 1157–1202 UT (Kennel et al., 1986).

downstream spectrum was strongly enhanced at frequencies of few 100 Hz, and the whistler mode was also excited (Kennel et al., 1986).

As discussed by Hartle and Wu (1973) and Sagdeev et al. (1986f), the pickup of new cometary ions in a perpendicular ring distribution could excite the ion loss cone instability with wave frequencies near the lower hybrid frequency of $f_{LH} \simeq 6$ to 12 Hz. The ICE E-field instrument, because of the limitation of the lowest frequency channel to 17.8 Hz, could not make direct measurements of the electrostatic lower

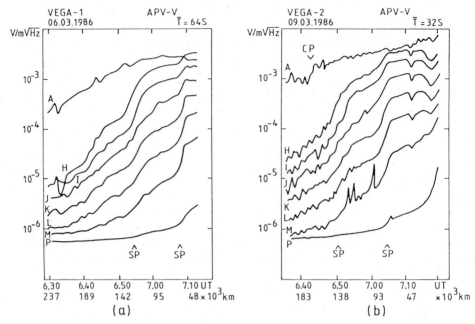

Fig. 5.37a, b. Electric field intensities measured by the filter banks on (a) Vega 1 (64 s average) and (b) Vega 2 (32 s average). The central frequencies of the filters are: A – 8 Hz, H – 860 Hz, I – 1.8 kHz, J – 3.7 kHz, K – 7.7 kHz, L – 15 kHz, M – 36 kHz, P – 203 kHz. The labels SP indicate the boundaries detected by the SP-1 experiment on Vega; and the label CP gives the position of the cometopause (Mogilevsky et al., 1987).

hybrid waves. However, an indirect identification using the observed low frequency magnetic turbulence has been attempted by Coroniti et al. (1986).

The high-frequency plasma wave analyzer (APV-V) on Vega 1 and 2 measured electric fields at frequencies in the bandwidth 8 to 300 kHz using the electric antenna (Grard et al., 1986). The field amplitude increases significantly at several locations, i.e., at 2×10^5 km, 1.2 to 1.5×10^5 km and 5 to 7 10^3 km in the coma of Comet Halley (see Fig. 5.37). A peak in the electric field spectra near 30 Hz and another peak at 200 to 400 Hz were detected in the cometocentric range of 7×10^4 to 1.5×10^4 km and 1.6 10^5 to 1.0 10^4 km, respectively (Mogilevsky et al., 1987). The APV-V experimenters interpreted the 30 Hz waves as electrostatic oscillations at frequencies around the lower hybrid frequency. The measurements of a pronounced peak at 15 Hz in localized regions by the Vega low-frequency plasma wave detector (APV-N) might also be indicative of lower hybrid wave activities (Klimov et al., 1986). The significance of the lower hybrid waves for electron and ion acceleration will be discussed in the next section.

5.6 Ion Acceleration

5.6.1 Energetic Ions in the Distant Coma

The first indication of ion acceleration in the vicinity of a comet was obtained during the ICE encounter with Comet Giacobini-Zinner when very intense fluxes of cometary ions were registered by the energetic particle detector (Hynds et al., 1986). As shown in Fig. 5.38, even at 10^6 km away from the comet, the ion flux at about 100 keV is quite appreciable. At distances beyond about 10^5 km from the nucleus, the ion anisotropy is very large (Fig. 5.39) and there is occasionally a good correlation between the ion intensity and the angle between the solar wind velocity and the interplanetary magnetic field (IMF). A simple interpretation of this effect is that the maximum speed of a new-borne ion, measured in the spacecraft frame, should be $2v_\perp$, i.e., a drift speed of $v_\perp$ plus an equal gyration speed. The maximum energy measured in the spacecraft frame is therefore

$$E_{\max} = \frac{1}{2} m_i (2v_\perp)^2 = 2m_i v_{sw}^2 \sin^2 \alpha. \tag{5.33}$$

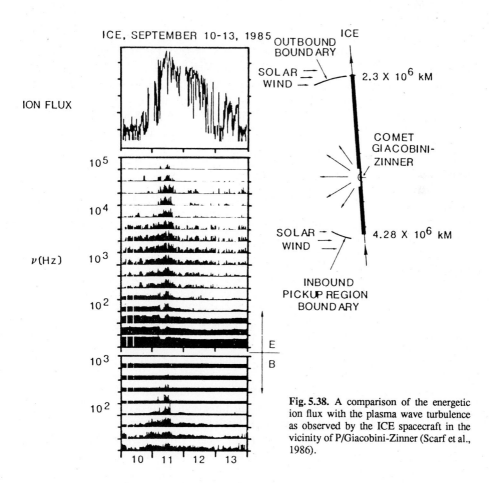

Fig. 5.38. A comparison of the energetic ion flux with the plasma wave turbulence as observed by the ICE spacecraft in the vicinity of P/Giacobini-Zinner (Scarf et al., 1986).

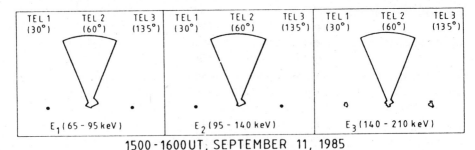

1500-1600UT, SEPTEMBER 11, 1985

Fig. 5.39. Anisotropies of energetic ions obtained from the energetic ion experiment on ICE at Comet Giacobini-Zinner: Sector plots from the three telescopes are shown for the period 1500–1600 UT on Sept. 11, for channels 1 to 3 (65–95, 95–140, 140–210 keV, respectively, for water-group ions) (Sanderson et al., 1986).

In the case of a quasi-perpendicular magnetic field, at $\alpha \simeq 90°$, $E = E_{max}$ for ions with a velocity vector pointing along the solar wind direction, and $E \ll E_{max}$ for particles moving in the opposite direction resulting in very high anisotropies (see Fig. 5.39). Note that because the energy spectra are very steep, a small change in the angle α could lead to a large variation in the energetic ion flux within the energy windows of the detectors.

Since $v_{sw} \simeq 400$ km/s, $E_{max} \simeq 60$ keV for water ions, with a cutoff in the energy distribution at this value. However, the measured energy spectra extended smoothly beyond 100 keV and even up to 0.5 MeV at P/Halley (Somogyi et al., 1986, 1990; McKenna-Lawlor et al., 1986). Significant ion acceleration must therefore have occurred. That such ion acceleration might take place near comet comae was first suggested by Amata and Formisano (1985) based on an analogy between diffusive shock acceleration at the Earth's bow shock and at the comet bow shock and noting the likely turbulent nature of the comet plasma.

Ion acceleration in the upstream region of the Earth's bow shock is mainly caused by diffusive shock acceleration, although direct reflections of particles at the shock also play a role. This process, which is a form of first-order Fermi acceleration, relies on the multiple random scattering of charged particles across the shock front separating "scattering centers" on the supersonic and the subsonic sides (Bell, 1978; Axford et al., 1977; Krimsky, 1977; Toptyghin, 1980). This is depicted in a qualitative manner in Fig. 5.40; the accelerated ion population builds up ahead of the bow shock with a characteristic length scale of $L_{DC} \simeq K_n/v_{sw}$ where K_n is the diffusion coefficient of the medium normal to the shock. Whereas $L_{DC} \simeq 5 \times 10^4$ km in the case of the Earth's bow shock (Lee, 1982), Amata and Formisano (1985) estimated that $L_{DC} \simeq 10^6$ km for the case of ion acceleration at the bow shock of Comet Halley.

Numerical experiments are helpful in illustrating the nature of a comet shock, which has a thickness on the order of a few comet ion gyroradii (Galeev and Lipatov, 1984; Omidi et al., 1986; see also Fig. 5.41). Thus, with an ambient magnetic field of 5 nT, the shock thickness of about three gyroradii of O^+ would be $3R_g = 3m_i v_{sw}/(q_i B) \simeq 4 \times 10^4$ km, i.e., a large fraction of the distance of the bow shock from the nucleus in the case of P/Giacobini-Zinner. This is perhaps one reason why the shock structure observed at Comet Giacobini-Zinner was not very well defined.

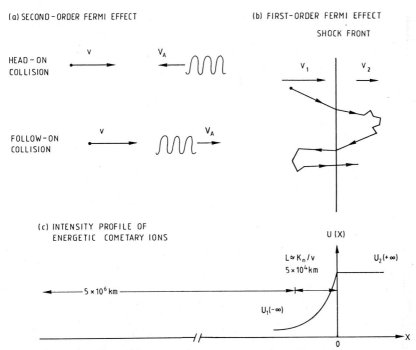

Fig. 5.40a–c. Schematic views of the acceleration process in the MHD turbulent regions of comet comae: (a) stochastic (second-order) Fermi acceleration at large distance from the comet; (b) (first-order) diffusive shock acceleration near the comet bow shock; (c) the intensity profile of the energetic ions resulting from diffusive shock acceleration.

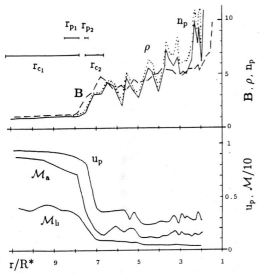

Fig. 5.41. Distribution of the mass density ρ, the number density n_p, the velocity v_p, the magnetic field B, the Alfvénic $\mathcal{M}_A$ and magnetosonic $\mathcal{M}_h$ Mach numbers along the Sun – nucleus line obtained as a result of numerical simulation. The proton gyroradius is r_p and the cometary ion gyroradius is r_c. From Galeev and Lipatov (1984).

In order to determine the importance of shock acceleration we should begin by comparing the expected energy spectrum with the observations. In the simplest situation, the basic process may be described in terms of one-dimensional diffusion–convection equations for the differential number density $U(E, x) = dn_i/dE$

$$\frac{\partial U}{\partial t} + \frac{\partial S}{\partial x} = -\frac{2}{3} v \frac{\partial^2}{\partial x \partial E}(EU), \tag{5.34}$$

where

$$S = v \left[U - \frac{2}{3} \frac{\partial}{\partial E}(EU) \right] - K_n \frac{\partial U}{\partial x}, \tag{5.35}$$

is the particle "flux" or "current" and K_n is the diffusion coefficient. The assumptions that the shock front can be approximated to be infinitely thin and the plasma velocity can be approximated as $v = v_1$ in $x < 0$, and $v = v_2$ in $x > 0$, with $v_1 > v_2$, lead to the following solution in $x < 0$, under steady conditions

$$U(x, E) = U_1(E) + [U_2(E) - U_1(E)] \exp(x/L_{\text{DC}}), \tag{5.36}$$

where U_2 is the number density at the position of the shock ($x = 0$) and $L_{\text{DC}} = K_n/v_1$ is the characteristic diffusion–convection length of the accelerated ions (see Fig. 5.40c). Furthermore, at the shock,

$$U_2(E) = (\epsilon - 1) \left(\frac{v_1}{v_2} \right) E^{-\epsilon} \int_{E_1}^{E} E^{\epsilon-1} U_1(E) dE, \tag{5.37}$$

where

$$\epsilon = 1 + \frac{3v_2}{2(v_1 - v_2)}, \tag{5.38}$$

and E_1 is the particle injection energy in the frame of the plasma ($E_1 = m_i v_1^2/2$). For a monoenergetic input spectrum ($E = E_1$), the diffusive shock acceleration leads to a power law energy spectrum for $U(E) \propto E^{-\epsilon}$ with index ϵ given by (5.38). Taking $v_1/v_2 = 1.4$, we find $\epsilon \simeq 4.8$ according to (5.38), whereas the corresponding spectral index of the energetic ions observed at P/Giacobini-Zinner is $\epsilon_1 \simeq 8.5$ upstream and $\epsilon_2 \simeq 6.5$ downstream. There could be several reasons for the discrepancy: First, the idealized conditions assumed, i.e., steady state and one-dimensional model, might not be applicable; second, the flow speed of the solar wind plasma is comparable to the velocity of the accelerated ions which violates a basic assumption in the equations of transport, (5.34 and 5.35), (Gleeson and Axford, 1967). Nevertheless, it is interesting to note that, in spite of being weak with $\mathcal{M} \simeq 2$, a comet bow shock may produce a discernable energization. The hardening of the energy spectra observed by ICE as it crossed the bow shock of P/Giacobini-Zinner (see Fig. 5.42) and similar effects seen at P/Halley could be a result of shock acceleration even if it is a secondary process.

The energization of cometary ions with energies larger than $100\,\text{keV}$ at large distances upstream ($R > 5 \times 10^6$ km) can have nothing to do with the above mentioned shock acceleration mechanisms. What could be of major importance in this region is the stochastic or second-order Fermi acceleration process. In recognition

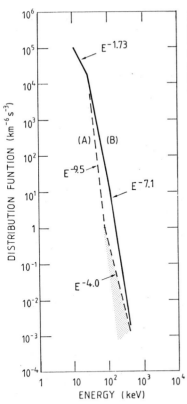

Fig. 5.42. Approximations to the observed distribution functions of energetic cometary ions in terms of power law distributions: (A) preshock and (B) post shock. The shaded region denotes the high-energy tail in the preshock energy spectrum (Ip, 1988). The curves are adopted from the energetic ion observations by the ICE spacecraft at Comet Giacobini-Zinner (Richardson et al., 1987).

of this possibility, Ip and Axford (1986) have formulated a simple model to check the validity of this assumption. By assuming that the mean free path for particles scattered by Alfvén waves is on the order of a few ion gyroradii, they showed that ions in the MeV range could be produced in the vicinity of the coma of Comet Halley. Further, more detailed theoretical work along this line has been carried out by Gribov et al. (1986), Isenberg (1987), Gombosi (1987b), and Barbosa (1989).

Subsequent publications by the experimental teams of energetic ion instruments on ICE (Richardson et al., 1987) and on Giotto (McKenna-Lawlor et al., 1987) have shown that there are limitations to the efficiency of the second-order Fermi acceleration upstream of the bow shock. First, the energy spectra of the energetic ions observed by ICE in the preshock region can be matched only when the scattering mean free path is on the order of 10 R_g, and about 30 R_g in the case of Comet Halley (see Ip and Axford, 1987b); the stochastic acceleration thus may be best described in the region of the weak scattering limit. Gombosi (1987b) has independently reached a similar conclusion using a self-consistent flow field model. The stochastic acceleration process can be expressed in terms of a Fokker-Planck equation for the differential number density assumed isotropic

$$\frac{\partial U}{\partial t} = D_o E \frac{\partial^2 U}{\partial E^2} + \frac{1}{2} D_o \frac{\partial U}{\partial E} - \frac{U}{\tau_D}, \tag{5.39}$$

where $D_0 = 3qBv^2/2$ is assumed to be a suitable form for the energy diffusion coefficient resulting from Alfvén wave scattering (Fisk, 1976) for ions with charge q. In (5.39), v is the particle velocity and τ_D is the diffusive escape time scale assuming a leaky box treatment is appropriate (see Barbosa, 1989). The new ions are subject to pitch-angle scattering by the plasma turbulence satisfying the following equation

$$\frac{\partial U'}{\partial t} = \frac{\partial}{\partial \mu}\left(D_{\mu\mu}\frac{\partial U'}{\partial \mu}\right),\tag{5.40}$$

where $U' = U'(t, \mu, E)$,

$$D_{\mu\mu} = D_\alpha(1 - \mu^2)[|\mu|^{\hat{q}} + b],\tag{5.41}$$

is the pitch-angle scattering coefficient ($\mu = \cos\theta$ with θ as the pitch angle) and D_α, $\hat{q}$ and b are constants. According to Jokipii (1971) $D_{\mu\mu}$ is related to the spatial diffusion coefficient, and $K_\parallel$, parallel to the average magnetic field, is given by the expression

$$K_\parallel = \frac{2\Lambda E}{3R_g qB} = \frac{v^2}{4}\int_{-1}^{1}\frac{d\mu(1 - \mu^2)^2}{D_{\mu\mu}},\tag{5.42}$$

where v is the particle speed. Thus, the time scale for pitch-angle scattering in the comet coma region can be used to infer the effective value of the spatial scattering mean free path Λ. Numerical integrations of (5.40) suggest that, for $\hat{q} = 2$ and $b = 0.01$ to 0.3, the pitch-angle isotropization time scale, τ_{pA}, is about 3 to 8 Λ/R_g s (see Fig. 5.43). On the basis of the partial shell structures in the velocity distributions

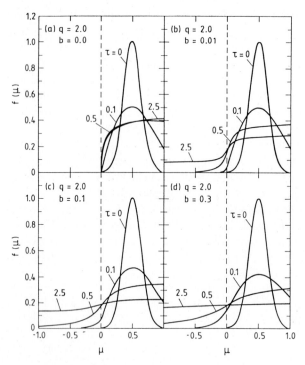

Fig. 5.43. The time evolution of the pitch-angle distribution for different values of $\hat{q}$ and b. Note that $\tau = D_\alpha t$ (Ip, 1988).

of pickup ions observed at Comet Halley ahead of the comet bow shock (Neugebauer et al., 1987), τ_{pA} may be estimated to be about 300 s for $\Delta B/B < 0.2$ and hence $\Lambda/R_g \gtrsim 30$ (Ip, 1988). Within 10^5 km of the comet bow shock, the magnetic field turbulence was observed to be enhanced such that $\Delta B/B \to O(1)$; thus, $\tau_{pA} \simeq 50$ s and $\Lambda/R_g \to 6$. This result is consistent with other estimates obtained by fitting the energy spectra.

The rapid isotropization of the cometary ions within 10^5 km of the comet bow shock may be a consequence of the chaotic trajectories characteristic of the "ion trapping" near the shock front shown in Fig. 5.41. Together with the pitch-angle scattering, the energy spectra were observed to harden (see Fig. 5.42) leading to a distinct enhancement of the energetic particle flux just ahead of and behind the bow shock. It is unresolved whether diffusive shock acceleration alone is sufficient to boost the particle energy, or stochastic acceleration with an enhanced scattering rate could be an important cause. A further complication is that, as discussed in Sect. 5.5, magnetosonic waves appear to account for a significant portion of the MHD turbulence in localized regions. Consequently, the transit time damping effect could contribute to ion acceleration (Glassmeier et al., 1987). As noted by Fisk (1976), transit time damping does not require oppositely propagating wave trains with similar magnitudes. This condition was not satisfied by the transverse Alfvén waves in the comae of Comets Halley and Giacobini-Zinner.

Other mechanisms under consideration include the lower hybrid turbulence and ion cyclotron wave absorption. In the comet environment, the observed lower hybrid turbulence (Scarf et al., 1986; Klimov et al., 1986) could be generated by the comet ion pickup process as well as by the bow shock formation. Buti and Lakhina (1987) suggested that these waves should play a role in ion acceleration on the basis of an estimate of the maximum energy achievable in such an acceleration process ($E_{max} \simeq 800$ keV for a wave amplitude of $\mathcal{E}_w \simeq 3$ mV/m). Note that in the case of electron beam excitation of lower hybrid waves, the wave vectors are nearly normal to the magnetic field, fulfilling the condition (Bryant et al., 1985)

$$\frac{k_{\parallel}}{k_{\perp}} < \left(\frac{m_e}{m_i}\right)^{1/2}, \qquad (5.43)$$

such that ions moving in resonance with the waves would satisfy

$$v_{\perp} \simeq v_b \left(\frac{k_{\parallel}}{k_{\perp}}\right) < v_b \left(\frac{m_e}{m_i}\right)^{1/2}, \qquad (5.44)$$

where v_b is the velocity of the electron beam, m_e the electron mass and m_i the ion mass. In this way, the accelerated ions would have their maximum energy reaching the electron beam energy. The resultant energy diffusion coefficient can be written as (Retterer et al., 1983)

$$D_{\perp} \simeq \frac{q^2}{4\pi\epsilon_0 m_i} \frac{|\mathcal{E}_w|^2}{\omega_i}. \qquad (5.45)$$

Thus, given finite time (the transit time through a distance of 10^6 km is on the order of 10^4 s in the outer coma), the energy gain from this form of stochastic acceleration would be

$$\Delta E \simeq \frac{q^2 \mathcal{E}_w^2 \Delta t}{4\pi\epsilon_0 m_i \omega_i}, \tag{5.46}$$

For $\mathcal{E}_w \simeq 3\,\mathrm{mV/m}$, $\Delta E \simeq 700\,\mathrm{keV}$; the lower hybrid turbulence thus could be of importance in particle acceleration if the effect is not limited to a very localized region as defined by the magnetic field configuration.

5.6.2 Energetic Electron Acceleration

As discussed in Sect. 5.5.2, the detection of intense electron plasma oscillations and ion acoustic waves at Comets Giacobini-Zinner and Halley (Scarf et al., 1986) suggests that electron heating and acceleration should take place. For example, in localized regions where the electron number densities were observed to be enhanced in the coma of P/Giacobini-Zinner, the electron velocity distributions were flat-topped. This may be related to electron heating via the ion acoustic wave instability (Dum et al., 1974). If ω_r and k_r are the frequency and wave vector of the linearly most unstable mode with $v_r = \omega_r/(2k_r)$, ion acoustic wave heating limits the electron heating effect to electrons with velocities v_e less than the resonant velocity v_r and no high energy tail on the electron distribution is expected to form; hence the flat-topped distribution. It is interesting to note that the electron velocity distributions observed in the spikes at P/Giacobini-Zinner (see Fig. 5.31) bear some similarities to the electron populations observed at weak interplanetary shocks (Thomsen et al., 1986). In numerical simulations of electron heating at collisionless shocks caused by the cross-field streaming instability, Winske et al. (1985b) found that electron heating occurs in directions both parallel and perpendicular to the magnetic field. The electron velocity distributions have a flat-topped form as a result of the condition $v_e < v_r$.

In contrast, lower hybrid wave heating should produce just the opposite effect with the formation of a high energy tail for $v_e \geq 2\,v_{th}$, where v_{th} is the thermal velocity (Wu et al., 1981). Thus lower hybrid wave acceleration could, in principle, be quite effective in generating a flux of energetic electrons capable of ionizing neutrals. However, for the electron impact ionization rate to be comparable with or faster than the photoionization rate, the number density of "hot" electrons with energy larger than 100 eV must be more than $10^6\,\mathrm{m}^{-3}$. The analysis by Sagdeev et al. (1986f) showed that lower hybrid wave instability should generate suprathermal electrons with a number density $\leq 10^{-6}n_e$, where n_e is the number density of the ambient electron population. Since n_e is less than $10^{10}\,\mathrm{m}^{-3}$ in the comae of Comets Giacobini-Zinner and Halley, electron impact ionization from this mechanism should be relatively small. This assessment is in effect consistent with observations from the electron experiments.

The extreme form of electron acceleration is found in the critical velocity ionization effect (Alfvén, 1954; Formisano et al., 1982; Galeev and Lipatov, 1984; Haerendel, 1986). Laboratory experiments as well as artificial gas releases in the Earth's ionosphere have shown that anomalous ionization can be produced when the relative motion between the neutral gas and the ambient plasma flow, normal to the magnetic field, exceed a certain critical value,

$$v_c = \left(\frac{2\eta\phi_i}{m_i}\right)^{1/2},$$ (5.47)

where ϕ_i is the ionization potential of the neutral gas molecules or atoms, m_i is the ion mass, and η a constant with a maximum value of 2/3 (Galeev and Lipatov, 1984; Formisano et al., 1982). The lower hybrid wave instability has often been invoked to be the mechanism coupling the kinetic energy of the new-borne ions to the electrons (Möbius et al., 1987). However, since the plasma speed is less than 20 km/s within 6 to 8 10^4 km from the nucleus of P/Halley, the condition for the critical velocity ionization effect was not fulfilled in this comet, even with $\eta = 2/3$.

5.6.3 Energetic Ions in the Inner Coma

As Giotto approached Comet Halley, hot comet ions with energies of the order of 100 eV disappeared at cometocentric distances $R < 6 \times 10^4$ km. This effect can be understood in terms of losses resulting from charge exchange with the neutral coma (Gombosi, 1987b; Ip, 1989). However, an intense flux of hot ions was observed in the vicinity of the contact surface by several of the plasma instruments (Johnstone et al., 1986; Goldstein et al., 1987). It has been suggested that these hot ions have their origin in the charge exchange process at larger distances, $R > 6 \times 10^4$ km. Since the motion of energetic neutral particles is not influenced by the magnetic field, they will simply follow ballistic trajectories until they are scattered in collisions with other molecules. The effects of collisions are important only at a distance of about 10^4 km (Eviatar et al., 1989), thus the energetic neutrals would be thermalized near the contact surface if they were injected towards the inner coma. Most of these particles in fact would not be able to penetrate much closer than a few thousand km from the comet nucleus as a consequence of collisions that result in a fraction of the energetic neutrals being reionized and hence appearing as hot ions (Eviatar et al., 1989).

An alternative explanation involves magnetic field reconnection in the inner coma. As shown in Fig. 5.44, the geometry of the diamagnetic cavity implies the presence of magnetic X-lines on the tailward side. The maximum ion energy that might be derived from reconnection is

$$\phi = V_A B_t L_t,$$ (5.48)

where V_A is the Alfvén speed in the tail, B_t the tail field, and L_t the effective width of the tail. With $B_t = 60$ nT, Alfvén speed $\simeq 10$ km/s, and $L_t \simeq 10^4$ km, we obtain $\phi \simeq 6$ keV. Energetic ions might therefore be generated this way *in situ*, but their presence would be sporadic.

In the context of reconnection, ion acceleration in the tail is suggested by the Giotto observations since no sharp directional changes of the magnetic fields near the contact surface were reported. On the other hand, near closest approach to Comet Halley, at the distance of 10^4 km, the magnetometer experiment on Vega 1 detected large variations in the magnetic field direction. In association with these magnetic

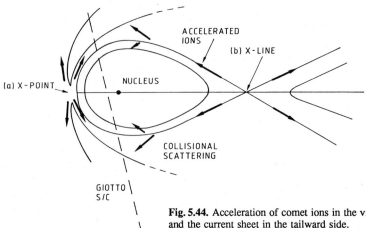

Fig. 5.44. Acceleration of comet ions in the vicinity of the X-line and the current sheet in the tailward side.

field changes, bursts of hot ions with energies reaching 200 to 600 eV were seen by the plasma analyzer (Gringauz et al., 1986a). This phenomenon has been interpreted by Verigin et al. (1987) as the result of front side reconnection in the manner suggested by Niedner and Brandt (1978) (see Fig. 5.44).

Finally, these hot ions, if confined within the magnetic field-free cavity, could contribute significantly to the pressure balance. Indeed, using an average number density of $n_i \simeq 10^8 \, \mathrm{m}^{-3}$ and a thermal energy of $kT_i \simeq 100 \, \mathrm{eV}$ (Eviatar et al., 1989), the corresponding thermal pressure would be more than sufficient to produce a magnetic field increment of about 60 nT at the sharp boundary of the ionopause. However, the exact origin of this hot ion population is still unclear at this time.

5.6.4 Structure of the Bow Shock

It was pointed out by Sagdeev et al. (1986f) that as a consequence of the presence of pickup ions, which dominate the plasma pressure despite their relatively small contribution to the mass density and momentum flux, the comet bow shock should have a structure similar to that of a cosmic ray shock (Axford et al., 1977). The pickup ions in this case play the role of cosmic rays and in both cases the nonthermal component is coupled to the background plasma via scattering by the turbulence which in turn is linked to the presence of this component.

A model of the comet bow shock has been discussed by Zank (1990) based on earlier work on the associated cosmic ray problem (Zank, 1988). It is assumed that the medium can be described by equations representing conservation of mass, momentum, and energy for the combined plasma consisting of solar wind and pickup ions, but neglecting the comet ion mass flux and mass density. In addition, a diffusive transport equation for the cometary ions is adopted with the diffusion coefficient assumed to be constant. In the presence of a magnetic field, a form of Ohm's law is required, and this is taken to be the generalized form allowing for the Hall current. For weak shocks, which are appropriate in these circumstances, it is found that the complete system of equations can be reduced to a single nonlinear equation in the form of a mixed Burgers-Korteweg-de Vries equation. In terms of normalized

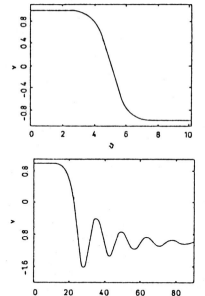

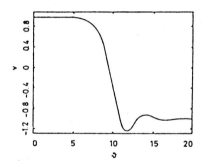

Fig. 5.45. The structure of cometary ion modified shocks for different values of ($\hat{\epsilon}$ = 0.01, 1.0, 10.0). The velocity v is in units of $v_\infty^- - v_\infty^+$, where v_∞^- and v_∞^+ are the upstream and downstream velocities of the plasma flow, respectively. The variable ϑ is a parameter relating the ratio of the solar wind plasma pressure to the pressure of the cometary heavy ions. (Zank, 1988).

variables, the traveling wave solutions of this equation satisfy

$$\hat{\epsilon}\frac{d^2v}{dx^2} + \frac{dv}{dx} = v^2 - 1, \qquad (5.49)$$

with boundary conditions $v(-\infty) = +1$, $v(+\infty) = -1$, and with the parameter $\hat{\epsilon}$ being determined by the Alfvén wave speed, the sound speed, the diffusion coefficient, etc.

If $\hat{\epsilon} < 1/8$, the downstream state of the plasma flow is completely smooth, whereas if $\hat{\epsilon} > 1/8$, there exists a damped oscillatory structure behind the shock (see Fig. 5.45). In the case of Comet Halley, assuming the gyroradius limit for the scattering mean free path, which is included in the diffusion coefficient, has an upper limit of 1, one might expect that an oscillatory structure be present. It is not clear that this was the case at the time of the Giotto encounter; during the Vega 1 encounter there was unfortunately a data gap at the critical time. On the basis of this theoretical work, there is some need for the data to be examined more closely. It should be noted that the recent numerical calculation by Sauer et al. (1990) has also shown that several new features (i.e., non-stationary wavy structures) could take place in a multifluid description of cometary shock formation.

6. Orbital Distribution of Comets

Jan H. Oort

6.1 Introduction

Comets can be classified by their orbits as long-period and short-period comets, according to whether their periods are longer or shorter than two hundred years. The long-period comets have revolution times ranging from two hundred years to 10^7 y. The short-period comets are concentrated in a narrow interval; 60% have periods between 5 and 6.5 y.

The two classes also differ greatly in their orbital inclinations. The orbits of the long-period comets are oriented essentially at random, while the short-period orbits are strongly concentrated to the ecliptic, their average inclination being about 13°. A similar difference exists for the argument of perihelion, ω, the angle between perihelion and the ascending node, which for the short-period comets is strongly peaked near 0° and 180°, about 90° from that of Jupiter, while there is no such anisotropy for the long-period comets.

At first sight it might seem that the long-period comets came from outside the Solar System, but this can evidently not be so. For though they usually reach very large distances, far outside the orbits of the most distant planets, no hyperbolic orbits have ever been found.

The best available data on long-period comets are those of the 55 orbits selected by Marsden et al. (1978) and two updates for eleven more recent comets with orbits of similar quality (Everhart and Marsden, 1983, 1987). The complete list contains 66 orbits, almost three times the number on which the original discussion of the comet cloud in 1950 was based (Oort, 1950). It is presented in Table 6.1 and Fig. 6.1. The reciprocal semimajor axes, $(1/a)_{\mathrm{orig}}$, refer to the "initial" orbits, before comets enter the perturbing influence of the planets. In order to avoid the effects of nongravitational forces, caused by insolation and subsequent asymmetric mass ejection from a rotating nucleus, comets coming close to the Sun have been excluded from the sample. Marsden et al. (1978) had shown that this effect can be quite appreciable in orbits with small perihelion distances, but that it is practically negligible if the perihelion distance q exceeds 1.8 AU. The orbits in the sample were restricted to $q > 1.5$ AU. They are arranged in order of increasing q.

The abscissa in Fig. 6.1 is the reciprocal of the semimajor axis, $(1/a)_{\mathrm{orig}}$. This is a measure of the orbital energy per unit mass, $M_\odot G/a$, where $M_\odot$ is the solar mass and G is the universal gravitaional constant. Units for $1/a$ are $10^{-5}\,\mathrm{AU}^{-1}$. Beside the initial values of $1/a$, Table 6.1 contains those for the osculating orbits with their mean errors. The last column shows $(1/a)_{\mathrm{fut}}$ for the orbits in which comets leave the planetary system when they are beyond their perturbing influence. Table 6.2 gives

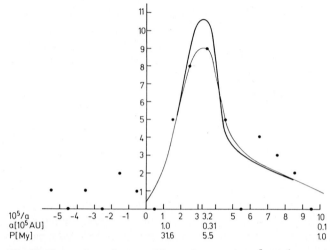

Fig. 6.1. The numbers of comet orbits per interval of $10^{-5}\,\mathrm{AU}^{-1}$ vs. $(1/a)_{\mathrm{orig}}$, in units of $10^{-5}\,\mathrm{AU}^{-1}$. The abscissa also shows the semimajor axis in units of 10^5 AU and the period in My. The data are from Table 6.1. The heavy curve represents the distribution corrected for accidental errors.

Table 6.1. Original and future orbits of comets with $q > 1.500$ AU ^{a}

Comet	q (AU)	i (°)	$(1/a)^b_{\mathrm{osc}}$	$(1/a)^b_{\mathrm{orig}}$	$(1/a)^b_{\mathrm{fut}}$
1948 II	1.500	77.5	− 724 ± 8	+ 28	+ 53
1888 V	1.528	56.3	+ 5571 ± 34	+5626	+ 5786
1925 VII	1.566	49.3	− 244 ± 12	+ 24	− 322
1975 XII	1.604	91.6	+ 1590 ± 6	+1964	+ 1545
1959 I	1.628	61.3	+ 82 ± 6	+ 76	+ 244
1925 III	1.633	27.0	+ 2990 ± 8	+3778	+ 4521
1932 VII	1.647	78.4	− 365 ± 15	− 56	− 327
1981 II	1.657	331.3	+ 619 ± 9	+ 581	+ 1045
1953 I	1.665	59.1	+ 2457 ± 6	+3277	+ 3323
1917 III	1.686	25.7	+ 353 ± 3	+ 17	+ 763
1968 I	1.697	129.3	+ 500 ± 7	+ 842	+ 704
1898 VII	1.702	69.9	− 549 ± 7	+ 68	− 709
1970 III	1.719	86.3	+ 509 ± 4	+ 555	+ 896
1946 I	1.724	72.8	− 678 ± 3	− 13	+ 373
1937 IV	1.734	41.6	− 79 ± 8	+ 62	+ 1400
1889 I	1.815	166.4	− 677 ± 3	+ 48	− 573
1890 II	1.908	120.6	− 139 ± 10	+ 89	+ 126
1910 IV	1.948	121.1	+ 104 ± 11	+ 474	+ 288
1892 II	1.971	89.7	− 188 ± 14	+ 846	+ 513
1907 I	2.052	141.7	− 489 ± 19	+ 25	− 290
1949 IV	2.058	105.8	+ 659 ± 7	+ 735	+ 957
1930 IV	2.079	72.0	− 252 ± 13	+ 524	− 44
1948 V	2.107	92.9	− 373 ± 4	+ 34	+ 31
1962 VIII	2.133	153.3	+ 4889 ± 1	+4935	+ 5403

Table 6.1 (continued).

Comet	q (AU)	i (°)	$(1/a)_{osc}^{b}$	$(1/a)_{orig}^{b}$	$(1/a)_{fut}^{b}$
1973 II	2.147	141.9	− 474 ± 6	+ 320	− 28
1944 IV	2.226	95.0	− 937 ± 13	+ 18	− 522
1983 XVI	2.255	20.6	− 87 ± 11	− 18	+ 402
1889 II	2.256	163.8	+ 81 ± 30	+ 933	+ 1268
1922 II	2.259	51.5	− 334 ± 9	+ 21	− 523
1898 VIII	2.285	22.5	+ 216 ± 36	− 71	+ 620
1932 VI	2.314	125.0	− 619 ± 3	+ 45	− 240
1947 I	2.408	108.2	− 393 ± 5	− 1	+ 26
1945 I	2.411	17.3	+ 2696 ± 27	+3058	+ 2385
1972 VIII	2.511	138.6	− 200 ± 13	+ 49	+ 362
1949 I	2.517	130.3	+ 257 ± 5	+ 498	+ 581
1950 I	2.553	131.4	− 279 ± 5	+ 263	+ 537
1951 I	2.572	144.2	− 479 ± 4	+ 37	+ 270
1904 I	2.708	125.1	− 496 ± 3	+ 227	+ 515
1903 II	2.774	43.9	− 218 ± 9	+ 26	− 488
1947 VI	2.828	97.3	− 372 ± 9	+ 234	− 233
1975 VIII	3.011	50.6	+ 132 ± 3	+ 36	+ 517
1947 VIII	3.261	155.1	− 699 ± 10	+ 34	+ 225
1983 XII	3.318	134.7	− 563 ± 8	+ 78	− 161
1905 IV	3.340	4.3	− 452 ± 8	+ 28	− 519
1983 XV	3.345	137.6	− 11 ± 13	+ 591	+ 273
1982 I	3.364	1.7	−17036 ± 2	+ 30	−16012
1927 IV	3.684	87.7	+ 494 ± 3	+ 623	+ 1087
1914 III	3.747	71.0	− 878 ± 20	+ 27	− 1
1973 IX	3.842	108.1	− 167 ± 14	+ 71	− 109
1955 VI	3.870	100.4	− 132 ± 3	+ 42	+ 252
1984 V	4.000	89.3	+ 121 ± 13	+ 35	− 17
1936 I	4.043	66.1	− 506 ± 6	+ 19	− 281
1956 I	4.077	79.6	− 1145 ± 3	+ 39	− 228
1942 VIII	4.113	172.5	− 774 ± 13	− 34	− 282
1925 VI	4.181	146.7	− 582 ± 9	+ 35	− 128
1959 X	4.267	125.5	− 207 ± 11	+ 40	− 2
1972 IX	4.276	79.4	− 1471 ± 7	+ 69	− 603
1957 VI	4.447	33.2	− 617 ± 5	+ 17	+ 148
1954 V	4.496	123.9	− 621 ± 16	+ 82	− 21
1981 XV	4.743	115.3	− 139 ± 13	+ 142	− 12
1973 X	4.812	137.4	− 7 ± 19	+ 536	+ 512
1972 XII	4.861	113.1	+ 19 ± 4	+ 476	+ 356
1977 IX	5.606	116.9	− 529 ± 10	+ 33	− 102
1976 IX	5.857	86.6	− 671 ± 24	+ 37	+ 189
1974 XII	6.019	60.9	− 656 ± 15	+ 11	+ 569
1975 II	6.881	112.0	− 296 ± 16	+ 68	+ 65

[a] From Marsden et al. (1978) and Everhart and Marsden (1983, 1987)
[b] In units of 10^{-6} AU^{-1}

Table 6.2. Counts of orbits with $(1/a)_{\text{orig}} > 10^{-5}\,\text{AU}^{-1}$

$10^5/a$ (AU^{-1})	Number per interval
1– 10	36
10– 20	1
20– 30	3
30– 40	1
40– 50	3
50– 60	5
60– 70	1
70– 80	1
80– 90	2
90–100	1
>100	6

the counts of orbits with $(1/a)_{\text{orig}} > 10^{-5}\,\text{AU}^{-1}$. The 36 orbits of the first entry in this table are plotted in Fig. 6.1.

The figure shows two remarkable phenomena: First, the very large distance to which the orbits extend. Table 6.1 and Fig. 6.1 contain comets with reliable orbital major axes up to 10^5 AU. At this distance there is a fairly steep outer boundary. The second remarkable feature is the very narrow peak at $1/a = 3.2 \times 10^{-5}\,\text{AU}^{-1}$, with a width of only $2 \times 10^{-5}\,\text{AU}^{-1}$. This width is not much larger than twice the mean error of $1/a$ as given by the orbit computations. The average mean error is $\pm 0.8 \times 10^{-5}\,\text{AU}^{-1}$. A rough correction for the effect of these accidental errors has been applied. The corrected distribution is shown by the heavy curve. The narrow peak is surprising since a single passage through the inner planetary system produces an average dispersion in $1/a$ of $\pm 35 \times 10^{-5}\,\text{AU}^{-1}$. It appears therefore that the comets in this part of the figure could not have previously passed through the planetary system. They have therefore been called "new" comets. Their aphelia populate a shell between about 0.2 and 0.7×10^5 AU from the Sun. This narrow peak is a strong constraint for theories of the origin of comets. In view of this constraint it is doubtful whether any model that involves an origin outside of the present Solar System, at $r > 30$ AU, is tenable.

In Table 6.1 are five comets with appreciably negative values of $(1/a)_{\text{orig}}$. At our request, Marsden has reinvestigated the calculations leading to these values. He concludes that two cases are probably and two other cases are possibly the result of nongravitational forces. There then remains one comet, 1942 VIII, with $q = 4.113$ AU that is enigmatic; something peculiar must have happened to this comet.

6.2 Origin and Dynamical Evolution

It is plausible to think that comets originated at the birth of the Solar System, in much the same way and in the same regions where the planets were formed, and that they were transported to their present large distances through perturbations by Jupiter and the other large planets. The Solar System must have begun as a condensation of the interstellar medium. It is likely that the medium had some angular momentum. Most of the condensation contracted into a compact nucleus, the proto-Sun, shedding most of its angular momentum into the surrounding medium that formed a fast rotating disk. Presumably the disk developed into rings in which the planets were formed by processes that are still poorly understood.

It is plausible that the interstellar dust particles began to grow rapidly when the medium was compressed into disks and rings and that they may have conglomerated into larger bodies. It is believed that these became the building blocks for the planets. They were therefore called "planetesimals." Probably only a small fraction of them took an actual part in the formation of the planets, with the others continuing to move in separate orbits.

It is tempting to think that these stray planetesimals were the forerunners of comets. The view that comet nuclei are conglomerates of planetesimals has been strongly advocated by Greenberg. It is supported in particular by the very low albedo shown by the spacecraft observations of the nucleus of Halley's comet (see, e.g., Greenberg 1982, 1990).

The proto-planets, in which many particles of the rotating rings were combined, would be likely to move in nearly circular paths, whereas the orbits of the particles that escaped from the planet-forming region presumably deviated from circles, thus preventing their being absorbed by the proto-planets. However, they must have been subjected to strong perturbations by the large planets and thereby have gradually been diffused outward. Successive perturbations will ultimately have thrown most of them out of the Solar System – most, but not all. For about 10% the outward diffusion would have been halted when their orbits had grown to a size of about 3×10^4 AU; i.e., an aphelion distance $Q \simeq 2a$, corresponding to $1/a = 0.00006 \, \text{AU}^{-1}$. At this size a new type of perturbation begins to take over, viz. perturbations exerted by other stars and tidal forces of the Galaxy. In contrast to the planetary forces, which influence the major axes only, the galactic forces can also cause changes in q and the inclination, i, and thereby withdraw the orbits from the influence of Jupiter and Saturn. Released from the large planetary perturbations, the planetesimals are "trapped" in a region extending from 1.7×10^4 to about 10^5 AU, where their orbits are virtually safe from further disturbances. The "proto-comets" will accumulate in this reservoir that is now usually called the "Oort cloud" around the Solar System.

There would thus be two classes of proto-comets that have survived from the time of birth of the Solar System: Those that formed in the trans-Neptunian protosolar disk and those that were transferred by planetary and stellar perturbations (or galactic tides) to the Oort cloud between approximately 2×10^4 and 10^5 AU, where they were likewise unaffected by perturbations from the planets.

The large cloud has a sharp inner edge. Numerical calculations show that the number of proto-comets captured in it becomes negligible for $a < 1.7 \times 10^4$ AU. The cloud will therefore not contain many comets with a below this limit. This is confirmed in a striking way by the steep drop in Fig. 6.1 at $1/a = 0.00006 \, \text{AU}^{-1}$. The cloud will be thoroughly "shaken up" by the outside perturbations. Initially the comets would have been concentrated to the ecliptic. But at the present epoch all traces of this must have been effaced. This is amply confirmed by Figs. 6.2 and 6.3.

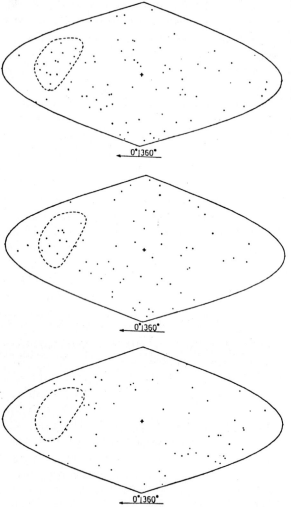

Fig. 6.2. Distribution of aphelion directions of "new" ($a > 10^4$ AU) comets (*top*), a between 400 and 10^4 AU (*middle*) and a between 40 and 400 AU (*bottom*) in ecliptic coordinates. The distribution is fairly isotropic. There is no tendency to be concentrated towards the ecliptic, nor is there a distinct relation with galactic coordinates. There appears to be some clustering on a small scale for "new" comets. A striking example has been indicated by a dashed curve (Biermann et al., 1983; Lüst, 1984).

The formation of a vast reservoir of comets of exactly the dimensions observed might thus have been predicted as a natural consequence of the interplay between planetary and stellar perturbations. Most of the objects will either have escaped or been transferred to this reservoir in the early stages of the Solar System's existence. It must be assumed that during this transfer the comets were able to retain the frozen gases of which they largely consist.

6.3 Time Scales

Numerical integrations of known orbits and extensive computer simulations have shown that the average of the change (without regard to sign) in the orbital energy, $1/a$, occurring at a passage through the central planetary system is ± 0.0005

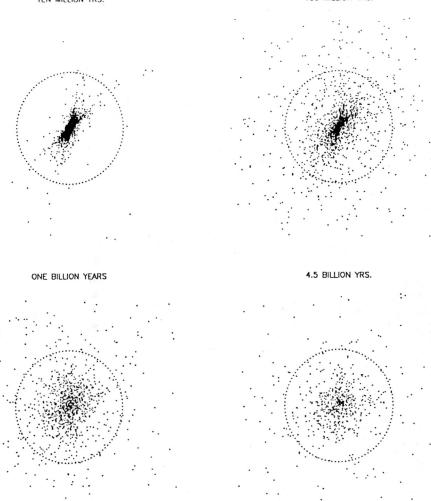

TEN MILLION YRS.

100 MILLION YRS.

ONE BILLION YEARS

4.5 BILLION YRS.

Fig. 6.3. Distribution of comets in space for a model discussed by Duncan et al. (1987), projected onto a plane for several different times. This plane is parallel to the galactic plane. The dotted circle corresponds to a radius of 20000 AU indicating the inner edge of the classical Oort cloud.

– positive and negative changes being about equally frequent. After n passages $(1/a)|_n \simeq 0.0005n^{1/2}$. The planetesimals will therefore require some 10^6 revolutions, or a few million years – a small fraction of the age of the Solar System – to get from their strongly bound initial orbits into orbits with $1/a = 0.00006$, from which they escape into the large comet cloud.

The "cloud" will continue to evolve by the action of the outside forces. The stellar perturbations gave the cloud its spherical shape and randomized it.

It should be noted that the constituents of the cloud are dark bodies, which become visible only when perturbations bring them into orbits passing close enough to the Sun to become observable as comets. It is their gas and dust emissions that make them observable so that we can study the shape and dimensions of the large cloud.

The cloud is also evolving by the escape of members through its outer as well as its inner surface. The latter are the more interesting, because they are observed in the tail of Fig. 6.1, stretching to large values of $1/a$. The decline of the number per unit interval of $1/a$ can in principle inform us about the rate at which comets gradually diminish in brightness or dissolve. It is of interest to see how far the observed evaporation and rupture rate can account for the decrease in the number density toward larger values of $1/a$. Very preliminary discussions show a fair agreement (see, e.g., Oort 1950).

6.4 The Outer Part of the Presolar Nebula

While the proto-comets are likely to have been expelled by planetary perturbations from the inner part of the presolar nebula, those formed in the region beyond Neptune must have survived. As was first pointed out by Kuiper (1951), it is therefore probable that there is a considerable swarm of comets with perihelia larger than about 30 AU. These will be almost free from perturbations and therefore long-lived. As we shall see in the following, they are likely to be the source of the short-period comets.

Compactly summarized: If we assume that comets, or their precursors, were born in the same regions as the planets, a large fraction will automatically evolve within about a million years into a cloud that has the three characteristic properties of the long-period comets: (1) Orbits extending to roughly 10^5 AU, (2) isotropy of orbital orientations, and (3) sharp inner edge at 1.7×10^4 AU. Those formed beyond the outer planets will remain semipermanently as a thin swarm outside of Neptune's orbit.

6.5 The Distribution of the Reciprocal Semimajor Axes

A rough indication of the general distribution of the semimajor axes is shown in Table 6.3, which gives the number of comets per interval of 0.00005 AU^{-1} in $1/a$. The numbers observed are added in parentheses. If there had been no losses by deterioration, the numbers per unit interval of $1/a$ should have been constant. Actually, comets do get lost through evaporation and through splitting and ensuing disintegration. Though no adequate discussion has yet been given, it seems probable that the decrease in frequency shown in Table 6.3 can be accounted for by the progressive deterioration of the comets. For a more extensive discussion the reader is referred to Oort (1950).

Table 6.3. Observed distribution of $1/a$ values

$1/a$ (AU^{-1})	P (y)	Number per interval of 0.00005 AU^{-1}	
0.00000–0.00050	> 90000	3.9	(24)
0.00050–0.00100	30000–90000	0.7	(3)
0.00100–0.00200	11000–30000	0.20	(7)
0.00200–0.00400	3950–11000	0.18	(8)
0.00400–0.01000	1000– 3950	0.058	(12)
0.01000–0.02000	354– 1000	0.023	(16)
0.02000–0.04000	126– 354	0.007	(10)

6.6 Other Theories for the Origin and Evolution

There have been various other theories about the origin of comets. Several are based on the supposition that comets formed in the far outer region of the solar nebula. They assume that the growth of bodies of the size of cometary nuclei at the low densities found there would have been made possible by some still undefined mechanism.

Most valuable among recent articles are those containing extensive numerical integrations of orbits in the large cloud. With regard to the latter the reader is referred to the article on "The formation and extent of the Solar System comet cloud" by Duncan et al. (1987). In their figure, which is reproduced here as Fig. 6.3, they give a suggestive representation of the evolution of the distribution of the comets in their model in the course of time. It should be noted that their initial distribution differed from that described in the present overview in that it was concentrated towards the center.

6.7 Short-Period Comets

Short-period comets are intimately related to Jupiter, as is particularly evident from the striking concentration of their aphelia toward Jupiter's orbit. This shows convincingly that a large fraction must have been captured by Jupiter. The evidence for this capture has recently been discussed extensively by Duncan et al. (1988) and Quinn et al. (1990). They showed that the capture theory does not only explain the concentration of the aphelia, but also the remarkable distribution of the arguments of perihelion, ω, with its two well-defined maxima at 0° and 180°. Figure 6.3, taken from Duncan et al. (1987), shows how strikingly these maxima are reproduced by their simulated orbits. Most important is their proof that the captured comets cannot have come from the Oort cloud, because the orbits in this cloud are randomly inclined while the short-period comets have strikingly low inclinations: 73% have $i < 15°$ while only 3% are inclined more than 45°. The captures must have been made from a thin swarm of comets born in the outermost regions of the solar nebula, the probable existence of which was initially pointed out by Kuiper (1951).

The problem of the origin of the short-period comets has been discussed by a number of astronomers. The reader is referred to Duncan et al. (1988) for an extensive review of former articles.

6.8 Differences Between Long- and Short-Period Comets

The intrinsic differences between the long-period and short-period comets were investigated by Oort and Schmidt (1951). They found that the long-period comets had a pronounced tendency for strong continuum spectra, especially at large distances from the Sun. This has been attributed to the presence of dust entrained by the gases from a very volatile icy component. Oort and Schmidt also found that the absolute brightness of long-period comets increased more slowly with decreasing heliocentric distance, r, than that of the short-period comets. This result was based on an investigation of cometary brightness estimates collected by Schmidt (1951). He presented the brightness as $m_o = A + Br^{1/2}$, where m_o is the magnitude reduced to 1 AU geocentric distance. In this formula, originally introduced by Levin (1943b) $B = 1.086L/R_oT$, where R_o is the gas constant, T is the temperature, and L is the latent heat of sublimation. (In the formula originally used by Levin L was the heat of desorption. For a more modern theory see Chap. 2, Sect. 2.2.2.) The difference between the long- and short-period comets has become more pronounced with the recent orbital data; for the long-period comets they yield an average value for $B = 2.8 \pm 0.3$, while for the short-period comets $B = 8.6 \pm 0.7$. Thus, the comets with small values of $1/a$ appear to have much lower values for the latent heat of sublimation than the short-period comets, confirming the above speculation on the behavior of the continuum spectrum.

By far the strongest evidence that the nuclei of the long-period comets are covered by an outer layer of extremely volatile matter comes from the narrowness of the peak near $1/a = 0.00003\,\mathrm{AU}^{-1}$ in Fig. 6.1. If the nuclei in this maximum were permanent objects it is impossible to explain its narrowness; for, after their perihelion passage, the planetary perturbations would have produced a dispersion in $1/a$ of ± 30 units, which is not observed. We must conclude that the the comets we now observe at $1/a < 0.00003\,\mathrm{AU}^{-1}$ do not return at a comparable brightness. Apparently they evaporate to such an extent during their first perihelion passage that their absolute brightness decreases by at least several magnitudes. This is plausible if the nucleus had an outer layer of highly volatile, frozen interstellar gases.

7. Comet Formation and Evolution

Hans Rickman and Walter F. Huebner

7.1 Introduction

In this chapter we first review the status of the hypotheses for comet formation and evolution as they were perceived before the spacecraft missions to Halley's comet. Since the missions gave the first observations of a nucleus, it is easily understood that the images of the nucleus will have the largest impact on modifying these hypotheses. Also a much more detailed composition of the frozen gases in the nucleus could be determined than was possible from ground-based observations and the dust size distribution and composition brought totally unexpected results. Thus the P/Halley data provide new constraints for the hypotheses of comet origin and evolution.

Comets have played an important role in cosmogony. Their own origin is usually studied within the framework of a particular cosmogonical model and is closely linked to the origin of the Solar System. The physical details and the sequence for formation of the Solar System and of comets are still open questions. However, there are a number of hypotheses that can now be confronted with conclusions based on the study of interstellar clouds as precursor models for a solar or presolar nebula and on cometary physics, chemistry, and orbital dynamics. To the extent that a choice can be made between different scenarios for the formation of comets, this also helps to judge the validity of different cosmogonical models.

The study of the evolution of comet nuclei is complicated. Comets form one of the few classes of celestial bodies in which various states of evolution might be observed on individual objects, but the evidence is often biased by observational selection, and the causes of the observed evolutionary effects are sometimes not uniquely identifiable. Extrapolation of these trends forward or backward in time is prone to error. The question of past evolution is of particular interest since it relates data from presently observed comets to the original, ancient comet population. If extrapolation backward in time were possible, then it would provide powerful clues about the probable mode of their origin. Alternatively, it would be useful to have a clear understanding of what structural or chemical modifications have occurred even in the most youthful comets observed.

At the heart of the discussion of cometary origin and evolution are two concepts that have been described in preceding chapters: The nucleus (with coma and tails as subsidiary phenomena) and the Oort cloud. These represent the basic observational facts, the foundations for the hypotheses. Changes in these concepts will influence our understanding of comet origin and evolution.

7.1.1 The Nucleus

The nucleus, as conceived by Whipple (1950), is a solid body consisting of a conglomerate of frozen gases and refractory dust grains with a broad size distribution.
Despite the serious problems encountered when trying to establish the composition of the ice from Earth-bound spectroscopic observations of the coma (see Chap.
3, Sect. 3.4), the idea of a water-dominated structure has been of use ever since
Whipple introduced the concept of a solid nucleus. The main reason was the simple
observation that the onset of cometary activity before perihelion passage and the termination afterward generally occur at heliocentric distances of 2.5 to 3 AU. This is
consistent with the evaporation of water ice. However, the existence of more remote
activity in some comets and the spectacular CO^+ tails of comets Morehouse (1908
III) and Humason (1962 VIII) have continued to stimulate debate about the role of
more volatile substances, such as carbon monoxide and dioxide, and whether they
are related to cometary evolution. Even from rough estimates of the gas production
rates necessary to explain the amounts of gas and dust observed in the comae of
comets at a heliocentric distance $r \sim 1$ AU, it has been evident that the typical mean
radius of an active nucleus must be at least 1 km and in extreme cases at least 10 km.

Thus, conceptually, the origin of comets is equivalent to the formation of
kilometer-sized bodies consisting of ice–dust mixtures. Where could this have occurred? The place must have been cold enough to allow the existence of frozen water.
Thus if comets are considered to be original members of the Solar System, only the
region in the vicinity of Jupiter's orbit and beyond guarantees permanent water ice
(water ice is apparently also an important constituent of Ganymede and Callisto).
The idea of a remote place for the origin of comets has indeed been commonly
accepted for some time.

How remote is the place of origin? This is where the consensus ends. Among
different hypotheses relating cometary origin to the birth of the Solar System, one
finds formation regions ranging from Jupiter's orbit via Saturn, Uranus, and Neptune
to a trans-Neptunian part of the solar nebula, all the way to satellite fragments of
the presolar nebula at even larger distances, or indeed different nebulae within the
same star-forming complex as that of the Solar System. The time of origin is closely
connected with the place of formation, either in the presolar nebula or after the
Sun had formed in the solar nebula. Cometary origins unrelated to that of the Solar
System, either in interstellar space or at the sites of other, more recent star-forming
events in galactic history, have also been proposed.

Observational stimulus to such discussions has come from indications about the
chemical composition of cometary material. Evidence for very volatile species has
been interpreted in terms of very low formation temperatures, i.e., large distances
from the proto-Sun. Difficulties in fitting the origin of carbon- and nitrogen-bearing
radicals in cometary comae with the mixture predicted by near-chemical equilibrium
condensation models of the solar nebula (mostly H_2O, NH_3, CH_4, CO_2, and CO) have
led to suggestions that cometary chemistry may be based on interstellar molecules.
Such an interstellar analogy does not necessarily mean that comet nuclei formed in
interstellar space. They could also have formed in the outer parts of the solar nebula
using essentially unaltered interstellar material. Depending on one's understanding

of the thermal evolution of the solar nebula, one may arrive at different conclusions about which formation site is compatible with the inferred formation temperatures, or how far out one must go to condense the volatile mantles of the interstellar grains without evaporating them again.

Equally important are the formation circumstances of the comet nuclei. Models based on the growth of ice mantles on grains, through sticking of molecules from the gas phase to the solid surfaces (condensation), require that the gas density be high enough to allow this growth to occur within a reasonable time scale. This requirement is independent of whether or not, e.g., gravitational instabilities in a rotationally flattened nebula of grains and gas played a critical role for the formation of comet nuclei. In order to have a high enough density, one cannot imagine the origin of comets to have occurred in too distant regions of the solar nebula. Traditionally, a compromise between the above arguments has tended to favor the Uranus–Neptune accretion zones.

However, the density requirement is dependent on the density structure of the solar nebula. Thus quite different conclusions appear as a result of the classical low-mass model, elaborated most notably by Safronov, as opposed to Cameron's very massive model. In the latter case, comet formation could have occurred at distances of many hundred AU from the proto-Sun. In a massive solar nebula the amount of condensible molecular material in one solar mass, i.e., about 10^{28} kg, may form comets. If the original mass of a comet was about 10^{14} kg, then a flattened population of about 10^{14} comets, concentrated toward the ecliptic plane, would arise.

7.1.2 The Comet Cloud

The comet cloud, as envisaged by Oort (1950), is a giant structure containing about 10^{11} comets; it surrounds the planetary system nearly isotropically out to distances of about 10^5 AU (see Chap. 6). The outer fringes are comparable to the distance of nearby stars. According to recent estimates, the number of comets in it may even be as high as 10^{12}. This cometary reservoir is considered to be a remnant from the formation of the Solar System. Thus, closely related to the origin of comets is the origin of the Oort cloud. Once comet nuclei were formed, how did they get into the isotropic distribution at these vast distances?

If *in situ* formation is disregarded because of the density argument, a dynamical transfer process is needed. On the assumption that comets originated in the planetary system, perhaps as representatives of the planetesimals that accreted as intermediate bodies in the formation of the giant planets and their satellites, the most natural transfer mechanism is provided by the orbital perturbations caused by these very same proto-planets. It then turns out that the perturbations by Jupiter and Saturn are so strong that a vast majority of their near-miss planetesimals would have been expelled from the Solar System with little chance of staying at its outskirts (Safronov, 1972). The ones expected to stay would be those perturbed to aphelion distances of about 10^4 AU, where the nearby stars of the solar star-forming complex would trap them by increasing their perihelion distances. Because of the smallness of the perturbations by Uranus and Neptune, such an outcome is quite possible. This has been yet another reason to associate comet formation with the growth of these planets.

However, this conclusion is based on the assumption that comets formed exclusively within the planetary system. If a massive solar nebula is considered, this assumption is not necessarily correct. On the other hand, how could comet nuclei, forming in a disk-like structure with dimensions of about 100 to 1000 AU, be transferred to hundred times larger distances and an isotropic distribution? An influential idea is based on a brief stage of significant mass loss from the solar nebula, allowing for an efficient transfer from nearly circular to nearly parabolic orbits. However, other possibilities have more recently come into consideration.

The Oort cloud may not be stable over the age of the Solar System because of perturbing encounters with Giant Molecular Clouds (GMCs), and thus it is perhaps not a remnant of the Solar System formation (Clube and Napier, 1982). One possibility to explain its indisputable existence is to imagine that the interactions between the Solar System and GMCs actually represent an exchange of comets so that, statistically speaking, as many comets are gained by the Oort cloud as are lost. If one believes that comet formation in the Solar System was restricted to the zones of planetary growth, one is faced with two alternatives. Either the Oort cloud is after all stable due to replenishment from the inner components of the cloud, or the present Oort cloud (and, by inference, the population of long-period comets as a whole) is, to a major extent, captured from other, more recent star-forming regions. Since there is no clear evidence for comets with hyperbolic orbits and since computations indicate that capture of such comets is a very unlikely event, an unstable Oort cloud would actually mean that its existence is hard to explain.

However, within the model of a disk-like population of comets having formed outside the planetary zone ("trans-Neptunian disk"), there is yet another possibility (Bailey, 1983). The initial mass loss from the solar nebula might have occurred in such a way that the comet disk was not drastically influenced. It may have survived and serves until the present time as a massive and dynamically active source of replenishment for the Oort cloud, joining onto its inner regions. The mechanisms of stripping and replenishment of the Oort cloud should be the same, i.e., exterior perturbations. Furthermore, as a result of the "stirring up" or "heating" of the orbital distribution of the trans-Neptunian disk, some transfer must also occur inward so that some comets become subject to the gravitational influence of the giant planets and can thus be captured into short-period orbits.

This then is an alternative to the traditional model of comet capture where the source for the observed comets is the flux of "new" comets continually arriving from the outer part of the Oort cloud. There are some indications that this traditional source may not be rich enough to explain the persistence of a population of short-period comets like the one we observe today. If so, support is lent to the existence of a second source situated anywhere from the outskirts of the planetary system to the inner core of the Oort cloud, but in principle this could involve primordial planetesimals from the Uranus–Neptune accretion zone as well as "intruders" from more remote regions. Also the question about a guarantee for the persistence of the present short-period comet population may need special attention if the source flux and thus the captured population varies with the varying galactic influences over time scales of 10 My to 100 My. Finally, there has been and still is debate about what is the typical lifetime of a short-period comet and, therefore, what is the required source flux.

7.1.3 Comet Evolution

The concept of comet lifetimes arises naturally from the observation that the mass of a comet nucleus cannot be so large as to allow losses of 10^{10} to 10^{11} kg per apparition (as a typical estimate for a reasonably bright short-period comet) during more than a minute fraction of the age of the Solar System. Observational verification is, however, difficult. The only clear piece of evidence concerns long-period comets, especially the new ones from the Oort cloud, where statistical conclusions can be drawn from the distribution of orbital energies. Apparently removal of a significant number of comets occurs already at the first approach to the Sun as evidenced by the near-parabolic peak of the energy distribution (see Chap. 6), but the reason for this "fading" is not clear. It may have to do with cosmic ray induced crusts formed on the surfaces of the nuclei during the previous 4.5 Gy. After that, a gradual loss of comets proceeds at a rate that implies a typical lifetime of several hundred revolutions.

The reason for this loss of comets is not clear either. Basically there are two mechanisms to consider. On the one hand the ice of the nucleus may be gradually consumed by sublimation until an end state is reached which may be a rapidly dispersing assemblage of meteoroids or, perhaps, an asteroidal body in case there is a silicate core at the center of the comet nucleus. On the other hand, refractory material may gradually accumulate on the surface of the nucleus, forming a dust mantle that terminates the gas production from the underlying ice by thermal insulation or choking. In this case, the end state is a defunct comet nucleus preserving its original bulk structure but deactivated by an insulating and relatively thin refractory crust. It is difficult to judge between the mass loss and dust mantle models, especially in the case of long-period comets. The frequently observed splitting of such comets is bound to accelerate the mass loss process and thus lends some support to the model of nucleus dispersal. However, the recent observations of a faint, dynamically old, long-period comet (IRAS-Araki-Alcock, 1983 VII) appear to fit very well with the predictions of the refractory mantle model.

The mechanisms of mass loss and dust mantle development should also be at work for short-period comets. In this case, there is substantial evidence that dust coverage does play a major role. The physical details of the processes are not well known. They may be sensitively coupled to the orbital evolution. Apparently the observations of brightness and nongravitational effects on short-period comets and of some low-activity comets and asteroids moving in cometary orbits fit well with the scenario of gradual dust coverage. Nonetheless, some objects pose specific problems, and it is quite likely that mass loss may be the critical factor for a significant number of short-period comets. In addition, the dust mantles so far predicted are so thin that the possibility of intermittent activity or rejuvenation by collisions with centimeter-sized meteoroids must be considered. In any case, some rough information on the typical lifetime of a short-period comet is available in the form of statistics of disappearance and dynamical captures; the result is about 500 revolutions. Fading of individual comets should then be hardly noticeable, and indeed such fading is just barely observed in a few of the most favorable cases.

In summary, we have sketched an outline of the ideas and theoretical concepts in recent discussions of the origin and evolution of comets. It is easily recognized that there is little certainty apart from a consensus regarding the basic concepts of the nucleus and the Oort comet cloud. There may be indications of a trans-Neptunian disk of comets as a remnant of a massive and extensive solar nebula, but clear-cut observational evidence is still lacking, and therefore the trans-Neptunian disk remains more or less hypothetical. Even though a possible analogy with the circumstellar disks of radii of about 100 to 1000 AU revealed by the Infrared Astronomical Satellite (IRAS) may be quite suggestive to the effect that these nearby stars might give a hint of what our Solar System would look like to an external observer, the question will not be settled until the disk surrounding our planetary system is either proved or disproved by observations. Likewise, there may be indications of gradual dust coverage playing a critical role in the advanced stages of comet evolution, but further observations such as, e.g., those feasible within the CRAF mission scenario are crucial for constraining the relevant models.

7.1.4 Recent Advances

What knowledge has the recent observational scrutiny of Comet Halley revealed? We have learned that a few isotope ratios are more or less terrestrial, which indicates that the origin of Halley's comet occurred along with the origin of the Solar System. We have also learned that the ice and the dust components of cometary material are not so easily distinguished as earlier believed and that the chemistry may bear a close resemblance to that inferred for interstellar grains (see Chap. 4, Sect. 4.9). The reduced abundance of the very volatile compounds CH_4, N_2, and CO in the nucleus suggests that the formation temperature may not be much below 50 K. The density of the nucleus appears to be only about 200 to 700 kg/m^3, i.e., the material of the nucleus appears to be very porous. Generally speaking, we now have more quantitative and more reliable evidence for a remote place of origin than earlier, although for only one special comet. However, a detailed interpretation of the data in terms of specific requirements on formation conditions remains to be made. For example, it is not obvious that present theories of the growth of comet nuclei by accretion are compatible with their low density, no matter how low the relative velocities are supposed to be. This does not mean that the model for accretion has to be wrong – only that it has to be weighed against the observational data and theoretical arguments inherent in the density estimate. It is obvious that we are still far from being able to specify the place of origin of Halley's comet – let alone that of comets in general.

Furthermore, we have learned that the nucleus of P/Halley is far from spherical. If its odd shape is primordial, it contains some valuable information on the formation processes, but it may also be strongly influenced by the previous sublimation history (depending on the past orbital evolution, which is not yet known). We have also learned that the gas and dust production by the nucleus is not an isotropic property, as most comet models have assumed, but that it is located in active regions covering about 10 to 20% of the surface area while the rest appears to be inert. Is this a manifestation of the gradual dust coverage already experienced during previous revolutions in the present orbit or is it a remnant from the cosmic ray irradiation to which the comet was exposed for 4.5 Gy? Is Comet Halley in a state of evolution by gradual dust coverage or is it simply consuming its mass via those active regions left open as gaps in a cosmic ray-driven crust? This important issue is not yet resolved and it seems fair to state that the observations of P/Halley have not solved the questions about comet evolution but rather complicated them.

In the remainder of this chapter we shall deal with the issues raised above in an attempt to clarify the situation with regard to observational data, expectations based on laboratory work, and theoretical developments. Obviously, no definitive answers will be found about the origin and evolution of comets, but perhaps an understanding will evolve for the relevant questions as they appear today, after the Halley exploration. As we shall see, much depends on the interpretation of the data and how well it is used to develop and constrain model calculations.

7.2 Chemical and Physical Clues from Observations

Clues about the origin of comets can be grouped into three categories: Chemical composition, physical properties of the nucleus, and orbital parameters. All observations are made on comets that have evolved to different degrees; to relate them to their origin, they must be extrapolated backward in time. Orbital evolution is specifically related to the place of origin; it was considered in Chap. 6 and will again be considered in Sect. 7.4. In this section we consider the chemical and physical clues with major emphasis on the composition of the frozen gases and the dust, the shape and density of the nucleus, and the material properties of the nucleus at very low temperatures.

Chemical composition, dust to gas ratio, surface structures, shape, density, and many other data from the missions to Comets Giacobini-Zinner and Halley and from most of the many Earth-based observations may be relevant only for short-period comets. All missions proposed for the next decades are also only to short-period comets. While presumably fresher, long-period comets would be more ideal for investigations to gain insight into the origin of comets, they too have undergone significant physical and chemical alteration since their formation. Examples of some changes are: Near perihelion the gases in the coma have been chemically altered by solar radiation and the solar wind; during the 4.5 Gy of their life in the outer Solar System the icy components of the nuclei have been modified by cosmic rays, UV radiation, solar and stellar winds, and heating caused by remnant internal radioactivity; and still earlier, during comet formation, selective condensation (fractionation), disproportionation, gas trapping, and diffusion have altered the relationship between the chemical abundances in the nebular region and the nascent nuclei. Thus, comets are not pristine in the sense that they provide a record of the nebular region in which they were formed. How then do comet missions help us to better understand the origin of comets and the early history of the Solar System?

The place, time, and environment of comet formation can be ascertained only through combining resources from many disciplines. Computer models for very slow physical and chemical processes over very long time spans, supplemented by laboratory determinations of material properties at very low temperatures, are required to deduce the conditions of the nebular region in which comets formed. Any successful model that simulates the evolution of comets from their origin through their period of "hibernation" in the outer Solar System to their entry into the inner planetary system must match the comet data from remote sensing and in particular the much more detailed data from *in situ* measurements made during comet missions. Successful execution of such a program is difficult, because the initial conditions must be guessed; running computer codes backward in time (e.g., time reversal of chemical kinetics) does not lead to a unique solution because it violates entropy. Figure 7.1 presents the interrelationship among (1) ground-based, airborne, rocket-borne, and *in situ* observations, (2) computer models, and (3) experiments to obtain needed data and to simulate cometary conditions.

The simple diagram presented in Fig. 7.1 does not indicate all of the details. For example, the label "condensation" also implies disproportionation and chemical fractionation, diffusion of gases is closely related to "gas trapping," and "grain coagulation and collisions" also includes all other processes of grain growth up to and including formation of subnuclei by accretion and aggregation. The chemical

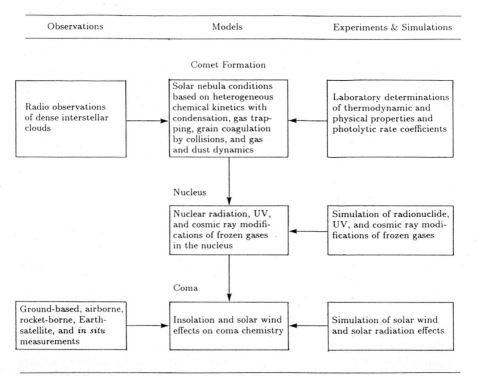

Fig. 7.1. A diagram relating observations, computer models, and laboratory experiments and ground- and space-based comet simulations.

and physical modifications of comet nuclei occurring in the Oort cloud (center of Fig. 7.1) also occur during the time of transit from the Oort cloud into the inner Solar System after a stellar perturbation. Some scenarios for the evolution of comet nuclei are discussed in Sect. 7.5.1. Comparisons must be made between model results and observations. The processes are iterative until agreement between models and observations is obtained.

7.2.1 Chemistry

Clues about the origin of comets based on chemistry include the molecular, elemental, and isotopic composition of the volatile and the refractory components in the nucleus. The division between volatile and refractory is somewhat arbitrary; in general we label a material to be volatile if it has a vapor pressure equal to or higher than that of H_2O. All other materials are refractories, including silicates, polymers, polycyclic aromatic hydrocarbons, and mixtures of many other complex organic molecules. The formation of the refractory and the volatile components may be closely connected, but it is also possible that they are only superficially related since they condensed at different temperatures and therefore at different times or different places. In the latter case, the refractory component may have existed in the presolar nebula in the form of simple grains that acted as condensation sites for the volatile component.

A particularly surprising result from the Halley missions is that the comet dust consists of three major components of about equal proportions: Silicates (as expected), particles rich on C, H, O, and N (CHON-particles), and mixed silicate–CHON particles (see Chap. 4). This composition is different from that of interplanetary dust, which consists mostly of anhydrous silicates. In some aspects it differs also from Greenberg's (1982) model for interstellar dust, which is exclusively of the core-mantle type. The presence of complex molecules, including polymers, in the coma of P/Halley may suggest a way out of this dilemma. Some polymers crystallize into fiber-like structures that have a high affinity for silicate or carbon surfaces. Submicrometer-sized grains of silicates, graphite, and other organic materials, including core-mantle particles, may be held together by polymers, forming very porous, fluffy, dust particles. At least one of the polymers, POM (polymerized formaldehyde) begins to disintegrate at about 400 K and totally disintegrates at 440 K; particles that reach that temperature (even higher grain temperatures were measured in Halley) will release the three observed types of grains. However, it is thought that the CHON particles formed before comet nuclei formed, possibly in the early stages of collapse of the protosolar nebula. The implication is that these dust particles were not heated above 400 K before they became part of Halley's nucleus. Thus comet formation took place in a region where even short-lived temperature fluctuations never reached this critical value. The composition of the grains is given in Table 3.2 in Chap. 3 and Table 4.6 in Chap. 4.

The volatile composition may be restricted similarly to an even lower temperature limit. If water condensed to form amorphous ice and remained as such, then the temperature cannot have exceeded 136 K even in a short pulse-like warming trend,

·otherwise a phase transition from amorphous to crystalline ice would have occurred. On a longer time scale of 1 My, the typically expected time of formation of comet nuclei, the ice is likely to crystallize even at 85 K (Schmitt et al., 1989). However, we lack evidence that amorphous ice exists in comets. Amorphous ice on grains in the interstellar medium may not have survived entry into the solar nebula. If it does exist, then most likely it will be found in the interior of nuclei of active comets, because the outer layers have been warmed above the transition temperature. In very large nuclei remnant radioactivity may have warmed even the interior to cause the phase transition to crystalline ice.

The composition of the volatiles in the nucleus is presented in Table 3.2 in Chap. 3. While the refractories showed an excess of C and N relative to cosmic abundances (i.e., carbonaceous chondrite abundances; see Table 4.6 in Chap. 4), the volatiles are deficient in these species relative to the solar composition. The combined composition results in elemental abundances very similar to solar values, except for the rare gases and H_2 that never condensed. The lowest condensation temperature of the gases is about 25 K. However, if gas trapping in the ice occurs, as measured in laboratory experiments and proposed by Bar-Nun et al. (1988), then some of the most volatile gases can be in comets at the 1% level even if comet formation occurred at about 50 K.

If there is no disproportionation and no gas trapping during condensation, the mixing ratio for condensed species i relative to species j can be calculated from the Clausius-Clapeyron equation and the equation of state for a perfect gas

$$\log \frac{n_i}{n_j} = \frac{L_i(T)}{L_j(T)} \left(\frac{1}{T} - \frac{1}{T_{ri}} \right) \bigg/ \left(\frac{1}{T} - \frac{1}{T_{rj}} \right), \tag{7.1}$$

where n is the number density, $L(T)$ is its latent heat of sublimation at temperature T, and T_r is the temperature at some reference pressure ($p_{ri} = p_{rj}$), usually 1 atmosphere.

The retention of ices on dust particle surfaces (condensation) at low temperatures and low densities was investigated by Yamamoto (1985). Gas molecules of species i with number density n_i and thermal velocity v_i at temperature T colliding with a grain having a radius a will heat the grain at a rate

$$\Gamma(n, T, T_d) = \pi a^2\, 2k(T - T_d) \sum_i n_i v_i \alpha_i(T), \tag{7.2}$$

where n is the total gas density, T_d is the dust grain temperature, $\alpha_i(T)$ is the temperature dependent accomodation coefficient for species i, and k is the Boltzmann constant. The dust grains cool by thermal emission at a rate

$$\Lambda(T_d) = 4\pi a^2 \epsilon \sigma(T_d^4 - T_b^4), \tag{7.3}$$

where T_b is the background radiation temperature, ϵ is the emissivity of the grain, and σ is the Stefan-Boltzmann constant. In steady state

$$\Gamma(n, T, T_d) = \Lambda(T_d). \tag{7.4}$$

As can be seen from Eqs. (7.2) to (7.4), a temperature difference may arise between gas and dust, with $T_d < T$ in general. While the most abundant molecules (H_2

and He) are responsible for the energy exchange between gas and dust, the heavier molecules (H_2O, CO_2, CH_4, etc.) are the ones that stick to the grain surfaces.

The sublimation (vaporization) rate, $z_+(T_d)$ per unit surface area on the ice mantle having molecular weight M and vapor pressure P is given by

$$z_+(T_d) = P(T_d) \cdot (2\pi k T_d M m_u)^{-1/2}, \tag{7.5}$$

where m_u is the unit atomic mass. On the other hand, the impinging rate, $z_-(T)$, per unit surface area for a species, whose abundance fraction in the gas phase in the nebula is f_i, is given by

$$z_-(T) = f_i n v_i / 4. \tag{7.6}$$

When $z_+(T_d) < z_-(T)$, net condensation occurs. The critical temperature at which the net sublimation rate equals zero is determined by

$$z_+(T_d) = z_-(T). \tag{7.7}$$

This corresponds to

$$f_i n k T / P(T_d) = (T/T_d)^{1/2}. \tag{7.8}$$

Based on these equations, Fig. 7.2 relates the temperatures of grains and several volatile species with plausible interstellar molecular abundances to the gas density. A very low abundance or the absence of the very volatile species of CO, CH_4, and N_2 would indicate that the temperature during comet formation was above 30 K.

The isotopic ratios for C, N, O, and S in volatiles and in refractories are the same as solar, within their limits of accuracy. On the other hand, the deuterium to hydrogen ratio is close to that of the Earth's oceans. The cosmogonical meaning

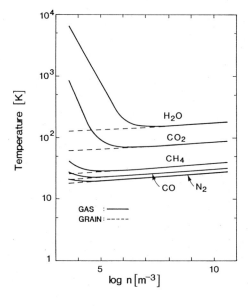

Fig. 7.2. Grain and critical sublimation temperatures of major candidate species in comet nuclei vs. gas density. Courtesy Yamamoto (1985).

of this ratio is not clear, because ion–molecule chemistry may have influenced this ratio through fractionation. In contrast to N, O, and S, for which the isotopic ratios in refractory samples show a modest amount of scatter around their mean values, the isotope ratio for $^{12}C/^{13}C$ presents such a wide range that the meaning of a mean value must be questioned. For volatiles, the $^{12}C/^{13}C$ ratio shows a much smaller amount of scatter even though the spectral analysis of $^{12}C^{13}C$ is complicated through the blending with NH_2 spectral lines. Recently Wyckoff et al. (1989) reported $^{12}C/^{13}C$ = 65±9 based on ^{12}CN and ^{13}CN observations of Comet Halley.

In terms of chemical compounds, H_2O is the most abundant constituent in comets. Everything else is at the level of a few percent or less. Less than half of the observed CO in the coma of Comet Halley comes directly from the volatile components of the nucleus, while the rest comes from the dust in the coma (Eberhardt et al., 1987). It is even possible that part of the CO coming from the volatile ices in the nucleus may be the result of dissociation of H_2CO (Boice et al., 1990) and, to a lesser extent, CO_2. The CO released from the dust may have its origin in the POM that holds the grains together. The production rates of CO and H_2CO from the distributed sources (probably dust) in the coma run close to parallel over distances of several thousand km, indicating a close relationship (Krankowsky, 1990). It is thus very likely that H_2O and CO, the two most abundant heavy molecules in the solar or presolar nebula condensed to form H_2O and CO_2 ice, part of which formed H_2CO through the action of cosmic radiation and during continued exposure to cosmic rays polymerized into POM.

The disproportionation of CO into CO_2 on condensation in the presence of cosmic or UV radiation may be a key in the chemistry of comet formation. Although laboratory experiments confirm that this process occurs (Sandford et al., 1988), its efficiency has not yet been established. Laboratory experiments by Pirronello et al. (1982) show that an icy mixture of CO_2 and H_2O irradiated by energetic ions, simulating cosmic rays, leads to formation of H_2CO. In an independent experiment, exposure of H_2CO to energetic ions (Gol'dyanskii, 1977) or UV radiation (Tsuda, 1961) leads to polymerization. The existence of POM in interstellar dust had been suggested by Wickramasinghe (1974, 1975) and in comets by Wickramasinghe and Vanýsek (1975) and Mendis and Wickramasinghe (1975).

To summarize, contrary to what is often implied, cometary matter is not pristine nebular material from which comet nuclei formed; but it is closer to it than any other type of matter accessible to us. The chemical composition of nuclei as derived by Delsemme (1988) is a guide, but the data from the P/Halley spacecraft missions revealed unexpected results that impose new constraints on the conditions for comet formation in a nebula. A new table of chemical abundances in comets has been prepared by Delsemme (1990). Computer modeling, supplemented by laboratory and space-release experiments and by ground-based, airborne, rocket-borne, Earth-satellite, and *in situ* observations of comets, provides the only means of deducing the chemical composition and physical properties of the nebular region in which comets originated. This type of analysis is a crucial step in expanding our knowledge about the early history of the Solar System and the surrounding Galaxy and propels advances of interdisciplinary character in space science, astrophysics, low temperature heterogeneous chemical kinetics, and formation of complex astrophysical molecules.

Isotopic abundances depend on the galactic evolution of elements and the individual history of comets. These ratios can be used as independent tests for theories of the general chemical evolution. For example, the chemical evolution of the Galaxy leads to enrichment of ^{13}C in interstellar matter. If comets were formed about 4.5 Gy ago, then cometary $^{12}C/^{13}C$ ratio should be slightly higher than that of present interstellar clouds. However, the cosmogonical meaning of isotope ratios is not clear, because ion–molecule chemistry may have influenced these ratios through fractionation (Vanýsek, 1987). Enrichment via the deuterium fractionation suggests that the chemical steady state is strongly sensitive to temperature. Thus, in principle, the observed D/H ratio may be used as a thermometer in the environment where comets formed. Deuterium becomes heavily fractionated in cold clouds, its abundance relative to hydrogen climbing as much as three orders of magnitude.

7.2.2 Shape of the Nucleus

The first and only comet for which we have pictures of the nucleus is Comet Halley. The detailed descriptions given in Chap. 2 will be analyzed here in regard to the origin of comets. The nucleus of P/Halley is twice as long as the diameter at its girth, i.e., it is $16 \times 8 \times 8$ km. Is this prolate shape a coincidence for the first comet nucleus that has been measured, or is this typical for comets? If it is typical, then this would be an important clue to the mechanisms of their formation. If, on the other hand, this aspect ratio is unique to Comet Halley, then we have observed one of the most unusual comet nuclei on our first mission! In that case the odd shaped nucleus could have resulted from erosion during its many passages through the inner Solar System, or it could be the result of a glancing collision of two nearly spherical bodies of about 10 km diameter. Such a collision would also be an important clue for the aggregation of a comet nucleus at the region of its formation. How did such a shape develop? There are four possible explanations: The shape arises from (1) erosion during the comet's exposure to insolation during its many perihelion passages, (2) splitting of the nucleus, (3) a glancing, inelastic collision of two nuclei of about 10 km diameter, or (4) its final stages of aggregation. We discuss the probabilities of these four processes below.

Erosion through evaporation of an icy body is largest at its subsolar point, unless that area happens to be covered by refractory material. However, it is not likely that a refractory area is always pointing to the Sun, unless the entire body is covered with refractory material. In that case, erosion is very slow. An area of ice evaporates at a rate $Z(\theta)$ [molecules per unit area and per unit time] when the surface normal is at an angle θ relative to the vector to the Sun. From energy balance at the surface of the nucleus and the Clausius-Clapeyron equation coupled to the perfect gas law, $Z(\theta)$ and the associated surface temperature $T(\theta)$ can be calculated (see Eqs. 2.3 and 2.4 in Chap. 2). At a heliocentric distance of 1 AU, $Z \simeq 2 \ 10^{22}$ molecules m^{-2} s^{-1} at the subsolar point. At small r, the right hand side of Eq. (2.3) is controlled by the evaporation term, while at large r it is controlled by the radiation term. The transition is between 2 and 3 AU. Thus in the region where the largest evaporation occurs, $Z \propto r^{-2}$. Integrating over that portion of the orbit where the comet is active (typically $r < r_m$, in AU), assuming optimum conditions ($\tau \simeq 0$, $\theta \simeq 0$, $A \simeq 0$), results in erosion of a thickness

$$|\Delta R| = \frac{M}{\rho N_0} \int_{t_1}^{t_2} Z(r)\, dt \simeq \frac{M}{\rho L} \int_{s_1}^{s_2} \frac{F_\odot}{v_\vartheta r^2}\, ds \simeq 2\frac{M}{\rho L} \int_{q}^{r_m} \frac{F_\odot}{\dot{r} r^2}\, dr,$$

$$\simeq 1.4 \times 10^7 \frac{M}{\rho L [q\,(1+e)]^{1/2}} \left(\arcsin \frac{r_m - q(1+e)}{e r_m} + \frac{\pi}{2} \right), \quad [m],$$

(7.9)

per revolution around the Sun (Huebner, 1967). Here ds is a path differential along the comet orbit, s_1 and s_2 are points on the orbit symmetrically spaced about perihelion, M is the molecular weight of the ice, L its heat of vaporization, ρ its density, v_ϑ the orbital velocity of the comet, e the orbital eccentricity, and q the perihelion distance in AU. For P/Halley ($q = 0.59$ AU, $e = 0.97$) assuming $r_m \simeq 2.5$ AU and water ice as the main constituent ($M = 18$ amu, $L \simeq 5 \times 10^4$ J/mol), $\Delta R \simeq 50$ m, for a low density of $\rho = 200$ kg/m^3. Orbital calculations of Halley's comet have shown that it has orbited the Sun at least 150 times without suffering a perturbation by a large planet that could have changed its perihelion distance (Carusi et al., 1987). Thus its diameter in the plane of the comet's orbit would be reduced by about 7.5 km. Assuming that the original diameter of the nucleus was at least 16 km (the present length of its major axis), then erosion could account for the observed width of the nucleus, but only in one dimension. However, the combined pictures from Vega and Giotto show that the width and the thickness are only about 8 km each. Thus, even under the extreme conditions assumed here, this is an unlikely explanation. Based on the mass loss required to account for the P/Halley meteor stream (McIntosh and Hajduk, 1983; Hughes, 1985), the comet may have orbited the Sun 1000 times. This would be a sufficient mass loss to explain the present shape of the nucleus of P/Halley, but it requires long-term stability with the spin axis remaining normal to the orbital plane and aligned with the long axis of the nucleus as the equator erodes. This is even more unlikely because then the maximum moment of inertia is not parallel to the spin axis and the spin orientation becomes unstable. Erosion is then distributed over many directions, which defeats the argument. Thus it is unlikely that erosion can explain the present shape of the nucleus.

Many comets have been observed to split. If a large comet splits into major fragments, then the largest fragments behave like individual comets that move in the same orbit as the original comet did. If Halley's comet fragmented into two parts of about equal mass, then this most likely occurred along the contour that looks closest to a straight line in the Giotto images (see Fig. 2.2 in Chap. 2). This line is about 10 km long and may have been longer before its ends were eroded. In that case we see now only about half of the original nucleus. Since we do not see the other half as an independent comet in the same orbit, it must have fragmented into many small pieces that have evaporated to the point that they are no longer observable. It is unlikely that one half should have disintegrated while the other half survived. Also, viewed statistically, splitting occurs mostly by breaking off small pieces; but it would be a coincidence that these pieces broke off along a straight line.

Inelastic collisions between large comets to form a double comet, as suggested by the shape of the P/Halley nucleus, must occur in the final phases of comet formation or later. If we assume the most optimistic case for such a collision, it must have occurred in the inner comet cloud. The probability for collision P_{coll} can be estimated from

$$P_{coll} = \frac{v\,\pi(2R)^2\,t}{2\pi r\,\Delta r\,\Delta z}.$$

(7.10)

Substituting optimistic values for the relative velocity $v \simeq 200$ m/s, the radius of each nucleus $R \simeq 5$ km, the time since formation of the comet cloud $t \simeq 4.5$ Gy $\simeq 1.5 \times 10^{17}$ s, the radius of the orbit $r \simeq 200$ AU $\simeq 3 \times 10^{10}$ km, the width of the comet ring $\Delta r \simeq 100$ AU $\simeq 1.5 \times 10^{10}$ km, and the thickness of the comet region in the ecliptic $\Delta z \simeq 10$ AU $\simeq 1.5 \times 10^{9}$ km, the probability for collision is 2×10^{-12}. Since there are about 10^{12} comets in the comet cloud (see Sect. 7.1.2), only about one collision occurred during the lifetime of the cloud. Stern (1988) has come to very similar conclusions. Thus Comet Halley would have to be a very unique object.

The final option is that an ellipsoidal shape with an aspect ratio of about 2 : 1 is a natural consequence of comet formation. If such a shape is to result as a consequence of rotation, then it is contrary to most large astronomical objects like stars and planets which have a bulge at the equator. However, these objects are or were fluids, while comet nuclei were always solids. Since there are many more small objects than big ones, the collision probability between them is greatly enhanced (Stern, 1988). One possible scenario, which has not yet been confirmed through model calculations, is that a large, slowly rotating body might have a reduced capacity for accreting small objects at its equator because of centrifugal forces. To have incoming material stick to the nucleus surface a minimum collision force might be needed. Since the low density of the nucleus demands very low collision speeds to begin with, the sticking probability at the equator may be too low. Such a body would thus accrete more effectively at the poles making them grow to produce the aspect ratio observed on Halley's comet. However, accretion must stop before rotational instability flips the body to maximize its moment of inertia. Consistent with this formation hypothesis is that some other comets may have a prolate shape (Sekanina, 1988a; Millis et al., 1988; Jewitt and Meech, 1988), but such claims must be viewed with caution; they are the result of ground-based observations and are not as definite as the spacecraft observations of P/Halley.

7.2.3 Structure and Density of the Nucleus

The images obtained with the Halley Multicolour Camera on the Giotto spacecraft show many unusual features on the nucleus (see Chap. 2). There are depressions several kilometers across and several hundred meters deep that are clearly measurable on the limb on the night side. A mountain on the night side of the terminator catches the sunlight; it is several kilometers long and could be several hundred meters high. There is a "finger" on the northern limb, followed by several smaller crater-like features on the day side of the terminator and a larger flat depression. A mountain-like bump on the anti-sunward side of the southern limb is several kilometers across.

Some of these characteristic structures, particularly the crater-like features, are probably caused by erosion. Although the mountain-like features are also affected by erosion, they are probably remnants of structural changes caused by collisions with smaller bodies during the accretion phase. If this proves to be the case, then the major accreting bodies were of the size of several hundred meters up to about a kilometer.

The mean density of a comet nucleus contains valuable information about the circumstances of its formation and early evolution. For instance, accretion of subnuclei may be expected to occur at different relative velocities in different regions of the solar nebula, and as a result, lower densities should arise at larger heliocentric distances. Formation of comet nuclei by local gravitational collapse in a disk of grains, small clumps, and subnuclei might imply even lower densities, since the material must agglomerate by collisions with a speed about equal to the escape velocity which is about 1 m/s for comets the size of P/Halley. Collisional compaction is then at its minimum but obviously also depends on the impact strength and compressibility of the subnuclei. Furthermore, the early stages of thermal evolution of the nucleus may be dominated by major heating events due to short-lived radioisotopes. Since these are concentrated in the refractory component of cometary material, the heating would be more pronounced in "dust-rich" comets that should tend to form relatively close to the proto-Sun. According to modeling scenarios, phase change of the H_2O ice (crystallization; Prialnik et al., 1987) and melting (Wallis, 1980; Irvine et al., 1980) might occur as a consequence of ^{26}Al heating. This would lead to a chemically differentiated nucleus and, in the case of melting, to a much higher density than the original structure if it had been very porous.

Comet densities have been estimated using two different methods. One of these is based on observed meteor densities (Verniani, 1969, 1973) which are interpreted as arising from evacuated, fluffy grains where the volatile component has been lost from the original structure (Greenberg, 1986b). Compensating for this lost component, according to the chemistry of the interstellar grain model (Greenberg, 1982), the result is a high porosity with a density less than 500 kg/m^3, even for the original cometary material. One must recognize that this is a local density referring to those layers of the comet nucleus from which the observed meteors originated. Furthermore the porosity thus found is only a microporosity which says nothing about the bulk constitution of the nucleus. Recently Greenberg and Hage (1990) analyzed the data on coma dust in Comet Halley in the same spirit and came to the conclusion that this dust is extremely porous, probably originating from a nuclear material with a density between 300 and 600 kg/m^3. This is a better indicator of the nuclear bulk density, since the dust particles sample the fresh material in active regions.

The mean density of the whole nucleus is found by independently estimating its mass and volume. Comet mass estimates are always based on a comparison of relative and absolute mass loss rates. For example, Hughes (1985) took the absolute mass loss per apparition of Comet P/Halley to be $\Delta M_{\oplus} = 2.8 \times 10^{11}$ kg based on absolute calibrations of comet magnitudes with water production rates (Newburn, 1981). Before the time of spacecraft imaging, the volume could be estimated from the visual photometric cross section assuming an albedo; Hughes found 440 km^3 using the albedo of 0.06–0.07, characteristic for the Moon and carbonaceous chondrites. Assuming a density of 500 kg/m^3 appeared reasonable at that time since this would imply a relative mass loss rate of $\Delta M_{\oplus}/M_{\oplus} \simeq 0.0013$, sufficiently small to explain the absence of observed secular fading. Furthermore, the mass of the Halley meteor stream (McIntosh and Hajduk, 1983) would imply an original mass of the nucleus seven times the present mass, which would mean that the comet is "middle-aged" (present radius about half the original radius) – again a reasonable expectation.

The most reliable mass estimate comes from the observed nongravitational effect in the orbital motion of a comet. This arises from the jet acceleration of the anisotropic outgassing of the nucleus. The force causing this acceleration can be expressed in terms of the gas production rate of the comet, so the nongravitational effect depends on the mass loss rate of gas in relation to the total mass of the nucleus. Combining this with the observed gas production curve, the mass can be determined. Using such an analysis, Wallis and Macpherson (1981) estimated from the typical values of the nongravitational parameter A_2 of short-period comets that the typical density should be less than $700\,kg/m^3$.

In general there are not enough data to estimate densities reliably for individual comets. However, there is one exception, i.e., Comet P/Halley. The mass of this nucleus, $M_{\circledcirc}$, is obtained by combining the expression for the jet acceleration

$$j = v m_u M Q / M_{\circledcirc} ,\tag{7.11}$$

with the expression for the main nongravitational effect, i.e., the perihelion delay

$$\Delta P = \frac{6\pi\,(1-e^2)^{1/2}}{n^2}\left(\frac{e}{p}\int_0^P j_r \sin f\, dt + \int_0^P \frac{j_t}{r}\, dt\right),\tag{7.12}$$

(Rickman, 1989). In these equations v is the average momentum transferred to the nucleus per unit mass of the sublimating gas; $m_u M$ is the average mass of the sublimating molecules of molecular weight M; m_u is the unit atomic mass, Q is the gas production rate from the nucleus; e, n, and p are the orbital eccentricity, mean motion (i.e., the time derivative of the mean anomaly or the mean angular velocity of the comet in its heliocentric motion), and semi-latus rectum, respectively; f is the true anomaly; r is the heliocentric distance; t is the time; P is the orbital period; and j_r and j_t are the radial and transverse components in the orbital plane of j, respectively.

The "nongravitational" mass determination thus involves three main steps: (1) Estimate $v = |v|$. The molecules leave the surface of the nucleus with a thermal (Maxwellian) velocity distribution characterized by the local temperature, T. By proper averaging over this distribution, for an assumed smooth surface one obtains a local vertical outflow velocity

$$v_0 = \left(\frac{2kT}{\pi m_u M}\right)^{1/2},\tag{7.13}$$

v is then obtained by a flux-weighted vectorial average around the nucleus over a spin period. Typically the sublimation flux distribution is expected to be strongly peaked and v is thus only slightly less than $v_0(T_{max})$. (2) Determine the average outgassing direction and thus the direction of j in the (radial, transverse, normal) reference frame. This amounts to an estimate of the thermal lag, given the spin properties of the nucleus and the activity distribution over the surface. For P/Halley the situation is simplified by the slow spin with an angular momentum axis far from the orbital plane, and by the likely existence of forced precession. Thus one may take the thermal lag to be predominantly in the orbital plane and put: $j_t = j \sin \eta$, where η is the lag angle. For a freely sublimating surface of an active spot that does not exchange heat laterally with its surroundings, η and its variation with r, can be estimated as a function of the unknown thermal inertia of the outermost surface layer. Near the Sun, very small values of 0°–5° are expected. (3) Specify the gas production curve, Q(t), with particular attention to its perihelion asymmetry. The reason is that the first integral in Eq. (7.12), i.e., the radial term, would vanish for a perfectly symmetric production curve but may dominate over the second term even with a moderate asymmetry. For P/Halley the still lingering uncertainty about the actual shape of the gas production curve (Rickman, 1989; Peale, 1989) is the main source of error in the mass estimate. The radial integral is indeed dominant, so the large uncertainty concerning the actual value of η, and thus the transverse integral, is less important.

The result of this analysis (Rickman, 1986, 1989) is a mass in the range (1.0–3.5)×10^{14} kg for the nucleus of P/Halley with a preferred value of about 1.4 × 10^{14} kg, corresponding to a density of 280 kg/m^3 with a possible range from 200 to 700 kg/m^3. Unfortunately the uncertainty of this density is still too large to draw detailed conclusions about the formation mechanism of the nucleus, but the object must have been assembled at very low speeds and the subnuclei must also have presented some resistance to collisional compression. Furthermore, it is evident that early heating events in P/Halley did not lead to large-scale melting or indeed any major structural modifications of the nucleus.

7.3 Formation of Comet Nuclei

We shall not attempt to give a complete review of the history and current status of research on this topic. The aspects not treated below were covered by Bailey et al. (1986).

There are three major hypotheses about the origin of comets. According to the first two hypotheses, comets were formed at about the same time as the Sun and the planets. After fragmentation and collapse of a molecular cloud, planets formed out of the gas and dust in the rotating disk that surrounds the proto-Sun. Small bodies that did not accrete onto the outer planets survived as comet nuclei. In this scenario comet nuclei are formed either in (1) the protoplanetary subnebulae of Jupiter and Saturn or (2) just beyond, in the solar nebula out to a distance of several 100 AU. The latter region shall be referred to as the trans-Neptunian region. The interstellar molecules were dissociated as they fell through the accretion shock in the solar nebula and were then reprocessed in the subnebulae and the trans-Neptunian region. The Oort cloud with its inner core arises naturally as a by-product of comet formation in the giant planet accretion zones (see Chap. 6), but it may need a comet population in the trans-Neptunian region as a source of replenishment for its long-term stability.

According to the third hypothesis, comet nuclei are aggregates of ice-covered interstellar dust. They are formed in parts of the presolar nebula from interstellar molecules, unaffected by the accretion shock. There are strong suggestions that comets also contain interstellar grains; the most important reasons are their high carbon content and their fluffiness. Greenberg (1982) had proposed a model for very fluffy interstellar grains containing radiation processed mantles that surround silicate cores. However, a first conclusion seems to indicate that the observed grains from Comet Halley do not clearly show the core–mantle structure predicted by his model. The requirement that interstellar grains be rich on carbon is common to all models of interstellar matter. As discussed in Chap. 4, about one third of the Halley particles are composed only of silicate material and clearly do not have organic mantles. A similar fraction of the particles shows no evidence of elements heavier than oxygen (with the possible exception of some sulfur), and they probably do not have cores. The rest of the particles do have compositions that are consistent with silicates and substantial organic components. Survival of bare grains and the formation of coreless organic particles are not predicted by the Greenberg model, but these difficulties could perhaps be reconciled as products of processing during or after formation of comet nuclei. However, as discussed in Sect. 7.2.1, the presence of very complex molecules and polymers suggests another particle structure that allows for the coexistence of all three types of grains. Some polymers have an affinity for carbon and silicates and can hold together pure silicates and other carbon and organic grains in a very porous structure. Since some polymers disintegrate at about 400 K, a temperature that grains in the coma of Comet Halley could easily reach, they can release all three types of the observed particles.

To investigate all three hypotheses, models appropriate for the protoplanetary subnebulae for Jupiter and Saturn, the outer solar nebula, and the presolar nebula (or a companion fragment of it) must be developed. Thus several options for the physical and chemical properties and thermodynamic conditions must be considered. In all cases the interplay between chemical and dynamical evolution with condensation and accumulation on grains, leading to aggregation and formation of small icy–dusty bodies, must be investigated.

However, there are problems in relating observations of comets to their mode of origin: (1) The place of origin and the mechanism of formation of comet nuclei are not known, (2) chemical models of dense interstellar clouds and the protoplanetary subnebulae of the giant planets are incomplete, (3) chemical modifications of comet nuclei in the Oort cloud are just beginning to be investigated, (4) physico-chemical models of comet comae are progressing rapidly but are still incomplete, and (5) all *in situ* measurements are on short-period comets; measurements on long-period comets can only be done by remote sensing and therefore are incomplete.

Of particular importance is the modeling effort interfacing items (1) and (2), which to-date has received very little attention. When coupled with observations of dense interstellar clouds and with ground-based and *in situ* observations of comets, such a model will provide a big step toward extending our knowledge in the exploration of space and beyond the Solar System into the surrounding Galaxy. It provides the bridge between space science and astrophysics.

7.3.1 Models for Giant Planet Subnebulae and the Solar Nebula

Prinn and Fegley (1989) developed models for the subnebulae out of which the giant planets formed. If comets also formed in these subnebulae they should contain CH_4 and NH_3 but no interstellar molecules such as H_2CO or S_2.

For the solar nebula two basic models exist: Cameron's model, in which the solar nebula is large and massive, and Safronov's model for a smaller, less massive nebula. However, a whole range of accretion-disk models have recently come into consideration by different authors, and depending on the choice of free parameters, the whole spectrum of mass and size may be considered. The temperature range from the orbit of Jupiter out to about 100 AU is 180 K to 18 K, based on Cameron's (1978) adiabatic model. For Hayashi's (1981) radiative steady state model the temperature range is somewhat narrower (about 130 K to 25 K). Cameron's model evolves with time during the infall with accretion to subsequent mass loss from the disk. The above values correspond to the end of the infall phase of the cloud when the density and temperature are near their peak values. Comets formed in the outer solar nebula will contain reprocessed molecules, i.e., CO and N_2, but very little or no CH_4, NH_3, H_2CO, or S_2.

7.3.2 Presolar Nebula

It is widely accepted that the presolar nebula must have been similar to moderately dense interstellar clouds. One difficulty with present models for interstellar cloud chemistry with condensation is that the condensation of heavy molecules on grains is so efficient and fast that the cloud would be depleted of heavy gas-phase species in less than 10^5 y, thereby contradicting observations. Gas-phase abundances of heavy molecules, much higher than predicted by models incorporating condensation, are observed in interstellar clouds that are believed to be much older than 10^5 y. Processes that return heavy molecules efficiently to the gas phase appear to be missing in the models. New mechanisms must be considered to increase the abundances of heavy molecules in the gas phase. Desorption by x-rays and heavy cosmic rays has been suggested by Léger et al. (1985) and Léger (1987), but it still needs to be applied in model calculations with detailed chemical kinetics. Comets formed in the presolar nebula will contain typical interstellar molecules including CO, N_2, H_2CO, and S_2.

Particles with high abundances of hydrogen, carbon, nitrogen, oxygen, and some sulfur, the so-called "CHON" particles, have been discovered in the coma of Comet Halley. These particles appear to be distributed throughout the nucleus of Halley's comet and therefore must have been present in the nebular region before comets formed. Since some components in CHON particles disintegrate only at temperatures of about 400 K, it is likely that they survived from the interstellar cloud. If CHON particles are present in interstellar clouds, they are also a likely source for some heavy molecules in the gas phase in these clouds.

7.3.3 Formation Mechanisms

The high abundance of frozen water in comets suggests that the nuclei formed at low temperatures. The very low abundance of very volatile materials such as helium, noble gases in general, molecular nitrogen, methane, and carbon monoxide, argues for a temperature that is not below 25 K. (Only about one-third of the observed CO in P/Halley comes directly from the nucleus.) The existence of CHON particles indicates that grains were not heated, that they may be of an interstellar character rather than of reprocessed interplanetary character. These conditions are also consistent with the very low density of comet nuclei. Thus the initial steps of comet formation must have started with the condensation of water and a few other gases on pre-existing grains. At the low temperatures, water would condense to form an amorphous ice, rather than crystalline ice. We must therefore expect that the ice in nascent comet nuclei will be amorphous. This is an important point for the evolution of comet nuclei because the phase transition from amorphous to crystalline ice occurs at 136 K and is exothermic (1.6 kJ/mol). Ice-covered grains evolved in an environment where cosmic rays and UV radiation caused minor heterogeneous reactions and possibly some polymerization. Polymerization of species relevant for comets is also exothermic (~ 2 kJ/mol). In places where density fluctuations enhanced collisions between grains, they coagulated at relatively low velocities forming fluffy, micrometer-sized and larger particles. High collision velocities (> 100 m/s) would evaporate some icy surfaces of the grains, reducing the probability to stick together. However, grains covered by organic materials (polymeric condensates) are much less sensitive to high-speed collisions than ice-covered particles. The processes of condensation, gas trapping (Bar-Nun et al., 1988), and grain evolution into particle clusters probably occurred simultaneously. Frozen gases may accumulate in the pores of the clusters.

There is little disagreement about the scenario up to this point, but there are widely different opinions about the cause for the density enhancements that made condensation and accretion possible. It is necessary to extend the models of Morfill et al. (1985) to investigate the chemical and physical properties of the various nebular regions that lead to comet formation. Comet formation is most easily understood if it occurs in a cool, gentle, but dense region. In first approximation, both the temperature and density of the nebular region are expected to decrease with increasing distance from the center. The temperature and density gradients are smaller in the plane of rotation than in the direction perpendicular to it. However, local fluctuations (enhancements caused by oscillations, turbulence, clumping, subnebulae that may lead to planet formation, etc.) may be possible. It is difficult to predict the location of the comet-forming region; additional criteria, such as the chemical species that condense, formation of complex molecules, trapping of gases, and dynamical effects, must be considered to identify the region. For example, presence of CH_4 and NH_3 in comets might suggest formation in a protoplanetary subnebula. Several options for the physical and chemical properties must be considered. Hence any development of a comet-formation model is a long-range program that needs to go through several phases, not only for the various conditions and processes that must be investigated, but also from the point of view of incorporating newly developed concepts for the basic physical processes.

Fragmentation and gravitational contraction of the parent cloud warmed the inner solar nebula, which subsequently cooled again. We already know that near-chemical equilibrium in which H_2O, CH_4, and NH_3 are the major constituents, CO, CO_2, HCN, and C_2H_2 important minor constituents, with most other chemical species present only in trace quantities does not satisfy the requirements. Lewis and Prinn (1980) had investigated the solar nebula to a distance of about 300 AU. Although condensation was treated only approximately for the major species, they found that conversion of CO to CH_4 and of N_2 to NH_3 was so slow relative to radial mixing and nebula evolution rates that the low-temperature equilibrium species were very underabundant. Thus most of the CO and N_2 dominated over other C and N compounds at the low temperatures. Also, comet models based on a near-equilibrium composition of the nucleus, cannot explain the abundances of species like CN, C_2, and C_3 which are observed in almost every comet (Mitchell et al., 1981; Biermann et al., 1982). Models based on heterogeneous chemical kinetics, with dynamic mixing, condensation, and gas trapping are needed. The heterogeneous processes require more basic research before they can be incorporated into comprehensive models. They should include phase changes in the formation and destruction of chemical species, in particular, changes from the gas phase to condensed phases (liquid and solid). Formation of monolayers on grains and disproportionation reactions will be important processes. Silicates may be good catalysts for the formation of some hydrocarbons including CH_4 and C_2H_4. Associated may be the production of complex molecules on grain surfaces. Cosmic and UV radiation appear to play important roles for polymerization.

Many different dynamical effects must be included in the description of the chemical properties of the system for comet formation. They can be grouped into two classes: Large scale and small scale (including microscopic) dynamics. Among the large scale effects are turbulent convection (Cabot et al., 1987a,b), motion of dust caused by radiation pressure that will change the local dust density (Hills, 1982; Hills and Sandford, 1983a,b), selective depletion of gas by radiation pressure, removal of gas by T Tauri winds, local vortices, oscillations of dust through the plane of rotation of the cloud (Biermann and Michel, 1978), etc. Among the small scale effects are coagulation of grains into fluffy clusters, growth of these clusters into clumps and small "snow balls", shattering in some high-speed collisions, sweep-up, etc.

The detailed chemical composition will vary with heliocentric distance and the relative abundances need to be determined in detail. Turbulent mixing will influence the relative abundances. Because dynamical effects lead to density enhancements, condensation, and coagulation, aggregation by sweep-up into comet nuclei may occur very rapidly. The advantage of comet formation in the intermediate solar nebula is that the comet nuclei are easily transferred into an inner comet belt (Weissman, 1985) and a flattened Oort cloud by planetary perturbations and can then be randomized into a spherical Oort cloud by stellar perturbations (Shoemaker and Wolfe, 1984; Remy and Mignard, 1985; Mignard and Remy, 1985; Duncan et al., 1987).

The outer solar nebula encompasses the space outside the region just considered. Dynamic mixing will influence the relative abundances that need to be calculated using chemical kinetics in a detailed model. In this region the densities are lower than in the intermediate solar nebula and formation of comet nuclei is slower. Radiation pressure may play an important role (Hills, 1982; Hills and Sandford, 1983a,b).

In models for the presolar nebula simulation of the dynamics for the collapsing cloud must be included. Existing models (Tscharnuter, 1980, and references therein) can be parameterized. Gas-phase chemical kinetics and gas–dust collision frequencies with accomodation coefficients (see Sect. 7.2.1) and sticking coefficients to calculate "condensation" and formation of ice mantles on grains are important. Fluid dynamic instabilities and radiation pressure may play an important role in the formation of comet nuclei.

Models for the fragments of the presolar nebula have many similarities with the models for the presolar nebula. The companion fragment is a smaller cloud that has separated from the original cloud. It is not big enough to form a star, but it can form comet nuclei. The physical processes for formation of comet nuclei have been investigated by Biermann and Michel (1978). This model should be investigated and developed further by including the gas-phase chemical kinetics and detailed "condensation" on grains. It has the appealing feature that stellar perturbations can easily extract comet nuclei from the companion fragment. The small number of these nuclei that will be captured by the main cloud of the presolar nebula will have a random (isotropic) distribution similar to that of the Oort cloud.

Models simulating conditions in different parts of the solar nebula may predict different chemical compositions of the condensed material which can then be compared with observations of comets. New clues may be obtained to differentiate between the possible places for the origin of comets, e.g., whether they are more similar to the original, interstellar composition, to a partially reprocessed, solar nebula composition, or to the composition appropriate for the protoplanetary subnebulae of Jupiter and Saturn. Models may also lead to criteria and discriminating clues for detection of subtle chemical differences between the origin of short-period and long-period comets. It is possible that comet formation is not localized to just one region; it may occur in several or all of the regions discussed here. Although comets have many compositional features in common, there are also differences, such as the dust to gas ratio and the relative abundances of some of the minor species. However, correlating these with orbital parameters has been unsuccessful so far because uncertainties in the abundances are too large.

Condensation of volatiles on grains proceeds very rapidly. Because of the very large number of grains, collisions are very frequent and the ice- and polymer-coated grains have a high probability for sticking together. Over a few thousand years, large particles in the nebula will oscillate through the midplane of the nebula under the influence of the gravitational field of the disk. The oscillation will be damped by friction and inelastic collisions, causing the particles to grow and to settle toward the midplane. Goldreich and Ward (1973) have shown that gravitational instabilities in such a dense layer will lead rapidly to formation of larger bodies in low eccentricity orbits. However, this concept has recently been questioned by Weidenschilling (1988) and Weidenschilling et al. (1989). Daniels and Hughes (1981), Donn and Rahe (1982), Donn and Hughes (1986), and Meakin and Donn (1988) have developed fractal and fluffy aggregate models for the formation of comet nuclei that show many of the aspects of low density and non-spherical shape and that are consistent with observations of Halley's nucleus.

7.3.4 Long-Term Processes

Planetary accretion is expected to proceed very slowly in the Uranus–Neptune region, but the relatively large quantities of nebular gases in these planets show that they grew to large sizes on a short time scale. Observationally, it seems that outside Neptune's orbit, near-ecliptic giant planets are lacking over a significant distance – say, out to at least 100 AU – and theoretically, as an outcome of accretion in the solar nebula, we don't expect such planets to have formed even further out (a distant brown dwarf companion to the Sun is not excluded but would have formed out of a separate subnebula).

However, comets might have formed in large numbers outside of Neptune's orbit; they should at least have started to accrete into larger bodies. The size spectrum of large comets should thus reflect an accretion process, their number decreasing rapidly with radius, but one can hardly predict the sizes of the largest bodies. Observationally, the 200-km diameter asteroid 2060 Chiron has recently turned out to be a comet (Tholen et al., 1988; Meech and Belton, 1989), proving the existence of nuclei much larger than those of comets that are usually observed. The dynamical properties of Chiron appear to be consistent with an origin in a trans-Neptunian population or the inner core of the Oort cloud and a future evolution into a Jupiter-crossing orbit where cometary activity might reach a much higher level. Also, the long-period comet of 1729 was visible to the naked eye for a long time in spite of a perihelion distance of ~ 4 AU, indicating a nucleus with a radius ~ 100 km to produce the observed amount of gas. The question whether Pluto's composition is icy, corresponding to a truly gigantic comet (nuclear radius ~ 1000 km), is complicated by the fact that the surface properties of Pluto and Charon seem to differ. A density of $\approx 2\,\mathrm{Mg/m^3}$ might be caused by gravitational compaction. The negative result of observational searches (Luu and Jewitt, 1988) indicates that if Pluto is indeed a giant comet nucleus, then it is among the largest ones – perhaps the largest.

Recent investigations (Sussman and Wisdom, 1988; Duncan et al., 1989) suggest that dynamical chaos can occur over time-scales ~ 1 Gy in that region of the Solar System as caused by planetary gravitation. Furthermore, there may be significant external influences building up over such intervals (see Sect. 7.4.1). Hence, initially near-circular in-plane orbits would likely have become dynamically stirred or "heated" so that accretion is hindered. Collisions will occur between subnuclei of comets. Some of these will probably lead to cratering, ablation, and shattering, while others will lead to the growth of the nuclei. This is equivalent to sweep-up of small objects, the size of a few centimeters up to a meter, by the larger bodies. Stern's (1988) results imply that collisions between large, km-sized bodies in the inner cometary cloud are extremely rare.

How could the cometary cloud have been formed? Gravitational scattering of icy planetesimals from the accretion zones of the giant planets is one possibility. Uranus and Neptune are most efficient in this respect. Their weak perturbations would make these bodies random-walk at a gentle pace outward to increasing aphelion distances. External influences would then take over, including the tidal field of the galactic disk (Sect. 7.4.1) and possibly nearby stars that formed along with the Sun out of the same cloud complex and remained in the solar vicinity over ~ 100 My. According to Duncan et al. (1987) this would occur as the semimajor axes reached several 1000 AU, and outward diffusion of the perihelia would lead to a comet cloud as depicted in Fig. 6.3 in Chap. 6. An alternative possibility is that a trans-Neptunian disk-shaped comet population is dynamically activated by the rare encounters between the Solar System and very massive perturbers in the shape of GMCs so as to act as a source of comets for all exterior regions.

7.4 Dynamical Evolution of Comets

Comets are obviously ephemeral phenomena. They develop their brilliant activity only while passing within several AU from the Sun, returning to anonymity as faint and tiny objects when they move out to larger distances. This leads to the expectation that the observed comets may represent only a minor fraction of a large population of such tiny objects, i.e., inactive nuclei, forever dwelling in the cold regions of interplanetary and transplanetary space beyond the asteroid belt. However, another aspect of the ephemeral nature of comets is equally obvious. The gas and dust observed in cometary comae and tails is not bound to the nuclei and thus lost from the comets. In relation to this mass loss, the material resources of a typical nucleus appear insufficient to guarantee substantial activity during more than 10^4 revolutions in Mars-crossing orbits. At least for short-period comets, whose orbital periods are about 100 y or less, the observable lifetimes are hence limited to a minute fraction of the age of the Solar System. Replenishment is necessary in order to keep their population in a steady state.

This replenishment may naturally be imagined as a consequence of the well-known instability of comet orbits. Even if we do not restrict our attention to short-period comets, we are always faced with a need for a dynamical mechanism transferring comets from a remote source into observable orbits, as explained in Chap. 6. Generally speaking, dynamical evolution thus appears as one of the most important features of comet physics, linking the presently observed comets to their place of storage or origin. By understanding this evolution, we may be able to trace the spatial distribution of comets and thus gain valuable insight into their mode of formation.

Figure 7.3 illustrates the classical picture of cometary dynamics by means of diagrams in which perihelion distance, q, is plotted against aphelion distance, Q, using a logarithmic scale. The distribution of observed comets is evident in the lower part of Fig. 7.3a ($q \lesssim 5$ AU); two distinct groupings appear immediately. One of these is formed by the "new" comets coming directly from the Oort cloud along nearly parabolic orbits. These are usually defined by an original, barycentric orbital energy parameter $1/a_{orig} < 10^{-4}$ AU^{-1}, which means that their previous aphelia were situated at $Q > 2 \times 10^4$ AU from the Sun. The orbital planes of these comets show no concentration to the ecliptic plane. There is a second group of comets at the lower left with aphelion distances in the range from 4 to 8 AU, which is called the Jupiter family. These comets show a strong dynamical link to the giant planet: They are characterized by low-inclination orbits with a certain tendency toward alignment

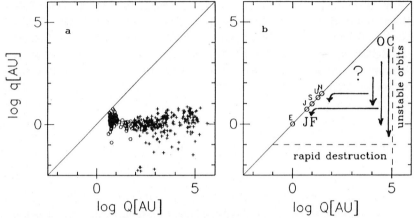

Fig. 7.3. (a) The distribution of comets observed up to 1987 over aphelion (Q) and perihelion (q) distances using logarithmic scales on the axes. Circles denote short-period comets and crosses denote long-period comets with non-parabolic orbits. For the latter the original orbits (before entry into the planetary system) have been used wherever available. The $q = Q$ line, below which all orbits must be situated, has been marked. (b) Basic transport routes in classical cometary dynamics illustrated in the same diagram. Vertical arrows denote stellar perturbations and horizontal arrows turning downward on the left denote planetary perturbations on low-inclination orbits. Planets have been marked by their initial letters on the $q = Q$ line. JF and OC indicate the locations of the Jupiter family and the classical Oort cloud, respectively. The question mark labels an orbital domain immune to all the considered transport mechanisms.

of the line of apsides with that of Jupiter's orbit. Furthermore, integration of their motions backward in time very often indicates close encounters with the planet, whereby strong orbital perturbations are experienced.

Let us briefly recall the remarkable properties of the concentration of new comets. Planetary perturbations, mostly caused by Jupiter, quickly scatter these comets over a much wider range of orbital energy than that corresponding to their previous aphelia. As a result, nearly half of them are expelled from the Solar System along hyperbolic orbits, and the remaining ones are mostly bound to reappear with much lower values of Q – much like those of the rest of the long-period comets that can be seen as a sparse background distribution to the left of the new comets in Fig. 7.3a. This background can thus be understood as arising from repeated passages of long-period comets; but to retain the concentration of near-parabolic comets we need a continuous infeed. Oort's idea that the infeed of new comets arises from the tendency toward an equilibrium distribution of angular momentum created by stellar perturbations has been most influential. As illustrated in Fig. 7.3b, this leads to the picture of the classical Oort cloud from which a steady transfer occurs toward perihelion in the inner planetary system to compensate for the dynamical loss of such comets by planetary perturbations.

In addition to this dynamical loss there is also a physical loss of long-period comets. For instance, many of the observed new comets certainly would not have been observable had they been making their second or third appearances. Otherwise the contrast between the height of the peak of new comets and the background level of the energy distribution would not have been as large as we observe. Furthermore, statistically the long-period comets must be dynamically older, i.e., have passed a larger number of perihelia near the Sun, the further we go to the left in Fig. 7.3b, and the decreasing number of such comets is most naturally interpreted by means of a physical loss mechanism limiting their observable lifetimes. It is not yet known whether the Halley-type short-period comets represent an extreme stage of such an orbital evolution, but this appears likely. Let us note that this type of cometary capture can in principle be aided by the action of nongravitational forces. These forces are weak compared to Jupiter's gravitational attraction. However, they may act coherently to produce a steady variation of orbital energy during a large number of revolutions, as sometimes evidenced by short-period comets with a long observational history, e.g., P/Halley. This is different from the gravitational perturbations which appear in a random way for long-period comets. Thus each perturbation is essentially independent of those experienced earlier and can be considered a random sample from a distribution of possible perturbations that extends over both positive and negative values. A succession of such perturbations produces a random walk along the orbital energy axis in contrast to the resolute march of nongravitational perturbations. The speed of this march may be high enough in some cases of low perihelion distance to contribute to the gradual decrease of aphelion distance (Weissman, 1979).

Clearly, the Jupiter family cannot be fit into the picture just described. The only reasonable way to explain this secondary concentration of observed comets is by invoking a different source. An evident indication about the nature of this source is the orbital affinity for Jupiter shown by these comets. We can leave aside those theories claiming a physical origin of comets on the giant planet or its satellites (Lagrange, 1814; Vsekhsvyatskij, 1966; Drobyshevskij, 1981) and concentrate on dynamical capture associated with close approaches. According to the classical ideas as indicated in Fig. 7.3b, this route leads from the classical Oort cloud by a separate branch. As the process of angular momentum transfer puts Oort cloud comets into orbits with perihelia near Jupiter's orbit, those moving at low inclinations with respect to the ecliptic plane are severely perturbed and tend to scatter over a wide range of orbital energy. This means that those which happen to experience predominantly decelerating perturbations are quickly transferred to the left in Fig. 7.3b, and the process of capture is relatively efficient. Toward the end of this process the comets arrive at low-eccentricity orbits in the Jupiter–Saturn region. Then, finally, during encounters with the planet, the comets may be captured into the observable Jupiter family via reversals of the apsidal line.

Thus, in conclusion we may say that in the canonical picture of cometary dynamics the ultimate source of all observable comets is the Oort cloud, but the observed long- and short-period comets represent two different branches of orbital evolution. The dynamical routes of this picture are limited to either the lower part of Fig. 7.3b ($q \lesssim 30\,\mathrm{AU}$) where planetary perturbations occur, or the right-hand part ($Q \gtrsim 10^4\,\mathrm{AU}$) where stellar perturbations are efficient. The question mark denotes an inaccessible domain characterized by orbits of low to moderate eccentricity with semimajor axes ~ 100 to $1000\,\mathrm{AU}$. If comets had formed in such orbits, they would not be able to get out. In the light of recent work one must conclude that the question mark still remains, but there are now a number of indications that the domain is more accessible than earlier believed, as we shall see in the following sections.

7.4.1 Infeed into the Planetary Region

When moving far away from the Sun, comets experience the gravitational attraction of masses outside the Solar System. They are thus dynamically influenced by our galactic environment. How can this gravitational environment be described? It is important to realize that the answer to this question is dependent on the time scale considered. We are dealing with an interplay of two effects: That of single objects passing in the vicinity of the Solar System and that of the myriads of objects perpetually present at larger distances. The further we extend our considered time interval, the closer encounters will have to be expected, or alternatively, the more massive and rare objects will be able to pass within a certain range. Let us write the mean encounter interval simply as

$$T_R = f_R^{-1} = (n\langle v \rangle \pi R^2)^{-1} ,\qquad(7.14)$$

for a limiting distance R, where n and $\langle v \rangle$ are the number density and mean encounter velocity, respectively, characterizing a certain class of objects. Figure 7.4 illustrates such a simple estimate of the mean encounter interval (reciprocal of encounter frequency f_R^{-1}) as a function of R.

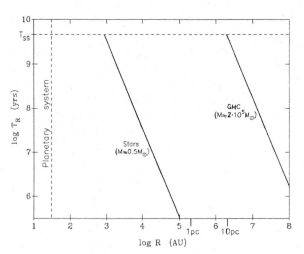

Fig. 7.4. The mean interval, T_R, between encounters as a function of R, the heliocentric distance, for two different kinds of galactic objects whose typical masses are indicated. Red dwarfs have been taken as representative of the stellar population. GMC stands for Giant Molecular Cloud complexes. The vertical dashed line indicates the approximate outer boundary of the planetary system. The horizontal dashed line gives T_{SS}, the present age of the Solar System.

Individual Stellar Perturbations. During a typical orbital period of several million years of an Oort cloud comet the most significant event to be expected is the passage of a typical red dwarf star ($\langle M_* \rangle \simeq 0.5\,M_\odot$) at a minimum distance of about 30000 AU from the Sun. This is bound to have a drastic effect on some Oort cloud comets that happen to be situated along the trajectory of the star and important perturbations will be suffered by most of the remaining comets in the cloud as well. However, a hypothetical population of comets at 100 to 1000 AU from the Sun would not be affected perceptibly by such a stellar passage.

Let us also estimate what may happen on the time scale of several hundred million years, characterizing the Sun's galactic orbit. According to Fig. 7.4, the closest stellar encounter to be expected is then at a minimum distance of about 3000 AU, so that some of the comets of the hypothetical trans-Neptunian population may be significantly perturbed. More far-reaching, however, is the expectation that during this interval a GMC will pass within 30 pc of the Sun (1 pc = 2.06265 × 10^5 AU). These structures appear to have masses of several 10^5 $M_\odot$, so even at such large distances their effects are bound to be important.

Thus, because of the importance of rare but drastic events, the effects of encounters with single objects may be totally different if considered on different time scales. The situation is not the same for the collective effects of the galactic disk. This imposes a tidal field in which the cometary orbits may exhibit large-scale variations over very long periods, so the question is only how much of this variation is contained within the time under consideration.

The effect of a stellar encounter on the orbit of an Oort cloud comet is usually computed by means of the impulse approximation. With respect to the Sun, the comet typically moves at a speed $\lesssim$ 100 m/s, while the star passes at least a hundred times more rapidly. Thus in a heliocentric frame of reference, the comet remains at radius vector r_c, and the star describes a rectilinear trajectory with a constant velocity $v_\star$, such that $r_\star = d_\odot + v_\star t$, where the time t is counted from the moment of closest approach of the star to the Sun (at radius vector $d_\odot$). One easily integrates the accelerations imparted to the comet and the Sun, and one thus finds the heliocentric impulse of the comet

$$\Delta v_{\&} = \frac{2GM_\star}{v_{\&}} \left(\frac{\hat{d}_{\&}}{d_{\&}} - \frac{\hat{d}_\odot}{d_\odot} \right) ,$$
(7.15)

where $d_{\&}$ and $d_\odot$ are the minimum distances of the star to the comet and to the Sun, respectively, and $\hat{d}_{\&}$ and $\hat{d}_\odot$ are unit vectors directed toward the respective points of closest approach.

For stellar passages through the Oort cloud one often approximates Eq. (7.15), neglecting the last term within the parentheses for comets in the immediate vicinity of the star track, and neglecting the first one for most other comets if the star makes a close approach to the Sun ($d_\odot \lesssim 10^4$ AU).

In general, for estimating the transfer rates of energy or angular momentum such approximations are reasonable as long as $d_\odot \lesssim a$, the semimajor axis of the cometary orbit. If θ is the angle between $r_{\&}$ and $d_\odot$, the impulse imparted to the comet by a more remote encounter can be approximated as

$$\Delta v_{\&} \simeq \frac{2GM_\star r_{\&} \cos \theta}{v_\star d_\odot^2} .$$
(7.16)

From Eq. (7.14) we realize that the closest stellar encounter expected during the interval T is at a distance $d_{min}(T) = (2\pi n_\star \langle v_\star \rangle T)^{-1/2}$. Substituting for T the orbital period $P = 2\pi (GM_\odot)^{-1/2} a^{3/2}$, we find the critical semimajor axis $a_{\&}$ for which $d_{min}(P_{\&}) = a_{\&}$

$$a_{\&}^{7/2} = \frac{(GM_\odot)^{1/2}}{4\pi^2 n_\star \langle v_\star \rangle} ,$$
(7.17)

i.e., $a_{\oslash} \simeq 24000\,\text{AU}$ using standard estimates of $n_\star = 0.1\,\text{pc}^{-3}$ and $\langle v_\star \rangle = 30\,\text{km/s}$. For $a < a_{\oslash}$ all stellar perturbations during one orbital period can be reasonably approximated by Eq. (7.16), while for $a > a_{\oslash}$ a more important contribution will be given by the close encounters.

The mean transfer of angular momentum J during one orbital revolution can be expressed by means of the variance σ_{J}^2 of the distribution of perturbations ΔJ. This can be estimated by adding the transverse impulses received from all stellar passages at each specific distance in a random-walk picture, and integrating the effect over all relevant passage distances. The result, as found by Bailey (1986a), is:

$$\sigma_{\text{J}}^2 = \frac{80\pi^2 \, (G\langle M_\star \rangle)^2 \, n_\star \, a^{7/2}}{3(GM_\odot)^{1/2}\langle v_\star \rangle} \cdot \begin{cases} 2\ln(1 + a/a_{\oslash}), & a > a_{\oslash}, \\ \left(a/a_{\oslash}\right)^2, & a < a_{\oslash}. \end{cases} \tag{7.18}$$

Thus we realize that the scatter of angular momentum per orbital period is a rapidly increasing function of semimajor axis (or aphelion distance for high-eccentricity orbits). The detailed numerical factors involved differ between different treatments of the problem, but the main structure of Eq. (7.18) remains unaffected. Figure 7.5 shows the scatter of perihelion distance (Δq) derived from Eq. (7.18) as a function of semimajor axis for $q = 15\,\text{AU}$ – the results can be translated into other values of q using the relation: $\Delta q \propto q^{1/2}$. Obviously the perihelion distance becomes drastically perturbed only at semimajor axes $a \gtrsim 30000\text{–}40000\,\text{AU}$.

The significance of this result becomes apparent when planetary perturbations are considered. The above-mentioned scatter of orbital energy due to planetary perturbations, $\sigma_{(1/a)}$, is a function of perihelion distance, and although each planet (with semimajor axis a_{p}) causes a peak at $q \simeq a_{\text{p}}$, the general trend is for $\sigma_{(1/a)}$ to decrease rapidly with q throughout the region of the giant planets (see Fig. 7.8 below). In comparison with the typical values of $(1/a)_0$ of new Oort cloud comets, or with the width $\Delta(1/a)_0$ of the interval within which they are contained, one finds that $\sigma_{(1/a)} \gg \Delta(1/a)_0$ for $q \lesssim 10\text{–}15\,\text{AU}$ (i.e., cometary orbits approaching or crossing

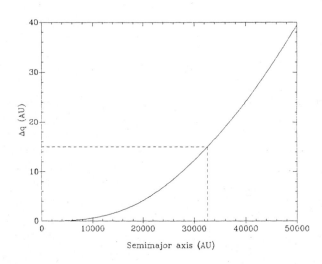

Fig. 7.5. Typical scatter, Δq, of the perihelion distance, q, for orbits with $q = 15\,\text{AU}$ and different semimajor axes, computed with the aid of Eq. (7.18). The dashed lines indicate the particular situation where $\Delta q = q$, thus separating relatively mild perturbations (*to the left*) from drastic ones (*to the right*).

the orbit of Saturn) while $\sigma_{(1/a)} < \Delta(1/a)_0$ for $q > 15\,\text{AU}$ (gravitational interactions limited to Uranus and Neptune). Thus if an Oort cloud comet penetrates to within about 15 AU of the Sun, it runs an overwhelming risk of being perturbed away from the cloud by Jupiter or Saturn. As a consequence the distribution of Oort cloud comets in angular momentum space may show a depletion in a central sphere of radius $J_0 = (2GM_\odot \cdot 15\,\text{AU})^{1/2}$ ("loss sphere"), and their distribution in velocity space at any particular heliocentric distance r may show a depletion within a "loss cylinder" along the radial axis of radius $(v_t)_0 = J_0/r$ (Hills, 1981).

The requirement in order for this loss cylinder to show no appreciable depletion of comets is that it can be replenished in the course of one orbital revolution. Considering stellar encounters as the mechanism of replenishment, Fig. 7.5 shows that filling the loss cylinder requires $a > 32000\,\text{AU}$. Let us call this encounter limit a_e. Recent estimates (Heisler and Tremaine, 1986; Bailey, 1986a; Torbett, 1986) are typically in the range of 30000 to 50000 AU for a_e. Thus one would have to conclude that for typical Oort cloud orbits ($a \simeq 25000\,\text{AU}$) the loss cylinder is far from being filled, and the observed orbits would represent a deep minimum in the true distribution of perihelion distances for Oort cloud comets. This effect has also been apparent in Monte Carlo simulations of the Oort cloud structure (Remy and Mignard, 1985; Weissman, 1985), thus contributing to an upward revision of the number of comets in the cloud to about 10^{12}.

Perturbations by the Galactic Disk. However, an important effect is missing from these Monte Carlo simulations and from the above picture of stellar perturbations, i.e., that of a quasi-continuous distribution of mass corresponding to the galactic disk. This can be modeled as a sequence of infinite, plane-parallel slabs where the density is constant within each slab. According to Bahcall (1984), the force law near the galactic midplane, perpendicular to this plane, can be written to very good approximation as

$$d^2\zeta/dt^2 = -8.4\ 10^{-30}\zeta, \tag{7.19}$$

where ζ is the distance from the plane, corresponding to a mass distribution where the density is independent of ζ, and the constant is in units of s^{-2}. In such a situation, an Oort cloud comet is accelerated with respect to the Sun due to the differential thickness of the disk, $\Delta\zeta$. We are obviously dealing with a compressional tidal field, and motion in the associated potential can be described in terms of an averaged Hamiltonian H_{av} as long as the perturbation to the Keplerian orbit per revolution remains small (Heisler and Tremaine, 1986). Defining $L = (GM_\odot a)^{1/2}$, $J = L(1 - e^2)^{1/2}$, and $J_\zeta = J \cos I$, where a, e, and I are the semimajor axis, eccentricity, and inclination with respect to the galactic equator, respectively. J is the total angular momentum, and J_ζ is its ζ component. If ω is the argument of perihelion of the orbit, again with the galactic equator as reference circle, this variable is canonically conjugate to J, so we have

$$\frac{dJ}{dt} = -\frac{\partial H_{av}}{\partial \omega} = -\frac{(5\pi G\rho_0)}{(GM_\odot)^2} L^2(L^2 - J^2)(1 - J_\zeta^2/J^2)\sin 2\omega. \tag{7.20}$$

The angular variables conjugate to L and J_ζ do not appear in H_{av}, so L and J_ζ

(like H_{av}) are integrals of motion. Thus the orbital energy, or semimajor axis, of the comet is not perturbed. The long-term motion can be described by means of trajectories in the (e, ω) plane which are either closed (ω libration around $\pi/2$ or $3\pi/2$) or open (circulation). The periods of such librations or circulations are at least ~ 0.1–$1.0\,\mathrm{Gy}$.

Of course Oort cloud comets are expected to deviate drastically from these trajectories because of the impulses received from random passing stars over about $100\,\mathrm{My}$ or more. Thus we may regard those impulses as redistributing comets not between different permanent Keplerian orbits but between different (e, ω) trajectories corresponding to different values of J_ζ and H_{av}. If this reshuffling is efficient, those trajectories leading into the loss cylinder with $q < 15\,\mathrm{AU}$ will be continuously repopulated. One may thus estimate the influx rate of comets into the loss cylinder using the phase space distribution function of the Oort cloud. There is a limiting semimajor axis a_t, corresponding to the tidal force of the galactic disk, which separates the regions where: (1) The change of angular momentum, $\Delta J = \int (dJ/dt)\,dt$, during one orbital revolution is much smaller than the radius J_0 of the loss sphere, and we may consider the loss sphere as empty ($a < a_t$) and (2) ΔJ is at least as large as J_0, in which case the loss sphere can be regarded as filled.

Generally, one finds that a_t is somewhat smaller than a_e although both values appear larger than the typical semimajor axes of Oort cloud comets. Hence the Oort cloud loss cylinder is more nearly filled if account is taken of the galactic tide than if only perturbations by stellar encounters are considered (Torbett, 1986; Fernández and Ip, 1987). This means that the number of comets in the Oort cloud is probably smaller than the estimate quoted above – i.e., closer to Oort's original value of about 10^{11}. The tidal influx rate dominates over the influx rate caused by stellar encounters, which can be derived from Eq. (7.19), at all semimajor axes of interest. It can be expressed, according to Bailey (1986a), per unit of semimajor axis, as

$$F_t(a) = 4\pi^2 f(a) J_0^2 \frac{GM_\odot}{2a^2} \cdot \begin{cases} \frac{20\pi\sqrt{2}}{3} \rho_0 a^{7/2} / [M_\odot (15\mathrm{AU})^{1/2}], & a < a_t, \\ 1, & a > a_t, \end{cases} \qquad (7.21)$$

where $f(a)$ is the phase space distribution function in terms of the semimajor axis. This function corresponds to the true energy distribution of the Oort cloud and is of course unknown. Different models of the comet cloud have been considered using power-law energy spectra with different indices (Bailey, 1983), and as a result, the energy spectrum of the cometary influx may have very different characteristics.

For observable comets, which have to penetrate deep inside the loss cylinder ($q \lesssim 5\,\mathrm{AU}$), one always obtains a peak near the limit of loss cylinder filling such that no comets are fed in from the cloud with $a \ll a_t$. Hence the characteristics of the observed peak of new comets are not very diagnostic of the true energy spectrum of the cloud, unfortunately. However, the infeed of comets to perihelia in the Uranus–Neptune region, near the limit of the loss zone, depends critically on the energy spectrum. Standard models, where $f(a) \simeq$ constant, have most of the comets arriving from the outer part of the cloud. On the other hand, for centrally condensed models of the cloud the integrated influx rate to the outer edge of the loss cylinder from the inner cloud regions by far outweighs the flux from the outer regions (Bailey, 1986a).

We shall see in Sect. 7.4.2 that there are indeed arguments in favor of the latter picture based on the problem of cometary capture. Furthermore, the apparent dynamical instability of the outer Oort cloud in the presence of GMC perturbations over the age of the Solar System (Bailey, 1986b) raises the need for a source of replenishment. The likelihood for this cometary halo to be captured from outside the Solar System appears too low to be seriously considered (Valtonen and Innanen, 1982), and thus a numerous population of comets in an inner core to this halo is an attractive possibility. However, there are also restrictions to the amount of central condensation of the comet cloud as we shall see.

The basic argument deals with the cratering rate of the terrestrial planets and the possibility that a substantial fraction of the short-period comets develop into asteroidal bodies capable of contributing to the cratering statistics. For further details we refer to Sect. 7.5.2. Active comets contribute very little to the present cratering flux (Bailey and Stagg, 1988), but dormant or deactivated comets in the disguise of Apollo–Amor asteroids may yield a highly significant population of impactors. This conclusion derives from the estimated present number of observable comets, but the picture would be modified in an important way if the Oort cloud had a dense inner core (Hills, 1981). At the rare but drastic perturbing events occurring at the typical frequencies of the Sun's galactic motion ($\sim 10^8$ y; see Fig. 7.4) the limit of loss-cone filling would temporarily penetrate to the densely populated inner core and thus the rate of passages of comets through the inner planetary region would increase dramatically (Fernández and Ip, 1987). If so, the terrestrial cratering rate might also show dramatic increases.

These hypothetical events are called "comet showers" and have indeed attracted attention in connection with the periodic phenomena of biological mass extinctions and episodes of intense cratering claimed to be revealed by the geologic record on Earth (for a review see Hut et al., 1987). An important point, however, is that the cometary showers cannot have been intense or frequent enough to have dominated the average cratering flux during the previous 3 Gy, which was apparently somewhat lower than the present flux. This and the apparent lack of extremely strong events in the terrestrial record (Stothers, 1988), put a constraint on the degree of central condensation of the Oort cloud.

7.4.2 Planetary Perturbations of Comet Orbits

When a comet enters into the planetary system, the gravitational attraction by the barycenter of the Sun and planets separates into a predominating force directed toward the Sun and perturbing forces due to the planets. Considering for simplicity one planet at a time, one can thus write the equation of heliocentric motion of the comet

$$\ddot{\boldsymbol{r}} + \frac{GM_\odot}{r^3}\boldsymbol{r} = -\nabla R_\mathrm{p}\,, \tag{7.22}$$

where the planetary perturbing function R_p can be considered as a time-dependent potential given by

$$R_{\mathrm{p}} = \frac{GM_{\mathrm{p}}}{r_{\mathrm{p}}} \cdot \left\{ \frac{1}{\Delta'} - \boldsymbol{r}' \cdot \boldsymbol{r}_{\mathrm{p}}' \right\}, \tag{7.23}$$

where M_{p} and r_{p} are the mass and heliocentric distance of the planet, respectively, Δ' is the comet–planet distance, and $\boldsymbol{r}$ and $\boldsymbol{r}_{\mathrm{p}}$ are the heliocentric position vectors of the comet and the planet, respectively. The primed quantities are expressed in units of r_{p}. Formally, the perturbation of orbital energy can thus be expressed as

$$-\frac{1}{2}GM_{\odot}\Delta(1/a) = \int_{r_1}^{r_2} \nabla R_{\mathrm{p}} \cdot d\boldsymbol{r} \quad , \tag{7.24}$$

and writing $R_{\mathrm{p}} = GM_{\mathrm{p}}R_{\mathrm{p}}'/r_{\mathrm{p}}$ in terms of a dimensionless perturbing function R_{p}' depending solely on the relative geometry we obtain

$$\Delta(1/a) = -2\frac{m_{\mathrm{p}}}{r_{\mathrm{p}}} \int_{r_1'}^{r_2'} \nabla' R' \cdot d\boldsymbol{r}'. \tag{7.25}$$

Here we have introduced $m_{\mathrm{p}} = M_{\mathrm{p}}/M_{\odot}$. The set of observed long-period comet apparitions provides a sample for which the relative geometry is reasonably similar regardless of which giant planet is considered, and Fig. 7.6 illustrates the distributions of energy perturbations for 392 long-period comets observed between 1801 and 1970, as computed by Everhart and Raghavan (1970). These sample distributions may all have parent distributions symmetric about zero, whose estimated dispersions are seen to vary approximately as $m_{\mathrm{p}}/r_{\mathrm{p}}$, as expected (Fig. 7.7). Furthermore, we see

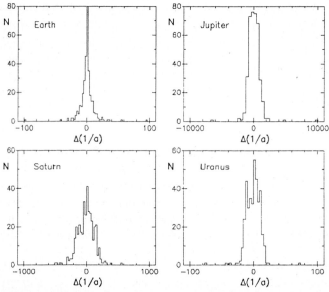

Fig. 7.6. Distributions of perturbations of inverse semimajor axis (proportional to the orbital energy) for 392 long-period comets observed between 1801 and 1970 (Everhart and Raghavan, 1970). Separate diagrams are shown for Earth, Jupiter, Saturn, and Uranus. Absolute numbers in each histogram box are given as ordinate, N. Note that the scale of the abscissa for Jupiter is 100 times and for Saturn 10 times that of Earth and Uranus.

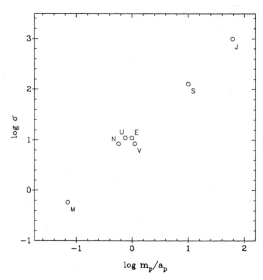

Fig. 7.7. Standard deviations of the distributions of energy perturbations given in Fig. 7.6, and of corresponding distributions for Venus, Mars, and Neptune, plotted vs. the ratio of planetary mass and orbital semimajor axis in Earth units. Each point is marked by the initial of the planet. The arrangement of data points along a line of slope 1 agrees with the conclusion drawn from Eq. (7.25); $r_p \approx a_p$ due to the low eccentricity of planetary orbits.

that (mainly because of Jupiter) σ is much larger than the typical barycentric values of $(1/a)_{\mathrm{orig}}$ of new Oort cloud comets before entry into the planetary system (see Chap. 6), as already mentioned. Thus nearly half of these comets are ejected from the Solar System along hyperbolic orbits and the rest is mostly brought into orbits with periods less than 1 My.

The orbital evolution of comets with periods less than 1 My is necessarily chaotic. A minute change in the value of $(1/a)_{\mathrm{fut}}$ (the reciprocal of the semimajor axis after exit from the planetary system) resulting from a minute change in the observed orbital parameters would probably lead to an entirely different geometry at the following passage through the planetary system, since the time of perihelion passage would easily change by many years. Thus subsequent perturbations $\Delta(1/a)$ are quasi-random, i.e., they may be considered as independent stochastic samples from the parent distribution of $\Delta(1/a)$. In this sense comets perform random walks on the energy axis and the outcome of this process depends basically on the characteristics of the $\Delta(1/a)$ distribution.

The dispersion $\sigma_{(1/a)}$ is of course the most important parameter of this distribution, but one should bear in mind that the shape deviates in a crucial way from a Gaussian. For instance, if we consider types of orbits with perihelion distances much larger than the planetary semimajor axis, no close encounters can occur and therefore the values of $|\Delta(1/a)|$ are strictly limited. In such cases the $\Delta(1/a)$ distribution differs from a Gaussian by the lack of tails. On the other hand, if comet orbits approaching or crossing the orbit of a planet are considered, we must take the close encounters into account. Some of their properties can be illustrated by using the concept of the "sphere of action" (Öpik, 1971) within which the motion of the comet is governed by the planetary attraction. A simple approximation to such motion is that of a hyperbolic deflection. Within this approximation we may take the heliocentric energy perturbation to vary as the deflection angle, i.e., inversely as the impact parameter. The tails of the perturbation distribution then correspond to the smallest

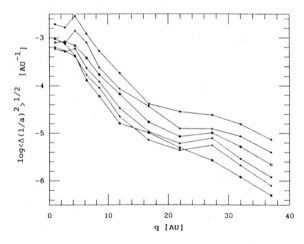

Fig. 7.8. Root-mean-square change of orbital energy plotted on a logarithmic scale as a function of perihelion distance q. The results are based on numerical simulations of gravitational scattering by the planets using a random sample of 3000 objects for each of the plotted symbols. The curves denote the six 30°-intervals of inclination. On the right hand side the curves fall from top to bottom in strict order of increasing inclination. From Duncan et al. (1987).

impact parameters p. From this it is easily seen that the relative frequency function $n(p) \propto p$ translates into a corresponding function $n(|\Delta(1/a)|) \propto |\Delta(1/a)|^{-3}$. We therefore have a distribution with tails that are much more important than those of a Gaussian curve. They give a significant and sometimes predominant contribution to the dispersion $\sigma_{(1/a)}$. Close encounters are thus of central importance for the planetary influence on the orbital evolution of comets.

Figure 7.8 shows $\sigma_{(1/a)}$ as a function of perihelion distance, q, and inclination, i, according to recent calculations by Duncan et al. (1987). We see that small perihelion distances (Jupiter-crossing orbits) and low inclinations are favorable to large energy perturbations. This has been realized since a long time and is of great importance for understanding the capture of comets into short-period orbits. Assuming for the time being that the classical Oort cloud is the only source population, such a capture requires a large change of $(1/a)$ from the initial value near zero. If this is to be accomplished by a random walk, an obvious problem is the presence of an "absorbing wall" at $1/a = 0$, a sink corresponding to hyperbolic ejections from the Solar System. Numerical simulations by Everhart (1972, 1977) showed that out of an initial population of comets entering the planetary system in nearly parabolic orbits the number of survivors decreased as $N^{-1/2}$ with the number of orbital revolutions, N. In order to traverse the range of orbital energy corresponding to capture in a random walk with typical step size of order $\sigma_{(1/a)}$, the necessary number of steps on the average varies as $\sigma_{(1/a)}^{-2}$. If we therefore define the capture efficiency to be the fraction of an initial population of near-parabolic comets that in due time is captured into short-period orbits, this quantity will vary as $\sigma_{(1/a)}$. The efficiencies of different giant planets in capturing comets from the Oort cloud should then compare as m_p/r_p – an expectation that is well borne out by the Monte Carlo simulations of Everhart (1977).

One can use these capture efficiencies as estimated from numerical simulations (Everhart, 1977; Fernández, 1980) together with the observed influx rate of Oort cloud comets to estimate a capture rate. Remarkably enough, this estimate is much smaller than the rate needed to keep the short-period comet population in a steady state (see Sect. 7.5.2 for a discussion of the observable lifetimes upon which this is based). Such a conclusion was preliminarily formulated by Joss (1973) and later substantiated by Fernández and Ip (1983a) and Bailey (1986a). Apparently we need an additional source in order to explain the capture of comets. Two alternatives may be considered: The inner core of the Oort cloud and a trans-Neptunian cometary disk with a pronounced concentration toward the ecliptic plane. Bailey (1986a) and Bailey and Stagg (1988) conclude that even a moderate central condensation of the Oort cloud may remove the discrepancy due to the large influx of comets from the inner core into the Uranus–Neptune zone. This influx occurs further from the hyperbolic ejection limit, so those planets might transmit a sufficient number of comets into the Jupiter–Saturn zone for capture into observable orbits. But indeed it is necessary to have only a modest central condensation since otherwise the steady-state number of short-period comets would be much larger than the presently estimated one! Apart from any possible consequences in terms of cratering flux (see Sect. 7.4.1), it would of course be unsatisfactory to conclude that the present short-period comet population is severely deficient relative to the average one. But the existence of a trans-Neptunian cometary disk is not ruled out by this argument, and suggestions of such a disk as a possible additional source of short-period comets have been made (Duncan et al., 1988).

The main problem is then to explain how comets can be continually fed in from this disk into Neptune-crossing orbits. If the size distribution is such that there exists a number of more or less Pluto-sized objects in the disk ($M_{\mathcal{O}} \gtrsim 10^{-4} M_{\oplus}$), then gravitational scattering by these might explain the infeed (Fernández, 1980). Alternatively, the Neptune crossers might be primordial, i.e., remnants of an original cometary population from the Neptune accretion zone (Fernández and Ip, 1983a). But dynamical studies (Fernández and Ip, 1983b; Duncan et al., 1988) indicate a somewhat too high removal rate of such objects to comply with outer Solar System cratering statistics (Shoemaker and Wolfe, 1982). The argument is that the number of Neptune crossers is reflected in the flux of comets transferred into the Jupiter–Saturn region and thus in the cratering rate at the satellites of these planets. This rate appears not to have evolved significantly during the last 3 Gy.

As yet there is no mechanism known to account for a continuing presence of Neptune crossers linked to a trans-Neptunian disk, but further investigations may still reveal a possibility of dynamical transfer. This appears as one of the major issues at present, since we need a complete picture of the infeed of comets into the outer planetary system from all possible exterior sources. With such a picture we may use the statistics of short-period comet orbits to draw conclusions about the structure of the sources. However, it is then also imperative to know with fair accuracy the transfer routes into the observable region and their relative efficiencies.

As already mentioned, this capture of comets in any case appears to occur via a multi-stage process where all the giant planets participate. This model was advocated a long time ago by Kazimirchak-Polonskaya (1972), but the details of the dynamics involved are only now beginning to be explored (e.g., Duncan et al., 1988; Stagg and Bailey, 1989).

In particular one seeks an interpretation of the inclination distribution of short-period comets. The latter is intimately connected with the orbital periods or aphelion distances, as shown in Figs. 7.9 and 7.10. The majority of short-period comets belong

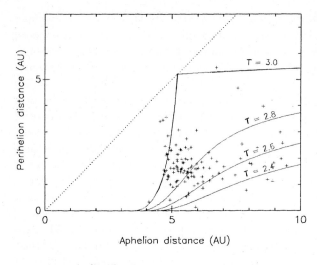

Fig. 7.9. The distribution of the Jupiter family comets observed up to 1987 over aphelion and perihelion distances. Each comet is marked by a cross. The dotted line, below which all real orbits are situated, corresponds to circular orbits, and the full-drawn curves indicate approximate dynamical flow lines for low-inclination orbits obtained from Eq. (7.26).

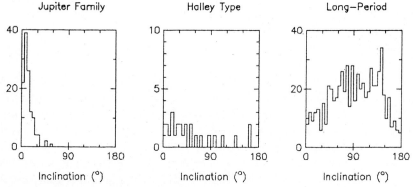

Fig. 7.10. The distributions of orbital inclination for three groups of comets: The Jupiter family is here defined by aphelion distance $Q < 10\,\text{AU}$, the Halley type short-period comets have $Q > 10\,\text{AU}$ and orbital period $P < 200\,\text{y}$, and the long-period comets have $P > 200\,\text{y}$ including parabolic and hyperbolic orbits. All comets observed up to 1987 are included.

to the so-called Jupiter family. They have aphelia close to Jupiter's mean distance from the Sun and predominately low inclinations (Fig. 7.10a). Actually the aphelion points of these comets are situated close to Jupiter's orbit as a rule. Thus they show a very strong kinematic link to this planet and orbital integrations indicate very unstable dynamical behavior due to major perturbations at close planetary encounters.

To understand the ensuing dynamical evolution it is instructive to make use of the Tisserand criterion

$$T = \frac{2a_J}{Q+q} + 2\sqrt{\frac{2Qq}{a_J(Q+q)}}\ \cos i\,,\tag{7.26}$$

where a_J is Jupiter's semimajor axis, and T (the Tisserand parameter) is an approximate form of the Jacobi integral neglecting potential energy terms that become large only when the comet approaches the Sun or Jupiter. Thus in between such encounters T remains quasi-constant because of the low eccentricity of Jupiter's orbit as long as perturbations by other planets remain small. Eq. (7.26) defines evolutionary surfaces in (Q, q, i) space and, from the point of view of Jupiter family comets, it is interesting to consider what may happen if i remains small ($\cos i \approx 1$) during the orbital evolution.

Figure 7.9 shows a few examples of the resulting evolutionary tracks in the (Q, q) plane along with the distribution of observed comet orbits. Obviously the orbital evolution is such as to connect the Jupiter family with low-inclination orbits whose perihelia are near Jupiter's orbit. The final stage of capture into the Jupiter family might hence be a drastic perturbation where the line of apsides is turned so that the new aphelion occurs close to the old perihelion. Indeed backward orbital integrations for Jupiter family comets show such perturbations to occur rather frequently (Carusi et al., 1985a,b). The pre-capture comets are in general unobservable since cometary activity is usually absent or very weak at Jupiter's distance from the Sun. They could in principle have any aphelion distances from this range upward, but the integrations have not yet revealed any direct capture from an orbit whose aphelion was far outside the planetary system.

The orbital evolution at this unobservable stage could hence either be a Jupiter-dominated random walk in aphelion distance starting from an Oort cloud orbit, or it could involve infeed from an even larger perihelion distance into a Jupiter-approaching orbit by the actions of Saturn, Uranus, and Neptune. According to the above-described efficiency argument the first scenario should be relatively unimportant and furthermore it is unclear whether it would explain the orbital distribution in Figs. 7.9 and 7.10. Everhart (1972) explored a wide range of combinations of q_0 and i_0 for initial Oort cloud orbits and found that practically only those with $4 < q_0 < 6$ AU and $i_0 < 9°$ could be captured into short-period orbits. This would indeed lead to good agreement with the observed orbital distribution. Duncan et al. (1988) challenged this conclusion, however, finding that capture from high-inclination or even retrograde orbits is quite likely, so that basically the Jupiter family would not stand out as clearly as observed from the rest of the short-period comet population. A definitive answer to this question would require detailed consideration of the observable lifetimes and their dependence on perihelion distance (see Sect. 7.5), but the observational basis for such an endeavor is as yet lacking.

As far as the multi-planet capture scenario is concerned, a detailed exploration of its dynamics will apparently help to clarify the structure of the source population. This is an important goal since direct observation of typical comet nuclei in the trans-Neptunian region or the inner core of the Oort cloud is very unlikely for a long time. In particular the question is still open as to whether the low inclinations of the Jupiter family comets demand a preference for low inclinations in the source population as well. A reliable answer to that question would yield an indication about the significance of the trans-Neptunian disk and thus an important constraint on solar nebula models.

7.5 Physical Evolution of Comet Nuclei

The state of a nucleus presently under observation is probably different from its nascent state in the solar nebula. Furthermore, in the future it will certainly be different from its present state. Comet nuclei are thus expected to evolve, at least over long time scales; indeed observations of different comets at the present time give some indications about the characteristics of this evolution. There is some analogy to the inference of stellar evolution from observations of a large sample of stars of various ages at a given time, e.g., using the analysis of their distribution over the HR diagram and in both cases an understanding of the underlying physical processes is obviously necessary. However, in the case of short-period comets one has a better possibility to see the evolution proceeding in individual objects. We will thus scrutinize the observed evolutionary effects in comets, but let us first concentrate on expected scenarios based on physical insight – both on theoretical grounds and stimulated by laboratory results. How should a comet nucleus evolve according to our understanding of a dusty snowball orbiting in interplanetary or transplanetary space?

We will attempt to describe some of the effects in chronological order along a typical route of orbital evolution as described in the previous section. It must be emphasized that an exhaustive study of global, quantitative models has not been made yet. Many more laboratory experiments are needed to limit the ranges of the physical parameters and new concepts need to be developed that could critically change the interpretations of the presently incomplete models.

7.5.1 Some Scenarios for Comet Evolution

The first effects to be expected are the heating by inelastic collisions, short-lived radioisotopes, and the liberation of latent energy by exothermic physico-chemical processes as outlined in Fig. 7.1. These heating events occur before, during, and after the formation of a comet nucleus. For example, the radionuclides were probably acquired already in the stage when the cometary material was dispersed as submicron grains, and the heat of their decay was then efficiently radiated away. The same holds for some of the exothermic processes such as polymerization. Thus the amount of energy available after the birth of the nucleus depends on the time required for the formation process.

One important consequence of this heating might be sublimation and recondensation or escape of volatiles. Therefore the entry of some very volatile species into the nucleus may have been prevented during the stage of formation, but in addition volatiles may be lost afterward as heat is released inside the nucleus. For a modest heating event (temperature increase $\Delta T \lesssim 10\,\mathrm{K}$), sublimation of volatiles is not very likely, but for a drastic event ($\Delta T \sim 100\,\mathrm{K}$), it would certainly occur. Heating by inelastic collisions leads to a temperature increase $\Delta T = v_{\mathrm{coll}}^2/(2C)$, where C is the heat capacity and v_{coll} is the relative speed of collision. For $v_{\mathrm{coll}} \simeq 100\,\mathrm{m/s}$ and ice temperature of $30\,\mathrm{K}$ the heat capacity is $C \simeq 2.3 \times 10^2\,\mathrm{J\,kg^{-1}\,K^{-1}}$ and $\Delta T \simeq 20\,\mathrm{K}$. At an ice temperature of $50\,\mathrm{K}$, $C \simeq 4.4\ 10^2\,\mathrm{J\,kg^{-1}}\ \mathrm{K^{-1}}$ and $\Delta T \simeq 10\,\mathrm{K}$.

The most important radionuclide to take into account is ^{26}Al, with a half-life of 7×10^5 y, which appears to have been produced in a supernova explosion shortly before the collapse of the protosolar cloud. If comet nuclei formed very quickly then they would contain enough ^{26}Al to undergo drastic heating during their first million years of existence. Thus even the H_2O ice might have melted or at least reached the transition temperature from amorphous to crystalline form (Wallis, 1980; Prialnik et al., 1987). The time scale of heat diffusion over the radius of the nucleus is $\sim 10^5$ y using the thermal diffusivity of compact, amorphous H_2O ice (Klinger, 1981, 1985) and probably $\sim 10^6$ y for a very porous material as far as bulk conductivity is concerned. If sublimation occurs, then gas-phase heat transport may allow for more efficient thermal diffusion (Smoluchowski, 1985).

What structure can we expect for a comet nucleus after these heating events? The ultimate source of heat in the form of radionuclides may be uniformly distributed throughout the interior of the nucleus, while the sink is caused by thermal radiation and thus situated at the surface. This leads to a temperature profile with a maximum at the center and a minimum on the surface and with a shape depending on the detailed ratio of the heat production and the diffusion time scales. As a consequence of this initial heating, additional heat may be given off by exothermic chemical reactions or phase transitions as the local temperature reaches a critical value. A few suggestions are: Recombination of ion molecule clusters (Shul'man, 1983) and reactions of free radicals formed in the icy mantles of the proto-cometary grains by UV photoreactions (Greenberg, 1982, 1986b).

Each such release of latent energy should eat its way out through the nucleus as long as the locally produced heat is sufficient to raise the untransformed material on the outside to the critical temperature. The process will stop when the transition or reaction reaches material that is too cold to respond to the heating. After all these modifications have occurred, the residence time of the comet in the Oort cloud is long enough for the whole nucleus to cool down to the equilibrium temperature of ~ 10 K, but the interior may have a frozen-in, layered, chemical structure as depicted in Fig. 7.11 for thermal events of different caliber. In particular, near the

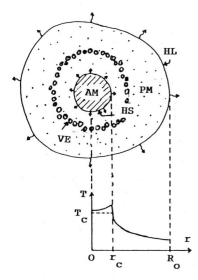

Fig. 7.11. Scenario for a nucleus with an internal heat source (HS) giving rise to chemically altered material (AM) with a release of latent heat. As long as the interior heat release is large relative to the thermal diffusion time scale, the alteration front moves outward. A typical temperature profile during this phase is sketched below the nucleus. As a result of the porous structure, volatiles may be driven out and recondense in a volatile enriched (VE) zone. Heat loss (HL) occurs at the surface of the still nearly pristine material (PM).

surface there may be a layer rich in recondensed volatiles that were sublimated in the interior parts during the heating events and transported by diffusion through the porous material (Whipple and Stefanik, 1966). However, the lack of evidence of [26]Al decay products in the data from the P/Halley missions rules out heating from that radionuclide in that comet (see Chap. 4).

Of course, there are also long-lived radioisotopes like [40]K, [232]Th, [235]U, and [238]U that can be expected to have been present in the silicate component of cometary grains, primordially in chondritic abundances. Their decay over time scales $\sim 1\,\mathrm{Gy}$ would heat the nucleus only insignificantly in comparison with the likely [26]Al heating for realistic values of the thermal diffusivity (Rickman, 1989), but in the possible absence of the latter, the long-term effect might still be large enough to have caused a chemical differentiation with outward migration of such volatiles as N_2 and CO. We are now considering time periods that comets spent in either one of the distant reservoirs that have been discussed. Closer to the Sun than 100 AU, solar radiation causes the equilibrium temperature of comet nuclei to be in excess of 30 K, but at much larger distances, characterizing the Oort cloud and its hypothetical inner core, radioactive heating might raise the central temperature significantly above this value. Heating events of limited extent are to be expected in the surface layers from the passage of luminous stars (Stern and Shull, 1988). A significant fraction of the comets is heated to temperatures of 25 to 30 K; this would imply the likely loss of such volatiles as N_2 and CO from the surface layers of the nucleus. A typical penetration depth of this heating can be estimated from the thermal skin depth

$$\Delta R = \frac{(K \Delta t_\mathrm{h})^{1/2}}{\pi \rho C}, \tag{7.27}$$

where K is the thermal conductivity, ρ is the density, C is the heat capacity and Δt_h is the time scale of the heating event. For the luminous stars that dominate the stellar heating one has $\Delta t_\mathrm{h} \gtrsim 10^4$ y. This means that a surface layer of several tens of meters is heated, even if a very low conductivity typical for porous, amorphous ice is assumed.

Thus it seems that if such volatile ices as N_2 and CO were indeed incorporated into comet nuclei, they may have sublimated to a large extent by internal heating events. Some of the gases thus produced can become trapped in the amorphous H_2O ice, but according to the experiments by Schmitt et al. (1989), this is limited to 7 to 10%, and the rest would have been driven out into a thin surface layer where temperatures were low enough for recondensation. Some of the comets that now arrive from the Oort cloud may thus have a surface enrichment of these volatiles, but in many cases those might have been lost during the heating by passing stars. Therefore observed abundances of CO or N_2 in a few objects apparently does not yet provide a very reliable formation thermometer. Inhomogeneities induced by collisions during nucleus formation – such as local heating, changes in the structure (e.g., porosity) that can influence heat capacity and heat flow, different chemical compositions (relative abundances of volatiles), and different dust-to-gas ratios in subbodies – have not been investigated yet.

A different effect is caused by energetic particle radiation. This starts at a very early stage because of the intense chromospheric activity of the proto-Sun during its

T Tauri phase (Strazzulla et al., 1983) and continues in the form of galactic cosmic rays during the entire period that a comet nucleus spends outside the heliosphere in the Oort cloud or the trans-Neptunian reservoir. Cosmic ray particles tend to deposit their energies at certain typical depths (measured as mass traversed per unit surface area, in kg/m^2) in the nucleus depending on their energies and the stopping power of the cometary material. Energy deposition occurs by nuclear reactions, relaxation caused by Coulomb scattering of the charged primary and secondary particles, ionization, and dissociation. When neutrons are produced, they can travel to larger depths.

Laboratory experiments have been performed by several groups to measure the stopping power of icy materials and to characterize the transformed material arising from chemical reactions involving the radiation-damaged molecules. It has thus been found that after 4.5 Gy of exposure to the typical cosmic ray spectrum of today's solar neighborhood the outermost 10^3 kg/m^2 receive an integrated dose high enough to cause profound chemical changes (Ryan and Draganić, 1986). The material so produced consists of volatile species as well as organic refractories; thus an unstable crust may be produced on a comet nucleus (Johnson et al., 1987). Upon approach to the Sun, the volatiles may sublimate at relatively large distances, and some of the refractory material may be torn off the surface as grains.

The irradiation-driven crust may cover the nucleus fairly uniformly because of the near isotropy of the cosmic radiation field and the rotation of the nucleus. On the average the expected thickness is $\sim 5\,$m for a density of $200\,kg/m^3$. Further investigations by laboratory work and theoretical modeling of the response of the irradiation-altered crust to solar heating are needed in order to specify the state of the nucleus after its capture from the Oort cloud.

Thus, unfortunately, it is not clear whether the next phase of evolution – that driven by insolation and surface heating – should act on a nucleus largely covered by an irradiation-driven crust of organic refractories or on one where nearly all this material has escaped during the first passages near the Sun. In the first case the nucleus would exhibit only spotty activity, while in the second case it would at least pass a stage of nearly isotropic outgassing ability.

Outgassing in response to insolation obviously implies mass loss and surface erosion, but there are secondary effects that may be very important. A thermal wave will penetrate toward the interior of the nucleus and along with this goes crystallization of the ice if this has remained amorphous in spite of the earlier interior heating. This implies the liberation of a significant amount of latent heat ($\sim 1.6\,kJ/mol$; formally allowing a temperature increase of up to 80 K for the heat capacity of water ice) and might thus have an important effect on the thermal evolution of the nucleus. What actually happens may depend on the chemical composition, as shown below. Let us first consider a case where none of the crystallization energy is consumed by liberation of trapped volatiles. The wave of heat release travels inward, triggered by external heating; this situation is fundamentally different from that of the early heating events with their outward propagation toward the continually cooled surface. According to numerical modeling by Prialnik and Bar-Nun (1987), the depth of the phase transition increases intermittently in distinct spurts of crystallization for a homogeneous spherical body of pure amorphous ice in the orbit of Comet P/Halley.

Their results were shown in Chap. 2, Fig. 2.14. An essential feature is the quasi-continuous recession of the surface caused by sublimation. Each time crystallization occurs, it spreads very rapidly downward since the amorphous ice next to the transition level receives part of the released heat, but it stops when the transition reaches material that is too cold. A period of slow evolution ensues during which heat gradually penetrates into the amorphous ice and the receding surface gradually comes closer to the transition level. As this happens, the temperature at that level, after a rapid initial decrease, rises again and reaches the transition temperature of 137 K for the formation of cubic ice as the depth is reduced to ~ 15 m. This initiates a new crystallization burst. A secular trend due to the progressive heating of the interior occurs in the sense that each new burst reaches a larger depth than the earlier ones. Consequently the intervals between bursts become longer.

In such an idealized case, where the surface remains icy, i.e., in a state of free sublimation at each perihelion passage, the comet nucleus would eventually crystallize completely. However, even for the model nucleus considered by Prialnik and Bar-Nun (1987) with an initial radius of only 2.5 km this final state does not occur until at least 90% of the initial mass has sublimated away.

Prialnik and Bar-Nun (1988) showed that a dusty ice nucleus may behave in a different way: First, the thermal insulation of a dust layer on the surface weakens the heat wave that triggers crystallization and, second, the dust consumes a large part of the latent heat released at the phase transition. Thus the crystallization bursts penetrate less deeply and occur more frequently than in the case of pure ice. It therefore appears that complete crystallization would require many thousands of revolutions for a nucleus starting out with at least the present size of P/Halley. A low density has a similar effect, so in a dust-rich, low-density comet the crystalline crust might remain at a nearly constant depth of ~ 15 m. Furthermore, if significant amounts of CO or CO_2 are trapped in the amorphous ice, those gases will escape upon crystallization (Schmitt et al., 1989); this requires an evaporation energy that may be important. With a gas content of 7 to 10%, the evaporation energy is 50 to 90% of the crystallization energy for the case of CO, implying that the process is still exothermic though much less so than in the case of pure H_2O ice (Schmitt et al., 1989). For the case of CO_2 all of the crystallization energy is consumed so that the process is isothermal. In such an extreme situation one would again expect that the phase transition on the average follows the recession of the surface so that its mean depth remains constant. This might represent an important source of cometary gases typically situated quite close to the surface. The depth would be expected to vary considerably in the course of an orbital revolution, and gas release would occur only during the interval of time when the level of crystallization moves inward. Thus the importance of those volatiles for the gas production of the comet might also vary considerably.

When planetary perturbations act in such a way as to capture a comet into a short-period orbit, the mean insolation of the nucleus surface per unit time increases statistically with time. As a result, the nucleus gradually warms up and one expects the temperature deep within the nucleus (the "internal temperature") to increase. Calculations of internal temperatures for various kinds of short-period comet orbits have been made repeatedly (Klinger, 1983; Kührt, 1984; McKay et al., 1986; Herman and Weissman, 1987) in an attempt to characterize the interior structure of such nuclei and the modifications they should have undergone since they were captured from the Oort cloud. However, simplifying assumptions are introduced, such as the neglect of the latent heat of crystallization (nuclei considered to be either crystalline or amorphous throughout) or the assumption of a thermal steady state except for an orbital heat flow in the surface layers. Thus the results have to be taken with care. Numerical simulations (Kührt, 1984; Herman and Weissman, 1987) as well as theoretical predictions (Klinger, 1981) indicate that the time scale for reaching a thermal steady state in a nucleus with a radius of more than 1 km is at least several thousand years. Short-period comet orbits are rarely stable over such intervals, and thus a layered structure may appear over some range of depth. In general the situation appears quite complicated and the actual interior temperatures are not yet known.

As a general though still tentative conclusion, it appears that the dynamically evolved comets and in particular the short-period comets should also be physically and chemically evolved with respect to their formation state (see Weissman, 1990). This expectation concerns their bulk structure, but specifically a very important chemical differentiation should occur in the uppermost surface layers. Heating from the surface may lead to preferential escape of the more volatile components and thus enrichment of the less volatile ones. Figure 7.12 shows a possible scenario involving the release of trapped volatiles at the H_2O phase transition, but above this level even the H_2O may become depleted relative to the silicates and carbonaceous material as the H_2O sublimates away. This means that a dust layer may form on the surface of the nucleus. In the presence of such a layer the outgassing flux would necessarily decrease with respect to the flux of free sublimation, since the layer would provide thermal insulation as well as an obstacle to the gas flow.

This idea offers an attractive explanation to the fact that short-period comets tend to be much fainter than long-period comets with similar perihelion distances. However, many aspects remain to be explored concerning the details. Considering first the problem of formation of the dust layer, it is natural to assume a continuous distribution of particle sizes. This distribution may encompass the distribution ob-

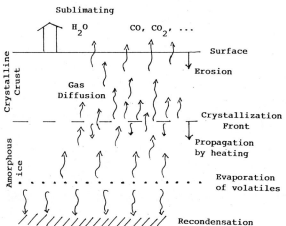

Fig. 7.12. Physical processes conjectured to occur near the surface of an active comet. Absorption of solar heat at the surface leads to H_2O sublimation and surface erosion as well as a thermal wave propagating downward. This implies crystallization of amorphous H_2O ice and the release of trapped volatiles like CO or CO_2. Even further down, the heating may lead to evaporation of frozen volatiles. Diffusion of gases causes recondensation at interior levels as well as an admixture of volatiles to the H_2O vapor leaving the nucleus. This scenario is basically the same as proposed by Schmitt and Klinger (1987).

served in the visible and infrared parts of the spectrum (see Chap. 4, Sect. 4.4), but it does not need to be the same. The observed distribution is a coma average that may be the result of fragmentation and evaporation of larger particles in the inner part of the coma. Thus the size distribution of particles on the surface of the nucleus may be weighted to larger particles than observed in the coma.

As discussed in Chap. 2 (Sect. 2.5.4) and in Chap. 4 (Sect. 4.4), there is a maximum size of particles that can be lifted from the surface of an active area on the nucleus, entrained by the escaping gas. All larger grains remain on the surface of the nucleus. Near Jupiter's orbit or further out the H_2O sublimation flux is so low that all the grains down to submicron size remain, but the rate of surface erosion is correspondingly low. As the comet comes closer to the Sun, successively larger grains are removed by the gas flow, but an increasing number of very large grains continue to be left behind. The critical particle radius reaches its maximum value near perihelion, so at each perihelion passage the nucleus surface is purged of most of the particles below this limiting size. A secular enrichment of the largest particles will take place with smaller grains that may stick to big particles and be sintered.

In such a situation, when more and more single grains reside on the surface, one may obviously expect that eventually a coherent dust layer will form. The criterion for this to occur is usually imagined in terms of a trapping mechanism (Shul'man, 1972), when the interstices between the particles become too small to allow escape of even smaller grains or when the surface gets so covered by particles that the effective gas flux and thus the maximum entrainable particle size is reduced considerably. The process of particle accumulation thus becomes self-accelerating and one can imagine a coherent layer to form very quickly.

The stability of such a dust layer depends on its weight and cohesion compared to the disruptive effect of gas pressure accumulating from below. In the first place we have to deal with continued sublimation of H_2O from the icy surface at the bottom of the dust layer, but explosion of volatile pockets further down may also be important (Prialnik and Bar-Nun, 1988). Alternative pictures of dust layer erosion by grains being entrained by the gas flow (Brin and Mendis, 1979) or broken into pieces and escaping from the surface as in the "friable-sponge" model (Horányi et al., 1984) appear less attractive after the realization that no dust emission seems to occur from the quiescent parts of the Halley nucleus (see Chap. 2) and that the abundant carbonaceous material may provide for efficient sticking and sintering of adjacent particles.

To evaluate the gas pressure difference across the dust layer one has to consider the mechanism of gas flow. Most of the models so far developed have assumed that gas flows freely through the layer; these models are of no concern here. Fanale and Salvail (1984) considered Knudsen flow with the porosity and tortuosity of the dust layer as model parameters, while Podolak and Herman (1985) and Rickman and Fernández (1986), following Shul'man (1981), introduced a model where the individual molecules diffuse outward with a characteristic mean free path set by the structure and the typical particle size of the dust layer. In an extreme case at the dust–ice interface, the local temperature may rise high enough to lift the dust layer off the surface.

What actually happens is as yet highly uncertain. Most investigations indicate that stability may be reached at a dust layer thickness of only a few cm and that by then the gas production of the comet has subsided almost completely so that further increase of the dust layer thickness is extremely slow. Thus it appears that a sublimation-driven surface dust layer, in contrast to the irradiation-driven layer of non-volatile material discussed above, should be very thin.

The long-term stability of such a dust layer appears uncertain for several reasons. For instance, a close encounter with Jupiter may suddenly bring the comet significantly closer to the Sun (decreases of the perihelion distance by $\gtrsim 0.5$ AU are quite common among Jupiter family comets; see Carusi et al., 1985a). If this occurs, the increased gas pressure might shatter the dust layer and expel it from the nucleus at least in the areas receiving maximum insolation. Similar results may arise from local enhancements of volatile ices or from local enrichments of the dust-to-gas ratio creating "hot spots." Thus the comet would be expected to brighten and show an increased nongravitational effect in the orbital motion. Afterwards a new layer would gradually form, and this would become thicker than the previous one, since the basic structure is now made up of larger grains. According to this picture short-period comets should exhibit low activity due to extensive dust coverage, but their level of activity should evolve in discontinuous jumps connected with the chaotic evolution of the perihelion distance. Statistically, a significant epoch in the physical evolution of the nucleus should be the overall minimum of the perihelion distance, after which the dust layer would be more stable than before.

Eventually the whole nucleus might be sealed off and we are thus faced with the dust-coverage model of extinction of short-period comets: The active lifetimes of the comets are limited by the growth of a dust layer on the surface of the nuclei, which prevents further sublimation of the ices. This is in contrast with the mass-loss model which is often considered in discussions of comet lifetimes: The end of the cometary activity would then result from exhausting the whole volatile inventory of the nucleus.

An important difference between these two models is that in the latter case the death of the comet is necessarily definitive, while the former scenario allows the possibility of rejuvenation. Active comets are known to split occasionally for as yet unknown reasons and if a dust-covered nucleus can split as well, its fragments would become active because of the exposure of fresh icy surfaces. However, to this hypothetical event we have to add another one which is certain to occur in due time: The nucleus is bound to collide with a meteoroid so that the dust layer is locally broken up and expelled by the cratering process and the ensuing surge of activity. The frequency of such events can be estimated for a nucleus of given size in a given orbit using an assumed spatial distribution of meteoroids, as done by Fernández (1990). In most cases the impacting particles are so small that only a very minor fraction of the nucleus surface is affected. The resulting activity may then be too weak to be observable as such, but debris would spread along the orbit of the inert-looking nucleus and might give rise to meteor showers on Earth, if the orbital configuration is favorable. More serious damage would be caused by interplanetary "boulders" of more than 10 m size, but such collisions appear to occur so rarely that most Jupiter family comets would be expelled by gravitational interaction with the planet before a collision takes place.

Thus a comet meeting the fate of mass loss would produce less gas and would thus fade as it gets smaller, while its nongravitational effect (varying as surface area/mass and thus inversely with effective radius; see Eq. 7.11 and 7.12) would become large and erratic. The comet might finish abruptly by total disintegration, and in any case no sizeable object would be left behind. All the refractory material initially present in the nucleus would be delivered to the zodiacal cloud. On the other hand, a comet meeting the fate of dust coverage would show a fading modified or interrupted by orbital perturbations, gradually entering a "dormant" state with sporadic, transient activity. The nongravitational effect would decrease to zero. An asteroid-like object would be left behind and might indeed be observed as an Earth-approaching asteroid. Only a minor fraction of the initial dust content of the nucleus would eventually contribute to the interplanetary dust cloud. In the next section we shall describe various ways to judge the relative importance of the two scenarios for cometary extinction using observational evidence.

7.5.2 Observed Evolutionary Effects

One may consider two different procedures for observing the effects of comet evolution upon repeated passages close to the Sun. The most straightforward approach is to compare different apparitions of the same short-period comet with regard to such quantities as the absolute magnitude, the nongravitational effect, or spectrophotometric indicators of coma composition (e.g., the ratio of continuum to emission band strengths interpreted as a dust-to-gas ratio). However, as we shall see, such comparisons are fraught with many difficulties and serious uncertainties plague the interpretation of the results even when reduced to the general trends of a large sample of comets. A different method is to use the orbital or dynamical properties to associate each cometary apparition with a dynamical age measuring the number of approaches to the Sun that the comet has experienced. One can then either look for correlations between the dynamical age and the above-mentioned observable quantities, or make inferences from a comparison of the observed and dynamically predicted distributions of comets with respect to dynamical age. The difficulties encountered with this method are on the one hand that the dynamical ages are sometimes very difficult to estimate, and on the other hand that there may be a wide primordial scatter in the observable quantities independent of evolutionary effects. For example, comet nuclei may have formed with a wide size distribution, and different formation circumstances may have led to different compositions.

Let us consider first the orbital energy distribution of long-period comets. As explained in Chap. 6, the existence of the narrow peak corresponding to new comets from the Oort cloud suggests that these comets fade considerably already in connection with the first passage near the Sun. Such fading means that a certain fraction of the comets becomes unobservable after their first apparition. Were the comets not to fade, the diffusive evolution of orbital energy $1/a$ caused by planetary perturbations over a long time would lead to a steady-state $1/a$ distribution given by the integral equation

$$n(1/a) = n_0(1/a) + \int_0^{(1/a)_{max}} n(\zeta)\varphi_{pl}[(1/a) - \zeta]\,d\zeta, \tag{7.28}$$

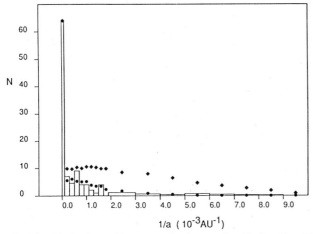

Fig. 7.13. Distributions of inverse semimajor axes for elliptic orbits of long-period comets. The histogram shows the number, N, of observed comets up to 1987 per interval of $10^{-4}\,\mathrm{AU}^{-1}$ in bins of varying widths from the parabolic limit down to $P = 1000\,\mathrm{y}$. The Oort peak shows a contrast of $\approx 10{:}1$ against the adjoining background. The results of two Monte Carlo simulations, for which we gratefully acknowledge Mats Lindgren of the Uppsala Astronomical Observatory, are also shown: The diamonds indicate a model with no physical decay, and the circles denote a model with a constant decay probability of 12 % per apparition. Both model results have been normalized to agree with the observed number in the Oort peak.

where n is the number of comets passing perihelion per unit time, n_o is the infeed rate from the Oort cloud, and $\varphi_\mathrm{pl}[\Delta(1/a)]$ is the probability density of planetary perturbations of orbital energy. A typical result is shown in Fig. 7.13, and it is easily seen that the model in question fails to represent the energy distribution of observed comets. Taking a probability p_d of random decay per orbital revolution into account, thus introducing a factor $(1 - p_\mathrm{d})$ into the integrand of Eq. (7.28), one can find a much better representation as also shown in Fig. 7.13. Such a conclusion was reached already by Oort (1950) and has been substantiated many times since then (e.g., Whipple, 1962; Shtejns, 1972; Weissman, 1979; Yabushita, 1983; Bailey, 1984). The exact amount of fading, or the precise decay probability, as a function of dynamical age is not known. Typically, however, the values found are larger than expected on the basis of physical understanding of comet nuclei, or judging from the evidence from short-period comets (see below); this is sometimes referred to as the "fading problem".

As an example, the best-fit model of Fig. 7.13 uses a decay probability of 0.12. Furthermore, Bailey (1984) considered a more elaborate evolution equation that included also the dependence on perihelion distance and the presence of losses caused by stellar perturbations at small $1/a$ values. He found a decay probability of 0.3 for $(1/a)$ values below approximately $4 \times 10^{-3}\,\mathrm{AU}^{-1}$. Weissman (1979) used a decay probability of 0.1 for new comets, decreasing smoothly with dynamical age, and obtained a satisfactory model of the $(1/a)$ distribution with a slightly larger box width. The tendency for the fading rate to decrease at larger orbital energies may be viewed either as a direct dependence on dynamical age such that comets fade more and more slowly as time progresses or as a selection effect (Weissman, 1979): Most comets do fade but some are immune, and the latter type becomes relatively more common as the former fade away.

The observed comet sample is not large enough to warrant a set of detailed analyses for different ranges of perihelion distance, but some indications are fairly obvious. For $q \gtrsim 2\,\mathrm{AU}$ the predominance of new comets over old ones appears even greater than that illustrated in Fig. 7.13, so the fading of large-q comets should be even more pronounced (Marsden et al., 1978). This is what one would expect if the fading is caused by the disappearance of a volatile-enriched surface layer (more volatile than H_2O) or the growth of a surface layer depleted in such volatiles (Marsden and Sekanina, 1973). The suggestion of such behavior dates far back in time (Oort and Schmidt, 1951), but positive evidence from physical observations is as yet lacking. Oort and Schmidt suggested that new comets have stronger continuum spectra, particularly at large distances, thus revealing a very volatile nucleus material whose evaporation leads to the formation of a grain coma. Donn (1977), however, did not find such an effect in his more exhaustive material. Kresák (1977) also failed to find physical differences between new and old comets if proper allowance for selection effects was made. The presence of particularly strong CO^+ emission observed in some comets (Festou 1984) might be indicative of a large amount of volatiles such as CO or CO_2, but this is not a common phenomenon in any group of comets (either new or old). Nonetheless, if the production of a coma consisting of large grains in the inbound range of heliocentric distance of 5 to 15 AU is indeed a characteristic property of new comets (Sekanina 1973), then one can infer the existence of a large gas flux at these remote distances and thus possibly a very volatile surface material on the nuclei of new comets. According to Whipple (1978b) new comets indeed tend to be brighter far from the Sun on the inbound branch of the orbit, as compared with old comets. This result comes from an analysis of photometric exponents n (see Eq. 7.29 below) showing that new comets on the inbound branch have particularly small values of n on the average. A similar study by Meisel and Morris (1976) was less conclusive. Whipple (1990) reanalyzed the Meisel and Morris data and concluded that long-period comets age persistently with increasing number of passages – i.e., not only from the first to the second one – in the sense that n grows larger on the average both before and after perihelion. Related to this is Kresák's (1977) conclusion that the extreme solar distances (r_{max}) of the first and last observations tend to be larger for the youngest long-period comets than for the older short-period comets. Whipple (1990) again found an indication of a persistent trend whereby r_{max} decreases on the average with increasing dynamical age. All these results favor the idea of substances more volatile than H_2O playing a rôle in triggering cometary activity which becomes less pronounced as comets grow older. A relatively direct observational indication comes from the analysis by Meech and Jewitt (1987) of photometric data of Comet Bowell (1982 I) at record heliocentric distances; they found that the gas production curve of this new Oort cloud comet is compatible with the latent heat of sublimation of, e.g., CO_2.

Marsden et al. (1978) also commented on the relative paucity of old comets with $q < 0.5$ AU, and Weissman (1979) found that in order to explain this an additional fading mechanism analogous to loss of volatiles but acting more rapidly would be appropriate. Thus the low-q comets would become unobservable after the sublimation of some 50 to 100 m of H_2O ice, and he suggested the build-up of a non-volatile crust as a possible mechanism. Modelling cometary decay as a random disappearance with a certain probability, Weissman (1979, 1980) identified the reason for such disappearance with physical disruption. Splitting has been observed in nearly 10% of the long-period comets although with no particular preference for new ones, but it occurs much less frequently for short-period comets (Weissman, 1980; Kresák, 1981). In most cases of the latter type the split comet has not disappeared, and there is even one case – that of Comets P/Neujmin 3 and P/Van Biesbroeck (Carusi et al., 1985b) – where two full-scale comets have resulted from such an event. For long-period comets the lack of groups or pairs (Kresák, 1982; Lindblad, 1985) was taken as support for identifying splitting with fatal disruption (Weissman, 1980), but this argument is weak and it appears that an acceptable physical mechanism for a decay probability at the level of 10% or more is yet to be proposed. Let us add to this discussion that in general terms the fading problem may have a different solution, as has sometimes been suggested (e.g., Marsden and Sekanina, 1973; Van Flandern, 1978; Yabushita, 1983; Bailey et al., 1986): The orbital energy distribution might not be that of a steady state, which means that the Oort cloud might not exist at all, or that we are seeing a temporary gust of comets sent in by one particular perturbing episode. From the dynamical arguments given above, however, neither of these possibilities appears attractive.

The intrinsic brightness of a comet is expressed in terms of an absolute magnitude referred to a standard distance from both the Earth and the Sun. Usually the apparent brightness is taken to vary as Δ^{-2}, where Δ is the distance from the Earth, and as r^{-n}, where r is the heliocentric distance and n (the photometric exponent) may be found by a least-squares fit to the observed magnitudes or given a standard value. The apparent magnitude, m, is generally given by

$$m = m_o + 5 \log \Delta + 2.5 \, n \, \log r + \Phi, \qquad (7.29)$$

where m_o is an absolute magnitude whose value is linked to the value of n. It represents the corrected magnitude which the comet would have if placed at $r = \Delta = 1$ AU. The corrected magnitude is thus $m - \Phi$, and the correction term Φ depends on the light scattering geometry (phase angle) and various phenomena affecting the magnitude estimates for extended objects (instrumental effects, sky brightness, solar and lunar elongations, personal corrections characterizing individual observers, etc.). Note that the values of Φ can be very large (see Kresák and Kresáková, 1987) and may thus introduce serious uncertainties into m_o.

Analyses of the bulk of photometric data with determinations of n and m_o have been performed for a large number of comet apparitions. Thus the photometric exponents are found to exhibit wide variations, especially for short-period comets. Nonetheless, for comparing absolute magnitudes of different apparitions, it is necessary to have a common definition. The most suitable value for such a purpose has been found to be $n = 4$, and the corresponding absolute magnitude is usually referred to as H_{10}. Most determinations of H_{10} are due to Vsekhsvyatskij and his colleagues (Vsekhsvyatskij, 1958, 1963, 1964, 1967; Vsekhsvyatskij and Il'ichishina, 1971). However, H_{10} is obviously of limited significance. It often provides a "temporary" fit to observed magnitudes over a limited period of time when the comet was well placed for observation and thus may not be representative of the whole apparition. Because of deviations from the $n = 4$ dependence and even departures from any unique dependence of brightness on r as found with light curves exhibiting perihelion asymmetry, the significance of H_{10} is further reduced. This is particularly true for comets with $q > 1$ AU (nearly half the long-period comets and most short-period comets), where H_{10} represents an extrapolation.

One should thus be aware of the uncertainties and possible biases influencing these data. Nevertheless it is important to know if there is an obvious difference between the H_{10} distributions of long-period and short-period comets, and secondly, if evolutionary trends can be seen comparing successive apparitions of short-period comets. For the first question the existence of time-dependent instrumental corrections (Kresák, 1974) is relatively unimportant since the samples of comets to be compared have been observed in parallel. Thus Hughes (1988) finds the cumulative numbers, N, of comets brighter than successive absolute magnitudes H_{10} as reproduced in Fig. 7.14, using the Vsekhsvyatskij data without corrections. The distribution function $N(H_{10})$ is seen to be exponential to a good approximation for both long-period (orbital period $P > 200$ y) and short-period comets (here defined by $P < 15$ y) up to a "knee" where the comets become faint enough for magnitude-dependent discovery incompleteness to take control of the curves. The index of the exponential distribution is very nearly the same for both groups of comets, and its value is very close to 2, so we may write

$$N(H_{10}) = N_{o} \cdot 2^{H_{10}} , \tag{7.30}$$

as a good fit to the magnitude distribution of long-period as well as short-period comets according to the Vsekhsvyatskij data. Donnison (1986) found a maximum-

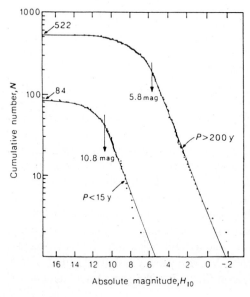

Fig. 7.14. The number of comets brighter than absolute magnitude H_{10} is plotted on a logarithmic scale as a function of H_{10}, separately for long-period ($P > 200$ y) and short-period comets ($P < 15$ y). It is evident that the same basic distribution law can be applied as a reasonable fit to both sets of comets, as indicated by the curves. The differences then reduce to: A difference in the total number on record such that the linear parts for bright comets are a factor 150 apart and a difference in the magnitude where break-away from linearity occurs, indicating that the short-period comet sample is complete to fainter magnitudes. From Hughes (1987).

likelihood estimate of 1.92 for the magnitude index of long-period comets, and Kresák and Kresáková (1987) found an index of 2.2 for short-period comets using absolute magnitudes derived in a different way (see below).

The level, N_0, of the magnitude distribution is about a factor 150 higher for long-period comets than for short-period comets; this difference is constant over the whole magnitude range where discoveries are complete. It is thus possible to adopt the view that short-period comets are evolved versions of long-period comets and that this evolution implies a certain amount of fading, ΔH. If this is so, then the ratio between the number of long-period comets and that of corresponding short-period comets is less than 150, since in order to compare like with like one would have to shift the short-period comet curve in Fig. 7.14 to the right by the amount ΔH. One should note that the concept of an evolutionary relationship between the observed long-period and short-period comets needs clarification: The real precursors of the short-period comets can hardly be the observed long-period comets but should have larger perihelion distances (see Sect. 7.4). One may nevertheless conjecture that at least in the beginning the physical properties are the same for long-period comets of all perihelion distances, so that in particular the observed magnitude distribution can be adopted for comets passing through all domains of the planetary system.

Observational selection naturally favors bright comets over faint ones, but short-period comets are also favored over long-period comets because they give us more chances of discovery. This explains why the knee position of short-period comets is 5^m fainter than that of long-period comets. But the similar slopes observed at brighter magnitudes, along with other statistical evidence (Hughes, 1988) indicates that a true magnitude distribution as given by Eq. (7.30) may indeed be assumed to hold for both types of comets. Returning to the problem of the origin of the short-period comets, this makes it possible to relate the above mentioned discrepancy of required versus dynamically predicted capture rates from the Oort cloud (see Sect. 7.4.2) to the fading ΔH. Depending on this magnitude displacement (see Fig. 7.14), one can obtain widely differing ratios between the flux of source comets and the number of captured comets. However, in order for the classical Oort cloud to be a viable source for short-period comets, they would have to brighten by one magnitude during capture ($\Delta H \simeq -1^m$). Of course this does not occur in reality.

Earlier in this section we have discussed fading and physical disruption of long-period comets. It is natural to ask whether such effects, inferred from the distribution of orbital energies, can be directly observed in the sample of short-period comets. Is fading seen when comparing different apparitions of the same comet? Are there well-documented cases of disappearance where a periodic comet has failed to return as an observable object?

A remarkable result concerning the first question was found by Vsekhsvyatskij (1958) using his compilation of H_{10} magnitudes. Drastic fading was seen in many cases as illustrated in Fig. 7.15. This result would indicate so short lifetimes that a capture origin of short-period comets would appear impossible, and this led Vsekhsvyatskij (1966, 1972) and Drobyshevskij (1981) to consider eruptions on the Jovian satellites as an alternative mode of origin, building on the old idea of Lagrange. However, the inferred rates of fading appear to be spurious. Extrapolations of these trends into the future can sometimes be proven untenable, as, e.g., for Comets

P/Pons–Winnecke   P/Grigg–Skjellerup

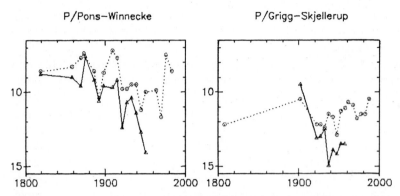

Fig. 7.15. Absolute magnitudes of comets P/Pons-Winnecke (left) and P/Grigg-Skjellerup (right) vs. time during the 19th and 20th centuries. Two different analyses are shown: H_{10} values from Vsekhsvyatskij (1958) by triangular symbols and full-drawn lines and H values from Kresák and Kresáková (1989) by circular symbols and dotted lines. Each symbol marks an observed apparition. Note that the rapid fading indicated by Vsekhsvyatskij's data is not borne out by Kresák and Kresáková's investigation.

P/Pons-Winnecke, P/Kopff and P/Faye which should have disappeared already before 1971 (Whipple and Douglas-Hamilton, 1966) but continue to be observed. As far as backward extrapolation is concerned, it is rarely meaningful because Jupiter family comets are very often discovered shortly after major reductions of the perihelion distance. However, there are cases of more stable orbits where the predicted brightness long before discovery is so high that the comet could hardly have been missed. A particularly well-known example is Comet P/Encke, discovered in 1786, where the lack of ancient observations implies an average fading rate well below 1^m per century (Kresák, 1985). Another example is Comet P/Grigg-Skjellerup (see Fig. 7.15) for which the observations during the interval 1902–1957 indicate a fading by 7^m per century – this comet was, however, recently identified (Kresák, 1987a) with Comet 1808 III, whose faintness limits the real fading rate to less than $0\overset{m}{.}5$ per century.

It would appear very strange if short-period comets should be fading at a particularly high rate during the 20th century, and equally strange is the general pattern shown in Fig. 7.16 from Kresák (1985). Judging from this summary of H_{10} values, even the brightest short-period comets appearing nowadays are fainter than the faintest ones recorded 200 years ago! Is thus the entire short-period comet population decaying so rapidly that with respect to the $H_{10} = 9^m$ level it has already disappeared? Are we continuing to observe short-period comets only because we can reach fainter objects with improving instrumentation? Probably not, the truth instead is that instrumental development has led to the assignment of fainter magnitudes to recent comets, as argued by Kresák (1974, 1985). With ever larger telescopes and increasing focal lengths the measured magnitudes have tended to concentrate only on the inner coma and nuclear condensation, while the outer coma has been ignored or even subtracted as part of the sky background. Moreover, significant fluctuations are superimposed on any underlying brightness trend. The brightness varies both intrinsically and as a result of orbital variations. Changing observing geometries

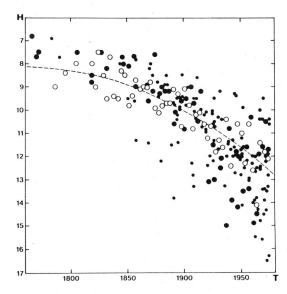

Fig. 7.16. Absolute magnitudes H_{10} for apparitions of short-period comets with $P < 20\,y$ according to Vsekhsvyatskij's data, plotted vs. time, T, in years A.D. Open circles denote comet P/Encke, filled circles denote all other comets with perihelion distance $q < 1\,AU$, and solid dots represent objects with $1 < q < 1.5\,AU$. The dashed curve is an approximate fit to the P/Encke data. From Kresák (1985).

are also important since better conditions imply brighter magnitude estimates. Thus, since the discovery apparition is usually a bright one, there is a bias in favor of rapid fading (Kresák, 1985, 1990).

Kresák and Kresáková (1987) used an empirically determined correction term Φ in Eq. (7.29), which grows rapidly toward fainter magnitudes, to derive an independent set of absolute magnitudes for all the short-period comet apparitions. Focusing on the brightest estimates, these are systematically nearly 2^m brighter than those derived by Vsekhsvyatskij, and they imply fading rates typically of order $0^m.1$ or less per century which are undiscernible in the irregular fluctuations. There are only five comets for which a secular fading appears significant: P/Encke, P/Faye, P/Brooks 2, P/Finlay, and P/Pons-Winnecke. Typical rates for these are one or several magnitudes per century.

In any case it is obvious that such large fading rates cannot be maintained for very long. The typical total lifetime would be only $\sim 100\,y$, and even if the above-mentioned comets were very bright upon capture, they could hardly survive for more than a few centuries. Moreover, as mentioned above the mean fading rate of P/Encke during the last millennia must have been significantly smaller, and for P/Finlay, Kresák (1987b) found that the lack of observations of its perihelion passage in 1827 means that the comet was fainter, not brighter, than at later apparitions. This is but one out of seven comets where missed apparitions of this type have been identified (Kresák, 1987b, 1989). The conclusion is that the aging of short-period comets is not a monotonous fading process but typically an irregular sequence of fading, dormancy and reactivation.

It thus appears that typical total lifetimes should be at least several hundred revolutions, as indicated by the small number of identifiable disappearances of short-period comets (Kresák, 1985) and the equally small number of net captures found over several centuries in catalogued dynamical evolutions (Fernández, 1985). The associated fading rate of a few hundredths of a magnitude per revolution is barely detectable in the absolute magnitude material, but consistent results have been obtained by Kresák and Kresáková (1989).

The observations just described relating to the evolution of short-period comets agree fairly well with the predictions of the dust coverage model, in particular as regards the irregular patterns. Taken together with the recent observations of a largely inert surface on the nucleus of Comet P/Halley and strong indications of a very low activity level for several other short-period comets (see Chap. 2), there is hardly any doubt that the surface enrichment of refractory material plays a major role in comet evolution. Further support for this view comes from the statistics of nongravitational effects. Consider a short-period comet that has just been captured by Jupiter or otherwise experienced a major reduction of its perihelion distance. Based on the dust coverage model such a comet might have a relatively fresh and icy surface, since much of the pre-existing dust may have been blown off. During the subsequent evolution a new dust layer should grow, and thus the outgassing rate decreases without any comparable decrease of the mass of the nucleus. Hence one expects a decrease of the nongravitational effect – possibly quite a rapid one – as time progresses from the last dynamical capture event experienced by the comet. This is exactly what comes out of the observed statistics (Rickman et al., 1987). It should be noted that a nucleus evolving only by mass loss (resulting only in a decrease of its surface area and assuming no change of its volatile mixing ratio) would behave in a radically different way: Its nongravitational effect would increase slowly and the rate of increase would grow as the perihelion distance decreases.

However, the question of the ultimate fate of the comets is to some extent different from that of the short-term evolution. In the mass loss scenario the exhaustion of all volatile resources necessarily puts a definitive end to the cometary activity, but in the dust coverage model there is always the possibility of rejuvenation of the comet by meteoroid impacts or nucleus splitting. Let us thus compare the typical total lifetime of 500–1000 revolutions indicated by the observations with the

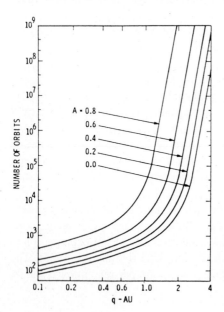

Fig. 7.17. The number of perihelion passages required for an icy sphere of 1 km initial radius to disappear by sublimation, as a function of perihelion distance q (plotted on the abscissa) and albedo A (five curves for different values of A). The assumed density is $600 \, kg/m^3$ and the assumed orbits are near-parabolic. From Weissman (1980).

expected number of revolutions required in order to consume the nucleus material by sublimation (the "sublimation lifetime"). Figure 7.17 shows Weissman's (1980) results for H_2O-ice sublimation applied to a homogeneous sphere of original radius $R_0 = 1$ km and density $\rho = 600$ kg m^{-3}. Considering recent evidence for low albedos of cometary nuclei, the bottom curve is the one most trustworthy. It can easily be shown that the sublimation lifetime τ_s for a sphere, where the fraction f_{fs} of the surface is free-sublimating and where the mass ratio of entrained dust grains to outflowing gas is χ, is

$$\tau_s = \frac{R_0 \rho}{(1 + \chi) S f_{fs}},$$

(7.31)

S denotes the gaseous mass loss per unit area of free-sublimating surface (integrated sublimation) over one apparition. The curves in Fig. 7.17 basically indicate the shape of the function $S(q)$.

Weissman used $\chi = 0$ and $f_{fs} = 1$. If we scale his results to a nucleus with $R_0 \sim 5$ km, $\rho \sim 300$ kg m^{-3}, $\chi \sim 1$ and $f_{fs} \simeq 1$ using Eq. (7.31), we obtain $\tau_s \simeq 1.2 \cdot \tau_s^{(W)}$. Thus for the average perihelion distance of 1.5 AU characterizing short-period comets $\tau_s \gtrsim 2000$ even if the whole nucleus stays freely sublimating until the end. For a more typical value of $f_{fs} \lesssim 0.1$ one finds values of τ_s far larger than the observationally indicated lifetime. There is hence some indication that the ultimate fate of most comets is choking off by dust rather than exhaustion of all volatiles.

As a consequence of this choking there should be a supply of objects that appear as asteroids but whose true nature is cometary. They should have orbits typical of the Jupiter family and would thus occasionally be observed as fast-moving, Earth-approaching asteroids. Such bodies, usually referred to as Apollo-Amor asteroids, are known in fairly large numbers. However, they tend to have orbits with aphelia far inside Jupiter's orbit ($Q \lesssim 4$ AU) and can mostly be understood as originating from the main asteroid belt via collisional break-up and transport of fragments into Earth-approaching orbits along chaotic routes (Wisdom, 1985; Wetherill, 1988). In recent times we have nevertheless seen the discovery of a significant number of Apollo-Amor objects in Jupiter-approaching orbits. Dynamical investigations by Hahn and Rickman (1985) have revealed several cases of typically cometary behavior. Hence, there is little doubt that a number of extinct comets in asteroidal disguise have actually been observed. In addition to the recent discoveries one may also mention the well-known singular asteroid 944 Hidalgo, which does not approach the Earth's orbit currently but which most probably has done so in the past (Kozai, 1979).

The number of observed objects is not yet large enough to allow any safe conclusion as to what fraction of the short-period comets do develop into asteroids, but apparently that fraction may be large (Rickman, 1985). We thus have ample reason to consider the formation of a refractory surface layer as the most important mechanism of cometary decay and disappearance. Furthermore, there is at least a possibility that the extinct Jupiter family comets also contribute a significant fraction of the usual Apollo-Amor asteroids, since very close encounters with the terrestrial planets or nongravitational effects during the active, cometary phase might decrease the aphelion distance far enough for gravitational decoupling from the giant planet to occur (see Weissman et al., 1990 for a general review of this topic).

7.6 Summary

It seems appropriate to end this chapter by summarizing the major problems as they appear today. These are of course the ones that hold the key to an improved understanding, if they can be tackled or partially solved in a satisfactory way.

Perhaps the most important problem relating to the advance of knowledge brought about by the exploration of Comet Halley is that these results pertain to a single comet and can at best provide suggestions about the properties of comets in general. These suggestions include: A porous nucleus of elongated shape in slow rotation covered on most parts by a layer of non-volatile and very dark material; a composition where H_2O ice with small admixtures of other volatiles is mixed intimately with refractory material, in which carbonaceous compounds are of essence; and a bulk elemental composition that is nearly solar except for a depletion in hydrogen and the noble gases. Further exploration of a large sample of comets by Earth- and satellite-based techniques and of a few comets by *in-situ* space probe studies is necessary as long as the ambition is to reach a general understanding of cometary formation and evolution.

Specifically, in terms of identifying the mechanisms whereby cometary material could have originated and the nuclei could have formed, it is necessary to broaden the modeling efforts that hitherto have concentrated on particular aspects. This is an ambitious goal, but in order to make use of the detailed chemical information that is becoming accessible for discriminating between various formation scenarios we must be able to make relevant predictions. For this purpose modeling of grain growth, evolution, and aggregation in the presolar and solar nebula has to be further developed in the light of laboratory results on, e.g., gas trapping in amorphous ice, polymerization, and other radiation effects.

As far as thermal and chemical evolution is concerned, major uncertainties remain as long as it cannot be specified in what fraction ^{26}Al entered into cometary material. In principle, by means of model calculations one should be able to predict what chemical structure results for a comet nucleus as a function of initial composition and physical structure. By checking such model results against observations of evolutionary trends in comets one might be able to constrain the initial conditions. At present we are far from this goal. The reason is partly that uncertainties remain regarding the behavior of the amorphous–crystalline phase transition in the presence of additional volatiles with gas diffusion, sublimation, and recondensation. Even more importantly, perhaps, the understanding of the evolution of surface phenomena on comet nuclei (non-volatile crusts vs. active spots) is still far from complete.

If we could characterize the dynamical transfer processes between various remote places of storage and observable comet orbits, we could make very valuable statements about the spatial distribution of comets, including the detailed structure of the Oort cloud with its inner core and a possible trans-Neptunian cometary disk.

However, although indications to this end are already appearing in the literature, much work remains to be done before a trustworthy picture can emerge. This involves an accurate modeling of the complex chaotic dynamics in the region of the giant planets and possibly beyond, but also a better understanding of the aging effects and observational selection effects operating in different kinds of short-period comet orbits. In addition, one needs to pay further attention to the possibility of time-variable phenomena, i.e., departures from steady state in the population of short-period comets.

8. Implications of Comet Research

Walter F. Huebner and Christopher P. McKay

8.1 Introduction

In this chapter we seek to summarize the progress and implications of comet research. We do not wish to leave the impression that comet data are sufficiently complete to draw unique conclusions, but the perspective we paint may be useful for assessing the current state. The attempted synthesis emphasizes not only new directions in thinking, but serves also as a guide for the future. However, the leap in knowledge gained by the spacecraft explorations of Comets Giacobini-Zinner and Halley and complemented by the Earth-bound observations of these comets is restrictive in the sense that the data, analyses, and interpretations pertain to only two comets and most of them to only one. The results can at best provide suggestions about the properties of comets in general, but even that may be a dangerous extrapolation. Ground-based observations of several comets show very different behaviors from that of P/Halley. Comets Morehouse and Humason are two extreme examples. Nevertheless, the P/Halley results hold the key to an improved understanding of comets if the limitations of the generalizations are kept in mind.

Comets are the most numerous objects in the Solar System. It is estimated that there are several times 10^{11} comet nuclei in the Oort cloud. A bright active comet also provides a spectacular display on the night sky. For these reasons comets are interesting objects in their own right that deserve to be investigated scientifically. However, when reasons are given why comets should be investigated and why they are interesting to science, the most commonly quoted objectives are the interdisciplinary information that comets reveal about (1) the early history of the Solar System, i.e., the chemical composition and thermodynamic conditions of the solar or presolar nebula region in which comets formed, (2) the formation of planetesimals from cometary subbodies, (3) the evolution of planets through comet impacts, i.e., the formation of oceans and atmospheres and the general enrichment of volatiles to the inner Solar System, and (4) the origins of life from the influx of comets on Earth. Two additional, but separate and equally important, objectives can be raised: (5) To understand processes that give rise to plasma structures in comets that might be identifiable from Earth so that comets can be used as probes of the interplanetary medium, particularly in regions where it is difficult and energetically expensive to make *in situ* measurements, e.g., close the Sun or out of the plane of the ecliptic, and (6) to use the coma–solar wind interaction region as a large plasma physics laboratory with no interference from wall effects. Rarely are results from comet investigations addressed in terms of these objectives in publications. It therefore appears prudent and appropriate to summarize the progress that has been made to meet these objectives from the modern observations and spacecraft encounters with comets.

The scientific details must be deduced from the progress that has been made on the physics and chemistry of comets. We will thus first consolidate and reiterate the new knowledge about comets in Sect. 8.2 and then direct our attention to the above objectives. The structure and composition of the nucleus are closely related to the early history of the Solar System and to the formation of planetesimals; they will be combined in Sect. 8.3. The formation of oceans and atmospheres and influx of organic matter are more closely related to the origins of life; they will be discussed in Sect. 8.4. Finally we will summarize and reiterate the relevant unsolved and new problems in Sect. 8.5.

8.2 Progress in Physics and Chemistry of Comets

There are, all will agree, no awards for originality for those who speak of the vast success of the comet missions and the increase in knowledge that they brought us. The results speak for themselves; they have been stunningly visible and are viewed with both wonder and delight. Here we review and restate the major results that have been described in detail in earlier chapters of this book. Basically, there are four broad areas of physics and chemistry of comets that relate to the above objectives: (1) The chemical composition of comet nuclei, (2) their physical structure, (3) the plasma structures in the coma, and (4) orbital theories.

8.2.1 Composition of the Nucleus

The frozen gases in the nucleus of Comet Halley consist mainly of water. It is now thought that the frozen gases of all short-period and many long-period comets are dominated by water at the level of about 80 to 90%. The most volatile gases such as hydrogen, nitrogen, methane, and the rare gases, are either absent or only present in trace amounts. Even the abundance of CO is much lower in the nucleus than coma observations suggest. About half to two-thirds of the CO appears to come from distributed sources in the coma, probably indirectly through the dissociation of complex molecules that are released from the dust. Closely related to the decay of these complex molecules may be H_2CO (Krankowsky, 1990), CH, and CH_2 (Boice et al., 1990). Even other species such as CN, NH, NH_2, C_2, and possibly C_3, may come, at least in part, from distributed sources, as the jet-like structures of these radicals in the coma suggest. The nucleus itself appears to be inhomogeneous, as indicated by the relatively small active areas that are responsible for the jet-like dust features and most, if not all, of the coma gas. The dust size distribution does not appear to have a lower limit; particles as small as 10^{-16} g were detected in large numbers. The abundance distribution of even smaller masses may continue without a cutoff down to molecular clusters. There is evidence that larger dust particles disintegrate into smaller particles in the coma. A size distribution of dust averaged over the coma may therefore underestimate the larger particles in the innermost coma within a few hundred kilometers above the nucleus and can certainly not be used to extrapolate to the distribution of dust sizes on the surface of the nucleus.

It is not clear whether all comets are dominated by water. Comets Morehouse (1908 III) and Humason (1962 VIII) showed very unusual activity in their comae and plasma tails (mostly CO^+) on time scales of less than a day. Both comets were dust-poor in the visible part of the spectrum. Comet Morehouse had a perihelion distance of 0.945 AU. Its eccentricity of 1.00067 does not mean that it was an interstellar comet; the hyperbolic orbit is most likely the result of nongravitational forces. Comet Humason had a perihelion distance of 2.133 AU and is on an elliptic orbit with a period of just under 3000 y. The active areas of these comets were certainly not dominated by water ice. A less well known, but unusual comet is Comet Stearns (1927 IV). It had a perihelion distance of 3.684 AU and showed C_2 Swan bands superimposed on a continuous spectrum, a stubby dust tail, and a narrow plasma tail. What is remarkable about the comet is that it was observed for four years, almost all of the time after perihelion (it was discovered only a few days before its perihelion passage), out to a distance of 11.5 AU. It is difficult to explain its activity at such large heliocentric distances if it is based on water-dominated ice. Comet Schuster (1975 II) behaved similarly. It had a perihelion distance of 6.881 AU, but was not discovered until a year after perihelion passage. It was then observed for nearly two years as it receded from the Sun. It is well to remember these unusual comets when generalizing the results from the P/Halley spacecraft encounters and observations.

One of the big surprises in the Halley encounters was the detection of large amounts of organic dust, the CHON particles. The idea of organic dust was not new (see, e.g., Greenberg, 1982), but its detection elevated it to much greater significance. CHON particles contain most of the carbon that was "missing" when comparisons were made of the coma composition with solar abundances. Not only do the CHON particles, as suggested by their name, contain carbon, hydrogen, oxygen, and nitrogen, they also contain some sulfur. It is believed that the disintegration of the CHON particles gives rise to the heavy molecules and contributes significantly to many light species such as CN, C_2, C_3, and probably NH and NH_2 that have been observed in the coma, some in the form of jets. Although the detection of very heavy molecules with molecular weight up to about 120 amu was not as surprising to some investigators as the detection of the organic dust, the abundances of these molecules were much higher than expected. The CHON particles and their disintegration products are also responsible for the IR emissions in the 3.2 μm range and very likely also for most of the CH and CH_n^+ ($n = 1$ to 3) detected in the coma. Some of the strongest evidence for this conjecture is the large range of CH in the comae of many comets. Since CH has a very short lifetime ($\tau \simeq 100\,$s), it must be continuously produced by distributed sources in the coma.

The detection of para- and ortho-water in IR spectra assumes a special importance. Because of its dominance in the frozen gases, water pervades the entire nucleus; thus the ratio of these two forms of water has the potential to open a new era of investigations about the formation temperature of comet nuclei. Investigations of two comets (Halley and Wilson) suggest formation temperatures of about 30 to 50 K, respectively. This technique of temperature determination must still pass the test of time.

The composition of comet dust deserves special attention because the comet missions investigated the first samples of dust that clearly come from a comet. The dust composition was found to be in about equal amounts organic (CHON), silicates, and mixed CHON and silicates. The silicates can be divided into two subgroups, those that are richer on iron and those that are richer on magnesium. The organic component can be divided into six categories: Those consisting primarily of (1) H and C, (2) H, C, and O, (3) H, C, and N, (4) H, C, N, and O, (5) C and O, and (6) H and O (Clark et al., 1987). The last two groups occur less frequently than the first four. The abundances of the elements heavier than and including Na in carbonaceous chondrites of type 1 (CI chondrites) and in the solar photosphere, normalized to the Si abundances, are the same over the enormous range of about seven orders of magnitude. However, the abundances of H, C, N, and O in CI chondrites are depleted relative to their abundance values in the Sun. This means that even though CI chondrites contain the least altered early Solar System material available for analysis in the laboratory, they did not retain some of the volatile chemical elements.

The relative abundances of the rock-forming elements (i.e., all elements heavier than oxygen) in P/Halley dust are within a factor of two of the CI chondrites, while hydrogen, oxygen, and especially carbon and nitrogen are highly enriched, i.e., less depleted in comet dust relative to CI chondrites. The abundance ratios of these elements in the comet dust approach the solar values. Thus the dust in Comet Halley is more primitive than CI chondrites; this testifies to the nearly pristine nature of comet dust. Concerning the apparent Fe deficiency and Si over-abundance in P/Halley dust relative to CI chondrites, these discrepancies may be caused by instrumental uncertainties and are probably not real.

The abundances of the elements H, C, N, and O in P/Halley, combined from coma gas and dust, were compared by Geiss (1987) and by Jessberger et al. (1988b) to the solar composition using two different methods (see Table 4.7 in Chap. 4). Results from these different approaches are in good agreement and show that nitrogen is underabundant by a factor of 3 and hydrogen is deficient by more than a factor of 700. However, the other elemental abundances are very similar to solar values. This finding corroborates the notion that comets are only slightly altered relics from the early Solar System.

Considering the mineralogy, the distribution of the Fe/(Fe + Mg) ratio in P/Halley silicate dust resembles more closely the distribution found in anhydrous interplanetary dust particles (IDPs). The distribution in these particles has a strong maximum at low values in addition to being spread over the whole scale. In contrast, in a layer lattice IDP the Fe/(Fe + Mg) ratio centers strongly around 0.5 and is very similar to the distribution in a carbonaceous chondrite matrix. The local maxima at about 0.3 and 0.5 in the P/Halley dust particle distribution may point to an admixture of layer silicates to the dominant anhydrous grain population, in accord with the results from infrared studies. However, grains rich in OH and poor in C (a signature of layer silicates) have not yet been identified in the comet dust spectra. The Fe/(Fe + Mg) distribution similar to that found in P/Halley and the anhydrous IDPs is very rare in meteoritic matrix material and clearly constitutes a major difference of bulk meteoritic and cometary compositions.

Comparing the ternary Mg-Si-Fe system, P/Halley dust again differs greatly from layer lattice silicate IDPs and matches better with anhydrous IDPs. The similarity is most obvious in the rather dense population of the near-pyroxene field and the extension to Fe-rich compositions. There is, however, a notable difference between comet dust and anhydrous IDPs: Anhydrous IDPs are practically devoid of samples below the olivine-Fe line, while in comet dust many Si-poor and Mg-rich grains occur. The distribution indicates a lack of both Fe-rich pyroxenes and Fe-rich olivines. The grains with compositions of Fe $\simeq$ 0, Mg $\simeq$ Si $\simeq$ 0.5 may indicate the presence of primary Mg-silicates. This is what one would expect for primitive Solar System material, i.e., silicates, that never had been exposed to high temperatures.

Assuming a solar ratio of C/Mg for the whole comet, i.e., gas and dust, and using the measured gas and dust compositions, the dust-to-gas mass ratio $\chi \simeq 1$ with an uncertainty of a factor of 2. This spread is much narrower than the range of values derived from ground-based measurements.

8.2.2 Structure of the Nucleus

In anticipation of the spacecraft encounters with Comets P/Giacobini-Zinner and P/Halley and other planned comet missions, many sophisticated theoretical models had been proposed for the nucleus. Several elaborate with imaginative detail about its structure, far beyond requirements to explain the observations. In spite of these efforts, the composition, the inner structure, and the evolution of comet nuclei remain very uncertain and will remain a field of justifiable debate (and sometimes unjustified ruminations) until the surface layer of a nucleus has been probed, e.g., by the CRAF penetrator, or a refrigerated, deep-frozen sample is returned for laboratory analyses, e.g., by the Rosetta mission.

Based primarily on spacecraft observations of Comet Halley, but also supported by radar, radio continuum, and IR measurements, the size of a nucleus is five to ten times larger in volume than estimates before the Halley missions indicated. The larger size of the nucleus, combined with the same response to the nongravitational forces acting on it (therefore the same mass), leads to a lower density. Although comet models always assumed a spherical nucleus for reasons of simplicity and ignorance about its actual shape, it was never believed that the nucleus would be a sphere. However, the aspect ratio of about 2:1 found for the nucleus of Comet Halley and now also suggested for several other comets (Sekanina, 1988a; Millis et al., 1988; Jewitt and Meech, 1989) is a much larger deviation from a spherical nucleus than had been anticipated. The surface morphology as deduced from Halley Multicolour Camera images suggests subunits of the nucleus that may have been as large as one to a few kilometers.

The surface area of the nucleus is about four times larger than had been thought. Consistent with the brightness measurements of the large nuclear surface is the very low albedo of about 4%. The surface temperature of P/Halley during the spacecraft encounters at 0.89 AU was about 330 K, compatible with a dark nucleus covered by refractory material rather than exposed ice. The temperature of sublimating ice would be only about 200 K. The surface of the nucleus must be inhomogeneous and covered by dark, refractory material with small patches of active areas that are also dark and are responsible for the gas and dust production. The low density, the jet-like activity, and the surface morphology strongly suggest an inhomogeneous structure throughout the entire nucleus.

8.2.3 Plasma Structures in Coma and Tail

The bow shock, the contact surface, the draping of the interplanetary magnetic field (IMF), and the magnetic cavity had been predicted before the comet encounters. New insight has been gained by the detailed, *in situ*, investigations of these plasma features and the region between them including the discovery of the cometopause. The new data are invaluable for the computer simulation of space plasmae. Plasma physics benefited through the development of models to simulate measurements of instabilities, waves, pickup ions, ring and shell distributions, energetic ions, and charge exchange processes with comet neutrals. Thus, the concept of a large plasma laboratory without wall effects has borne fruit.

Except for details concerning the two-fluid nature of the comet plasma flow, the formation of the cometopause, and the relative drift of the solar wind plasma and cometary ions, magnetohydrodynamic (MHD) simulations have been quite successful in providing a global view of the whole interaction process. This assessment is based on the work of Wegmann et al. (1987), in which the global properties of the plasma flow, magnetic field structures, plasma densities, and ion and electron temperatures are described using a three-dimensional model with extensive chemistry. They have also shown that stationary MHD models can be used effectively to analyze the data of magnetic field measurements.

Only one spacecraft, the International Cometary Explorer (ICE), passed through the onset of the ion tail region of a comet. Four major findings, based on its plasma observations of P/Giacobini-Zinner, are therefore of particular importance: (1) Except for the detection of the formation of a magnetosheath at the ion tail boundary, the model for field draping is basically confirmed. The magnetic field strength in the lobes of the ion tail is on the order of 60 nT. This relatively high field may be explained in terms of pressure balance at the tail boundary where the total external pressure of the cometary ions was as large as the solar wind ram pressure. (2) A thin plasma sheet with a total thickness of about 2000 km and a width of about 1.6×10^4 km was found at the center of the ion tail. The peak electron density was $n_e \simeq 6.5 \times 10^8$ m^{-3}. The energy distribution of the electrons can be approximated by a two-temperature model for the cometary electrons with $T_e \simeq 2.6 \times 10^4$ K and 1.1×10^5 K. The second value agrees with the excess energy of photoelectrons from H_2O produced by solar radiation (Bame et al. 1986). (A third electron temperature appears to be associated directly with the solar wind). (3) The plasma flow velocity gradually decreased to zero at the center of the ion tail. A significant amount of electron heating was detected between the ion tail and the bow shock. (4) The degree of magnetic field turbulence and plasma wave activity was seen to correlate with the production of ions from cometary neutrals in the solar wind. Outside the bow shock, the anisotropy of the energetic heavy ions was found to be very high. The ion flux became more isotropic inside the bow shock probably because of stronger pitch-angle scattering and lower flow speeds.

The P/Halley encounter on the subsolar side of the coma produced similar observations as those described in (3) and (4). Regarding the formation of the magnetosheath, first detected at P/Giacobini-Zinner, it is worth noting that during the inbound passage of the Giotto probe, a sharp magnetic field discontinuity was observed at about 1.3×10^5 km from the nucleus. The magnetic field strength increase by about 20 nT has led to the terminology "magnetic field pile-up region." This structure was not detected outbound nor was it observed by the Vega magnetometer experiments.

Even though the complete model of the physical processes involved in the formation of the ionospheric boundary layer still needs to be developed, several major features can be identified. The best signature of the contact discontinuity comes from magnetometer measurements in which the magnetic field strength was observed to decrease from a value of about 20 nT to zero over a distance of 25 km (Neubauer, 1988). The concept of a magnetic field-free cavity was not a new idea, but the stability of the interface raised new issues that will be discussed further in Sect. 8.5.4.

The plasma environment near a comet is extremely turbulent. During the ICE encounter with Comet Giacobini-Zinner very strong wave activity was detected, together with energetic particle fluxes, even at a distance of about 2×10^6 km from the nucleus. In addition, very strong, large-amplitude magnetic field variations were observed with $\Delta B/B \simeq 1$. The component of the magnetic field fluctuations along the average magnetic field direction was found to have a maximum in the wave power spectral densities with a peak at a period of 75 to 135 s. The 100 s waves observed at Comet Halley have somewhat different characteristics; they may correspond to ion cyclotron waves for cometary water-group molecules ionized and picked up in the solar wind. Similar fluctuations were also observed in the solar wind electron number density and the solar wind flow velocity. As described first by Wu and Davidson (1972) and Wu and Hartle (1974), the basic cause of the plasma turbulence is associated with free energy produced by pickup of cometary ions.

8.2.4 Orbital Theories

Calculations of orbital transfer of comets to and from the Oort cloud through perturbations by planets, stars, and giant molecular clouds have made significant progress during the past several decades. However, attempts to relate chemical differences, such as the relative abundances of C_2, C_3, CN, and NH_2 in the coma, dust-to-gas mass ratios, and cometary activity to short-period versus long-period comets has been mostly unsuccessful. In light of the new findings from the P/Halley missions this should receive more attention in the future. It may be the hard-to-detect minor and trace mother molecules that show most of the differences. For example, if some comets originated in the giant planet subnebulae, they would be richer on CH_4 and NH_3 than those originating in the main solar nebula or the presolar nebula.

Some specific aspects appear to be promising: (1) Although CO can only be observed in the brightest comets because of its small g-factors, a pattern of the variation in the relative abundance of CO (between 1 to more than 30% relative to H_2O) appears to be correlated with comets that are normally classed as dusty, i.e., have dust of a size that produces a continuum in the visible range of the spectrum. (2) There are some systematic trends in which CN and C_2 observations are correlated with heliocentric distance. The ratio of CN/C_2 increases at large distances. Compared with the variation of CO, most trace species seem to have a reasonably constant abundance relative to H_2O, as measured either with OH or $O(^1D)$ emission, varying only by small factors in either direction from the average for most comets. Also, the variation of CN is strongly correlated with the dust-to-gas ratio of a comet, which is consistent with the interpretation that the parent of CN observed in jet-like features is associated with the dust. This dust may, however, be too fine to scatter visible light. The variation of C_2 is comparable to that of CN but not well correlated with the dust; this variation might be a reflection of the varying efficiency of photolytic reactions in the coma rather than a reflection of the abundance of the parent. NH and NH_2 may also be related to CHON in dust particles.

These aspects must be viewed with caution. Counterexamples exist: Comet Humason showed very strong activity in its plasma tail, related to CO^+ abundance, but was relatively poor on visible dust. The CN jet-like features observed in P/Halley are not correlated with visible jet-like features in the dust continuum. The abundance of CN relative to OH, i.e., water, was normal in Comet Giacobini-Zinner, while C_2, C_3, and NH were depleted.

8.3 Progress on the History of the Early Solar System

The new findings from the Halley encounters should help attempts to reconstruct the chemical and thermodynamic conditions in the outer parts of the early Solar System, where physical and chemical processes must have been so gentle that chemical bonds predating the Sun could survive the formation of comets and indeed the solar nebula.

A basic objective is the identification of the sources of the most frequently observed radicals CN, C_2, C_3, NH, and NH_2. Wyckoff et al. (1988) and Magee-Sauer et al. (1989) have shown that the abundance of NH is less than 1% in the coma of P/Halley and the scale length of its parent molecule may not fit the dissociation lifetime of NH_3. Schleicher et al. (1987) find that NH, together with C_2 and C_3, is unusually low in P/Giacobini-Zinner, a comet with less than average dust-to-gas ratio. Allen et al. (1987) and Wegmann et al. (1987) concluded that NH_3 and CH_4 constitute a few percent of the frozen gases in the nucleus of P/Halley. However, their early analysis of the ion mass spectrometer data did not consider the release of NH_n- and CH_n-complexes from the CHON particles as possible sources of the ion radicals NH^+, NH_2^+, CH^+, CH_2^+, and CH_3^+ (see, e.g., Boice et al., 1990). It is now believed that CHON particles may be important contributors to the abundances of these ions. Thus the abundances of NH_3 and CH_4 may be much less than 1%. If this can be confirmed, then Comet Halley did not originate in the protoplanetary nebulae of Jupiter or Saturn.

Considering the chemical composition, we note that the most volatile species of the frozen gases in the nucleus, including CH_4 and N_2, are present at the level of a percent or less and that CO is present at about 5%. Most of the CO detected in the coma originates from distributed sources, presumably from the decay of complex molecules produced by thermal and possibly electrostatic disintegration of CHON particles. This suggests that the nebula region in which comets formed was never cold enough to condense the most volatile gases. However, as Bar-Nun et al. (1985, 1988) have shown (see Fig. 8.1), gases can also be trapped at low temperatures in amorphous water ice. About as much CH_4 would be trapped as CO. The lower abundance of CH_4 relative to CO suggests, as had been pointed out by Lewis and Prinn (1980), that CO conversion to CH_4 was very incomplete in the solar nebula. They also predicted that N_2 conversion to NH_3 would be very low. While less than 1% of NH_3 appears to be present in the frozen gases, the abundance of N_2 may not be very much higher. However, this would still be consistent with the expected abundance ratio of CO/N_2. There are several difficulties in assessing the abundance of nitrogen in the coma. In mass spectra N_2 occupies the same channel as CO and C_2H_4 and N is in the same channel as CH_2, a potentially abundant decay product of CHON particles. From astrophysical abundance considerations, CO can be expected to be more abundant than N_2. In optical spectra N_2 has not been identified and the strongest emission line of N is dominated by the wing of the very strong hydrogen Lyman-α line. Only N_2^+ has been identified in the plasma tail, but its intensity indicates a very low abundance. Analysis of the CHON particles indicates a low abundance of nitrogen relative to the solar abundance, but high compared to chondritic material.

Relative to CH_4 and CO, N_2 as well as rare gases are poorly incorporated in water ice. Although argon has not yet been detected in comets, N_2 has. Therefore, a search for Ar should be a target in the CRAF mission investigation. If the N_2 in comets dates back to the solar nebula, then the ratio of Ar/N_2 should be close to the cosmic value which is about 0.1.

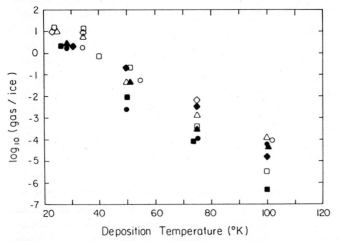

Fig. 8.1. The total amounts of trapped gases vs. deposition temperature of the (water vapor) gas mixtures. The following symbols indicate: ($\Diamond$) CH, ($\triangle$) CO, (o) N_2, and ($\square$) Ar. Open symbols represent the results of a 1 : 1 deposition ratio for each single gas relative to water vapor. Solid symbols represent the deposition of gas mixtures $H_2O : CH_4 : Ar : CO$ (or N_2) = 1.00 : 0.33 : 0.33 : 0.33. Each point is an average of at least two experiments. Some points were moved slightly on the temperature scale for clarity. From Bar-Nun et al. (1988).

Besides direct condensation and trapping (adsorption), volatiles can also be incorporated into water ice by clathration. Physical adsorption is particularly effective on amorphous ice at low temperatures depending on the degree of exposure of the volatile to the ice. Clathration, on the other hand, can occur at temperatures well above the pure condensation points of the volatiles. This is one of the reasons that it had been suggested by Delsemme and Swings (1952) as a possible mechanism in comets. A nearly complete conversion of ice to a clathrate depends on the ability of volatile gas molecules to collide with fresh icy surfaces. In the absence of such collisions, clathrate formation is extremely limited because of the limited number of cage sites available. A detailed investigation during the CRAF mission to Comet Kopff should improve our knowledge about this long-standing question of clathrate hydrates in comets as originally suggested by Delsemme and Swings (1952) and again by Delsemme and Wenger (1970).

Considering that comets formed in the outer parts of the solar or presolar nebula, it would be surprising if the elemental abundances of comets should differ from those of the Sun or carbonaceous (CI) chondrites. Based on the abundance estimates for the rock-forming elements Mg, Al, Si, Ca, Ti, and Fe of Breneman and Stone (1985) for the Sun and those of Anders and Ebihara (1982) for the "cosmic" abundance derived mostly from chondrites (which agree to within a factor of two), one would expect the abundance differences for these elements in comets not to be very different. Indeed, Table 4.6 in Chap. 4 shows that the abundances of all elements measured in Comet Halley, except hydrogen, carbon, and nitrogen, are within a factor of two of those of the CI chondrites (see also Delsemme, 1990). Except for hydrogen, nitrogen, and possibly iron, they are also within a factor of two the same as those for the Sun. In Comet Halley, hydrogen and nitrogen are considerably more abundant than in CI chondrites and much less abundant than in the Sun, while carbon is much more abundant in P/Halley than in CI chondrites, but less than a factor of two underabundant compared to the Sun. Iron is less than a factor of two underabundant in Comet Halley relative to CI chondrites, but underabundant by a factor of about 2.5 relative to the Sun. The underabundance of iron could, however, be instrumental.

The major abundance deviations are therefore in hydrogen, carbon, and nitrogen. Why are these abundances in Comet Halley, and by deduction in comets in general, closer to solar than to chondritic abundances? One possible explanation is that two different types of binding energies are responsible. For the Sun it is gravitational binding, while for comets it is chemical bonding. Hydrogen is so overabundant that only a small fraction can be bound in molecules other than H_2, which cannot be retained by chondrites or comets because of its volatility. However, gravity prevents it from escaping from the Sun. On the other hand, carbon is chemically one of the most active elements. It has already been pointed out that conversion of CO to CH_4 may have been quite incomplete, but some reactions between carbon, hydrogen, oxygen, and nitrogen did take place; possibly forming some complex molecules in the presence of UV or cosmic radiation. Oxygen, being one of the most abundant, chemically active elements can also combine with silicon to form very refractory silicates (enstatite and pyroxene are by far the most important sinks of oxygen), while nitrogen assumes the peculiar chemical form N_2 in interstellar clouds. Molecular nitrogen is hard to condense and transforms only very slowly to NH_3. Because of their high volatility, most of the CO, N_2, and the CH_4 that formed could not be retained by chondrites or by comets. Thus most of the oxygen is chemically bound in refractories, while carbon compounds are less refractory and nitrogen, mainly in the form of N_2, is least condensable. This would explain the underabundance of N_2 in P/Halley. Many of the less volatile carbon and nitrogen compounds could not survive on chondrites because their environmental temperature is higher than that of comets. On comets they could survive in part as frozen gases and in part as CHON particles, until the comet comes close to the Sun. When the comet releases the CHON dust, it is warmed by solar radiation and the CHON slowly disintegrates.

The water in P/Halley contains from five to ten times more deuterium than is found in interstellar clouds or in the protosolar nebula (i.e., as accreted in gaseous form from the protosolar nebula by Jupiter and Saturn). The D/H ratio is consistent with other Solar System hydrogen that was accreted in the form of volatile molecules and their ices such as in the Earth's ocean water, deuterated methane on Titan, or interplanetary dust. Since the D/H ratio in P/Halley was determined from water and that in Titan from methane, chemical fractionation could have influenced this ratio; but the effect was probably small. Thus the very similar ratio could mean that comets formed at the same heliocentric distance range where the Earth, Titan, and meteorites were formed. It could also mean that the Earth and other terrestrial planets acquired their volatile organics from comets. The latter explanation enjoys broader acceptance.

The isotope ratios of other elements in Comet Halley, mainly nitrogen, oxygen, and sulfur are consistent with solar values. Although the carbon isotopes also encompass the solar value, the ratio of $^{12}C/^{13}C$ shows such an enormously wide range that it poses some new questions that will be discussed further in Sect. 8.5.2.

The molecular composition of comets resembles that of dense interstellar clouds, strongly suggesting that comets formed in a gentle environment of the solar or presolar nebula in which the composition had not significantly changed during the collapse of the cloud. In such a cloud low temperatures lead to an unusual chemistry: Neutral molecules do not readily react with one another; the chemistry is therefore dominated by reactions between the neutral molecules and the trace amounts of ions that are produced by UV and cosmic radiation. As a consequence, in the gas phase nitrogen is found mainly in its molecular form, carbon in the form of carbon monoxide and some related molecules, and condensed matter is enriched in deuterium.

The density of a comet nucleus appears to be much lower than had been anticipated. It seems to be about $300 \, kg/m^3$, but could be as low as $100 \, kg/m^3$ or as high as $1 \, Mg/m^3$. The low density, odd shape, and surface morphology of the nucleus must reflect the aggregation and accumulation processes of comet nuclei. The extremely nonspherical shape of the nucleus and the morphology of the surface rule out accumulation solely of very small subnuclei of a meter or less in size. The models of Daniels and Hughes (1981) and Meakin and Donn (1988), which start out as fractal models of very small flakes and gradually change into the fluffy aggregate models in which larger clumps accumulate with some deformation, appear to be on the right track. Such computer simulations are very important to understand formation of comets and planetesimals. Deformation must occur during slow collisions and shattering may occur in high energy collisions. Not all of the shattered material may escape sideways, since the closing speed of the colliding blocks may be faster than the escape of some of the shattered material. Clearly, the collision velocity plays an important role. Compressed material must exist side-by-side with very loosely shattered material in "voids" where the solid bodies did not meet. In this manner one can explain the low density, the surface morphology, and the necessary deformation and associated local density enhancements during collisions.

The initial density of the colliding blocks must, however, also be very low. They themselves are products of collisions of smaller clumps. When small blocks collide, more of the material shattered during a collision can escape than in a collision between large cometary subbodies. This leads to the following three, alternative hypotheses. (1) Collisions are so slow that no shattering occurs and the density of the colliding blocks is not significantly increased, i.e., deformation is small. (2) Average collision speeds between small blocks do lead to shattering; only the blocks that collided with very low velocities survived. The shattered material may be accumulated again on the surfaces of larger, low-density blocks. (3) The frozen gases do not shatter, but the small blocks penetrate into each other during a collision at moderately high velocities of the order of 1 km/s and their density approximately doubles. In this case the initial densities must be extremely low so that the density after several collisions still does not approach the value $1 \, \text{Mg/m}^3$.

Unfortunately, no data for fluffy, very low temperature ice exists in order to test which of the three hypotheses is the most likely. Laboratory experiments must be conducted and models must be developed to test all three hypotheses. One guideline for collision velocities appears to emerge: If the temperatures obtained from the abundance ratio of ortho- to para-water are correct, then comets never exceeded a temperature of about 50 K. If the water condensed as amorphous ice, then its heat capacity at 30 K is $C \simeq 2.3 \times 10^2 \, \text{J kg}^{-1} \, \text{K}^{-1}$. Thus heat generated during collisions of icy clumps may have raised the temperature from about 30 K to not more than 50 K, i.e., $\Delta T \simeq 20 \, \text{K}$. This corresponds to a collision velocity $v_{\text{coll}} = (2C\Delta T)^{1/2} \simeq 100 \, \text{m/s}$.

Dust particles may be the nuclei for the condensation of the gases in dense interstellar clouds as well as in the presolar and solar nebula. They most likely existed in the interstellar cloud and survived the collapse in the outer parts of the presolar or solar nebula until comet formation. CHON particles could have formed under the action of cosmic radiation on the surface layer of comet nuclei in the Oort cloud. However, since CHON particles are still being released from Comet Halley, after most of its surface has been eroded to some depth during its many visits into the inner Solar System, it appears that they are also deep in the interior of the nucleus. Thus even the CHON particles must have existed before comets formed.

Hence the chemistry of comets provides us with a link between the Solar System and its ancient ancestor, an unnamed dark interstellar molecular cloud dispersed long ago. Comets are the most numerous class of bodies composed of bulk matter in the Solar System. In addition, this matter has the least changed composition since its aggregation.

8.4 Comets and the Origins of Life

In addition to the origins of life, this section also deals with the possible influence of comets on the evolution of atmospheres and oceans and the general enrichment of volatiles on the planets of the inner Solar System. This enrichment and the atmosphere and oceans are prerequisites for the origins and evolution of life.

8.4.1 The Origins of Life

One of the fundamental questions that can be addressed by studying comets is: What can comets tell us about the origins of life? Can the study of comets be an aid to understanding the relationship between the chemical and physical evolution of the Solar System and the appearance of life, one of the key goals of planetary science? In many ways the study of comets is the study of origins. Comets may provide crucial clues to the events leading to the origin of the Solar System, to the origin of the planets, and to the origin of life itself.

The standard hypothesis for the origin of life, as first outlined by Oparin (1924, 1938) and Haldane (1928) begins with the abiological production of organic materials. The motivation for organic material as the primal stuff of life is clear; all life on Earth is composed of organic material. A most interesting aspect of life is that only a relatively small number of the many possible organic molecules are used. Life is a small and particular subset of the possibilities of organic chemistry. All the different proteins are made from only 20 amino acids, the nucleic acids are largely made from only eight different nucleotides, and the polysaccharides are mostly made from a few simple sugars (see, e.g., Lehninger, 1975). The basic building blocks of life are few in number compared to the myriad possibilities of the carbon bond.

A demonstration of the significance of the biochemical unity of life, and a major advance in the development of the origin of life hypothesis, were the classic experiments of Miller (1953). Miller showed that a gaseous mixture of methane, water, and ammonia, when subjected to a high-voltage electric discharge produced organic molecules. More significantly, high yields of many of the important biogenic organic molecules are produced in Miller-type syntheses, including many of the essential amino acids. Thus, there seemed to be a natural mechanism for abiologically producing some of the needed structural components of life. As an aside, it has proven to be much more difficult to abiologically synthesize the nucleotides, leading to the suggestion that the metabolic structure of life evolved abiologically to be followed by the biological evolution of genetic codes (Dyson, 1985).

Initially it was suggested that a Miller synthesis of organic material could have operated in a reducing atmosphere on the early Earth. However, studies of the photochemistry of ammonia and methane have indicated that these molecules are photolyzed in the atmosphere on time scales measured in years or less (see, e.g., Levine, 1985a). Thus it is unlikely that they were major constituents of the early atmosphere. Furthermore, studies of the formation of the terrestrial planets, Venus, Earth, and Mars, have indicated that the juvenile atmosphere of the Earth was composed of CO_2, H_2O, and N_2 (see, e.g., Walker, 1977; Pollack and Yung, 1980; Levine, 1985a). Organic materials are not readily made, if at all, in such an atmosphere (Chang et al., 1983). Thus, while the early Earth appeared to have had an environment suitable for life to grow, this same environment may not have allowed for the *in situ* production of the primal biogenic organic material that may be necessary for life's origin.

In fact the inner Solar System seems bereft of organic material. However, as the distance from the Sun increases, the amount of organic material appears to increase. The asteroids of the outer belt, starting at about 2.7 AU, begin to show signs of significant organic material (Hartman et al., 1987). The giant planets are very reducing, being composed primarily of H_2, and organic material is common in the moons of these systems (Encrenaz, 1986; Simonelli et al., 1989).

This is one of the central problems of the origins of life: The organic material thought to be the building blocks of the first organisms is found in the outer Solar System, while the environments conducive to water-based chemistry and the growth of life are found in the inner Solar System. There does not appear to be a region of overlap. Thus one credible suggestion, first expressed by Oró (1961), for the origins of life would be to transport the biogenic organic material from the outer Solar System to the inner Solar System. This is the key procreative role that comets may have played and is the basis for the current interest of comets in the origins of life.

Recent spacecraft explorations of Comet Halley have confirmed earlier suggestions that comets are rich in organic material (see, e.g., Kissel and Krueger, 1987a) and have lent support to the concept, first suggested by Oró (1961), that comets may have played a role in the origins of life on Earth. The spacecraft observations of organic material have been limited to determinations of elemental composition, but an indication of the possible complexity of comet organics is the discovery of polymerized formaldehyde (Huebner, 1987).

In the following sections, we attempt to address the role of comets in transporting organic material to the early prebiological Earth. We begin by considering the nature of cometary organics and the possibility that this organic material would have survived entry into the early atmosphere. The strong shocks and high specific kinetic energy of the impactor may have effectively destroyed all the organic material in the incoming comet, unless the comet shattered into smaller fragments. However, it has been shown that organic synthesis of complex organics can occur in high energy shocks under reducing conditions. Therefore we also consider the possibility that the local conditions in the cometary entry shock could have been made sufficiently reducing by the organics of the entering comet to resynthesize organic material in the shock wave. Further we consider the amount of organic material that may have been imported by comets and the role that comets may have played in importing other volatiles essential for life, in particular water, to Earth.

8.4.2 Organic Material in Comets

Even before the spectacular results of the P/Halley flybys, there was ample evidence indicating that small molecular weight organic compounds are present in the coma of comets (see, e.g, Wykoff, 1982). Of particular interest was the detection of HCN by Huebner et al. (1974) and CH_3CN by Ulich and Conklin (1974). By inference these molecules imply the presence of more complex parent organic molecules.

Ground-based observations of comets taken over many years had indicated that the abundance of C was 25% of that expected based upon solar composition. The P/Halley results confirmed the suggestion (Delsemme, 1973, 1981, 1982) that the low carbon indication in the coma was caused by carbon being "hidden" in the dust. The most likely form for any carbon in the dust is organic material (suggested by Greenberg, 1982 and Delsemme, 1982), which is consistent with measurements taken during the P/Halley encounter (Kissel and Krueger, 1987a).

The primary evidence for the prevalence of organic material in comets comes from the dust particle impact mass spectrometers flown on the P/Halley missions. Dust particles impacted a silver foil at encounter velocities of about 70 km/s. The impact destroyed the particles and the elemental fragments were analyzed with a time-of-flight mass spectrometer. The dust that had been analyzed was categorized into several general types. Most numerous were grains that contained silicate materials (Clark et al., 1987; Langevin et al., 1987b; Jessberger et al., 1988). The second most numerous class of particles were characterized by high concentrations of the elements C, H, O, and N ("CHON" particles, Clark et al., 1987; Langevin et al., 1987b; Jessberger et al., 1988). Some CHON particles also contain small amounts of sulfur. Another class of organic particles were composed of H, C, and N and a final class was composed of H, C, and O (Clark et al., 1987). Together, these results indicate that about one-third of the mass of the dust and about 10% of the mass of the comet nucleus is composed of organic material of a refractory nature.

In a detailed interpretation of the results from the particle mass spectrometer, Kissel and Krueger (1987a) suggested that the dust particles were composed of a mineral core with an organic mantle, similar in many respects to the laboratory models proposed by Greenberg (1982, 1986a). The dominant organics in the mantle were inferred to be unsaturated hydrocarbons. Nitrogen-containing compounds, such as hydrocyanic acid (a possible precursor to CN that was observed in jets by A'Hearn et al., 1986a), acetonitrile, and propanenitrile, were also indicated. Included in this category were the biologically important pyrimidines and purines (in particular, the purine, adenine, was reported). No indication was found of oxygen-rich organics such as sugars or peptides. There was no detection of amino acids, suggesting that their concentrations are reduced by at least a factor of 30 compared to the pyrimidines and purines.

Kissel and Krueger (1987a) also reported the possible detection of phosphorus in the mineral core of the particles, but not in the organic fraction. This completes the list of the biogenic elements (H, C, N, O, P, and S) detected in comets.

Other results from the P/Halley encounter also suggest the presence of organic compounds. Schloerb et al. (1987) reported HCN at about 0.1% relative to water. Formaldehyde is of particular interest since its presence in comets was suggested by Oró (1961) based upon general considerations and by Delsemme (1981, 1982). It was detected in the monomer, H_2CO, by Combes et al. (1986, see also Moroz et al., 1987), Snyder et al. (1989), and Krankowsky (1990), and as the polymer polyoxymethylene (POM) by Huebner (1987). In addition, CN has been observed in jets in the coma (A'Hearn et al., 1986a). The density distribution within these jets indicates that the CN is the decay product of a more refractory parent molecule.

Table 8.1. Abundances of the Biogenic Elements in Atom Numbers Normalized to Si=1.0

Element	Solar	Comets	CI Chondrites
H	26600.	30.0	2.0
C	11.7	8.1	0.7
N	2.3	1.4	0.05
O	18.4	18.8	7.5
S	0.5	0.6	0.5
Si	1.0	1.0	1.0

As seen in Table 8.1 (adapted from Delsemme, 1988; a similar, but more extensive table has been prepared by Delsemme, 1990), the biogenic elements within the comet nucleus are close to solar abundance relative to Si. The notable exception is H. Similar results were obtained from the P/Halley encounter analysis (see Chap. 4). Hydrogen is depleted in comets by many orders of magnitude with respect to cosmic abundances. Indeed, the overall redox balance of cometary material may be slightly oxidizing ($O > H_2$). However, this does not necessarily imply that comets were formed under oxidizing conditions. In fact they probably formed under reducing conditions in hydrogen dominated outer regions of the solar nebula, but could not retain the highly volatile hydrogen. This is confirmed by the D/H isotopic ratios measured during the Halley encounters ($0.6 \times 10^{-4} <$ D/H $< 4.8 \times 10^{-4}$; Eberhardt et al., 1987a), which is in the range for hydrogen in Solar System objects that acquired their hydrogen as part of condensed phases, i.e., ices.

There are actually several ways in which organic material could have been incorporated into comets. During formation of the solar nebula, comets may have acquired organic material of interstellar origin directly. About 80 molecules have been detected in interstellar clouds (Irvine and Knacke, 1989) of which about 60 are organic. The main uncertainty with respect to direct incorporation of these organics into comets is the possibility that nebular temperatures exceeded the decomposition temperature of these compounds. This depends on the dynamics of the solar nebula and whether comets were formed within the Solar System and subsequently transported to larger orbits and the Oort cloud, or were formed at greater distances in the presolar nebula. Based upon the carbon budget of the outer Solar System and models of the formation of the outer planets, Simonelli et al.

(1989) have recently suggested that up to 10% of the carbon in the solar nebula, in the region beyond Jupiter, could have been organic material, with the rest being CO. The presence of CO (as opposed to CH_4) may indicate that the outer solar nebula was composed of essentially unaltered material from the interstellar medium. If this interpretation is correct, then this provides evidence that the organics in comets are remnants of interstellar organic material. Alternatively, organic material could also have been formed on small grains in the nebula itself due to a variety of reactions, including shock formation, UV radiation, and cosmic radiation (Wood and Chang, 1985). A third possible source of cometary organics is *in situ* production due to cosmic radiation.

The organic material in comets may share a similar origin as that observed in meteorites. The more volatile components, which would be lost during high-speed atmospheric entry of a large meteorite, would be retained in the cold nucleus of a comet. There is some evidence based on studies of stratospheric extraterrestrial particles that suggests this is the case (Brownlee, 1979, Fraundorf et al., 1982).

The formation (or decomposition) of organic material, as well as the diffusion of material through the nucleus, would result in characteristic isotope fractionations of the carbon atoms (as well as H, N and O). The primary source of isotope data for comets are imprecise ground-based determinations of the isotope ratio of the volatile carbon, as obtained from observations of the C_2 Swan bands. The data indicate a ratio of $^{12}C/^{13}C$ of $\sim 100 \pm 30$ (see, e.g., Stawikowski and Greenstein, 1964; Danks et al., 1975). Similar results have been reported by Šolc et al. (1987b) based on P/Halley data. This range of values brackets the terrestrial value of 89, but has such large uncertainties that it cannot be used to place useful constraints on the theory of organic production (e.g., Abell et al., 1981). More recently Wyckoff et al. (1989) reported $^{12}C/^{13}C = 65 \pm 9$ based on CN observations, which may indicate a systematic difference from terrestrial values. Clearly, an *in situ* determination is essential. Furthermore, the isotope ratios of the nonvolatile carbon may be more indicative of chemical processes.

The temperature under which comets condensed and the temperature history that they experienced before ejection into the Oort cloud (T < 10 K) would have determined the nature of the volatile compounds in the initial "pristine" comets of the cloud. Models of the subsequent evolution of comets in the Oort cloud suggest that the passage of nearby stars could cause many comets to reach temperatures of between 16–30 K (Stern and Shull, 1988). Most organic compounds would be preserved at these low temperatures. The low temperatures of the comet nucleus are not, in principle, a barrier to prebiotic chemistry. Organic synthesis can occur in carbon-rich ices (see, e.g, Thompson et al., 1988) and fairly complex organics can be produced. For example, Schwartz et al. (1982) have demonstrated adenine synthesis from HCN in ice. Furthermore, the low temperatures may actually prevent the thermal decomposition of complex organic molecules.

It is very unlikely that the comet nucleus presents a favorable milieu for these organic molecules to further react to form polymeric compounds and ultimately the structures of life such as membranes (Ponnamperuma and Ochiai, 1982; Lazcano-Araujo and Oró, 1981). Perhaps the most serious difficulty preventing further prebiotic evolution is the absence of liquid water. However, it has been suggested by Irvine et al. (1980, 1981) and Wallis (1980) that radiogenic heating of a comet nucleus by the decay of an initial burden of ^{26}Al could have resulted in the formation of a liquid water core inside of large ($R > 10$ km) comets. The existence of such a liquid core requires three uncertain conditions (Irvine et al., 1981): (1) The time scale for formation of comets must be shorter than the decay time of ^{26}Al, which is 7×10^5 y, (2) the cometary material must be a good insulator, and (3) the comet must be large. Given these conditions Irvine et al. (1981) suggest that the duration of liquid cores in comets could be from 0 to 10^9 y. Since the latter is comparable to the time from the formation of the Earth to the oldest definitive evidence for life at 3.5 Gy ago (see, e.g., Schopf, 1983), the possibility of the origins of life on comets is not entirely out of the question. In fact the high concentration of organics and the reducing conditions in a comet, coupled with the presence of liquid water, provide an environment that is actually closer to the conditions of the Miller synthesis than the early Earth would be, even under the conditions of a reducing atmosphere. One of the problems in the synthesis of biologically significant molecules in an early ocean on Earth is the polymerization of amino acids and nucleic acid monomers; the higher concentration of organics in a comet would aid this polymerization (Irvine et al., 1980). The continuation of any comet life past the period of a liquid water core is unlikely, given the essential requirement for liquid water for all life as we know it (see, e.g., Kushner, 1981).

Hoyle and Wickramasinghe (1981) and coworkers (see, e.g., Hoover et al., 1986) have presented an alternative theory in which comets play a central role in the cosmic cycle of life. Their model requires that the comet nucleus undergoes periodic melting due to various heat sources. Within the resulting solution, microbial life can thrive using the abiologically produced organics as nourishment. This model suggests that life is spread throughout the galaxy by comets carrying microorganisms, in particular, bacteria and viruses. However, there is no evidence to support this view. Hence, most researchers in the field have concluded that comets are probably not sites in which life can currently arise and even hardy microorganisms would not long survive (see, e.g., Bar-Nun et al., 1981).

8.4.3 Comet Transport of Volatiles

Direct observation of OH and H_2O^+ (Wyckoff, 1982) and the fact that most comets begin to form spectacular comae only at $r < 2.5$ AU indicate that water is the major component of the icy fraction of the nucleus. The P/Halley results provided further evidence that the major volatile constituent of a comet nucleus is water ($> 80\%$) with a smaller amount of CO (e.g., $CO/H_2O < 7\%$, Eberhardt et al., 1987b; $C/O < 10\%$, Steward, 1987) and little CO_2 ($< 3\%$, Combes et al., 1988; Moroz et al., 1987). Several comets, however, have exhibited light curves suggesting that CO_2 (or CO) ice may play a significant role in controlling the gas production rates on some "new" comets at distances greater than about 3 AU (A'Hearn and Cowan, 1980). The infamous light curve of Comet Kohoutek may be the result of high CO concentrations. Allen et al. (1987) and Wegmann et al. (1987) inferred a ratio of NH_3/H_2O of 1 to 2% and CH_4/H_2O of 2%. Boice et al. (1990), however, show that the ratio of CH_4/CO in the ice is less than 1% and may be close to zero. Eberhardt et al. (1987b) reported that the N_2/H_2O ratio was $< 10\%$.

Comets may have been an important source of volatiles during the formation of the terrestrial planets. Because the terrestrial planets formed in the inner, hotter regions of the solar nebula, they are low in volatile elements. For example, the C/Si ratio for the Earth is about 10^{-4} of the solar ratio. It is plausible that collisions with comets during the late accretionary stages of planet formation supplied much of the atmospheric CO_2, N_2, and water for Earth, Venus, and Mars.

Recently, several researchers have suggested that a considerable amount of water could have been accreted by the early Earth in the final stages of formation due to the interception of comets scattered from the vicinity of Neptune and Uranus (Chyba, 1987; Ip and Fernandez, 1988). Chyba (1987) has estimated that an ocean of water could have been accreted between 4.5 and 3.8 Gy ago if comets comprised more than 10% of the mass of the impacting population. Since the ratio of organic material to water based upon the P/Halley encounters is over 10%, an influx of this amount of water represents an enormous influx of organic material.

However, it is not clear that organic material within a comet nucleus impacting the Earth could survive. The typical impact velocities are of the order of 20 km/s or more (Sleep et al., 1989). The energy per unit mass contained in this kinetic energy is equivalent to 2×10^8 J/kg, more than enough to completely pyrolyze organic material. These numbers would suggest that virtually no cometary organic material would survive entry into the early atmosphere. However, there are at least three possibilities that argue against complete loss of the incoming organic material. First, as has been pointed out by numerous researchers, the highly porous fragile cometary nucleus is likely to fragment upon entry (see, e.g., Greenberg, 1986a; Overbeck et al., 1989). The mid-sized and small particles can be slowed down by atmospheric drag to a sufficient extent that their organic burden remains intact (Anders, 1989). Second, comets are known to exude a significant fraction of their mass as dust. This dust is known to be rich in organics. As these small dust particles impact the Earth they are gently slowed down by atmospheric drag and do not reach pyrolytic temperatures. Dust of this nature is currently collected in the atmosphere (see, e.g., Brownlee, 1979). During the final stages of the formation of the Solar System, when the scattering of comets by Neptune and Uranus occurred, the rate of infall of this dust may have been many orders of magnitude larger.

A third possible way in which organic material can be supplied by comet impacts would be as the result of shock resynthesis of organic material from the reducing mixture that results from the destruction of the initial cometary organics. The elemental composition of comets suggests that the composition of the wake from the ablation of material upon entry will be reducing. High-temperature shock synthesis, in a reducing atmosphere, of complex organic molecules, including solid polymeric material, is well documented (see, e.g., McKay et al., 1988; Scattergood et al., 1988).

8.4.4 The Role of Comets

The evidence for life on Earth dates back 3.5 Gy (Schopf, 1983). By this time life appears to have had a fair amount of biological sophistication, indicating that the origins of life may have been much earlier. Thus life may have developed within the first few hundred million years after the accretion of the Earth. Only those comets that deposited organic matter, volatiles, or possibly even spores before 3.5 Gy ago could have played a contributing role in the formation of life on Earth. The distribution of the flux of Earth-crossing comets during this early period is uncertain but was probably much higher than at the present time.

Recently, Clark (1988) has suggested a rather unique way in which comets could have contributed organic material in high concentrations to the surface of the Earth. His scenario postulates a grazing impact of a comet that transverses a sufficient amount of the atmosphere to be slowed down to terminal velocity. This "soft" impact would leave a coherent "pond" of organic-rich material on the surface. Such a pond, processed by rain, would then serve as a site for the origins of life. The origins of life based upon such a primeval procreative comet pond would suggest that life is a chance event of very low probability.

If, as seems likely, comets did indeed play a role in the origins of life by transporting organic material and volatiles to the early Earth, then similar events should have transpired on Venus and Mars. The record of the events 3.5 Gy ago are obscured on Earth and probably on Venus as well. However, over half of the surface of Mars dates back to the late bombardment that may have included the impact of numerous comets. Thus, Mars may very well retain the best record of the effects of comets on the early stages of the terrestrial planet and the role that cometary organics may have played in the origins of life. Even though there may be no life on Mars today, it may hold the best record of the events leading to the origins of life (McKay, 1986) and may help us understand the role that comets played in the origins of life on Earth.

Did comets play a role in the origins of life? The question remains unanswered. However, results from the recent explorations of Comet Halley have lent support to a generally positive answer and have helped refine the questions for future space missions. Only with future missions that provide detailed *in situ* analysis of cometary materials can we begin to understand their contribution to the organic seed of early life and to the volatiles that nourished this early life.

8.5 Unsolved and New Problems

We have already compiled a list of topics and questions that new comet missions like CRAF and Rosetta will explore (see Chap. 1, Sect. 1.7). Here we reiterate some of these topics and introduce a few new ones.

The spacecraft flybys of Comets Giacobini-Zinner and Halley were highly successful. They confirmed many untested theories and provided additional clues for competing models, but without leading to a clear preference. They also raised new questions. Based on this new knowledge, research efforts will now be much better focused than was possible before the encounters. Moreover, we should be better equipped to interpret Earth-based observations; a statement that still needs to be tested and verified over the next several years.

8.5.1 Nucleus Rotation and Surface

Beginning with the nucleus, we recognize immediately that the rotational state of P/Halley still needs to be resolved. If the long-period variability of coma bright-ness is caused by free precession, then the motion of the nucleus is very difficult to analyze. Since precession may be expected to be a common phenomenon of comets, interpretation of brightness variability from ground-based observations will require even greater care and effort.

We have seen that the gas and dust production of a comet is not an isotropic property, as most comet models had to assume for lack of guiding details, but that it is confined to active regions covering about 10 to 20% of the surface area while the remainder appears to be inert. The gas production of most short-period comets is only 1 to 10% of that of P/Halley but the sizes of the nuclei, such as those of the Comets IRAS-Araki-Alcock, Neujmin 1, and Arend-Rigaux, do not seem to be much smaller than that of P/Halley. Therefore the total active area of these comets must be smaller, i.e., the relative fraction of their active area to their total surface area must be at the 1% level or less. Is this a manifestation of the gradual dust coverage already experienced during previous revolutions in the present orbit or is it a remnant from the cosmic ray irradiation to which the comet was exposed for 4.5 Gy? Is Comet Halley in a state of evolution by gradual dust coverage or is it simply consuming its mass via those active regions left open as gaps in a crust processed by cosmic rays? What about long-period comets such as Comets Morehouse and Humason? Are they giant comets with many active areas, or do they not have inert areas, or are they indeed of a different, more volatile composition?

Models of comet nuclei must be able to account for the small active ar-eas. They cannot be randomly distributed at one place on the surface in one apparition of the comet and completely redistributed in the next apparition, un-less volatile inclusions in the nucleus are of a size that typically last for just one apparition. Models that choke the nucleus activity by accumulating debris (regolith) on its surface, thereby reducing the amount of solar energy reaching the volatiles, are probably not correct. There must be a process that operates efficiently from underneath an inert surface. The idea of local inhomogeneities, where more volatiles are present and released when the nucleus is heated during a new perihelion passage, becomes increasingly more attractive.

In view of the above, it is questionable whether sublimation alone properly describes comet activity. The gas production, however, cannot be much lower than that from ice, otherwise the active fractions of the surface would not sustain the observed coma production. No visible distinctions between active and inactive surfaces could be found on the spacecraft images. From the point of view of the physics of the nucleus, the details of the sublimation process and the question of what makes an area active will provide essential insight into the nature of comet nuclei. Why is only such a small part active? How can this activity be sustained for many revolutions? This important issue is not yet resolved, and it seems fair to state that the observations of Comet Halley have not solved the questions about comet evolution but rather complicated them. Such are the vicissitudes of exploring uncharted regions.

8.5.2 Chemical Abundances in the Coma and the Nucleus

Chemical abundances in the coma are related to the composition of the nucleus, but the abundances of volatiles have been altered by gas-phase reactions, distributed sources of gas originating from evaporation and decay of organic dust, and heterogenous reactions on dust. In the innermost coma there may also be some recondensation of gases as they cool by the expansion of the coma. A fundamental question that is still not answered is the complete inventory of molecular species and their abundances in the coma and in the nucleus. The evidence seems to favor abundances of NH_3 and CH_4 of less than 1%, if they are present at all. It is not yet clear whether NH_3 is the parent of the widely observed NH_2 and NH, nor is it clear whether they even share a common parent. Most data are consistent, with both NH and NH_2 coming from NH_3, but the data also seem to suggest a much wider variation in the abundance of NH_2 than in the abundance of NH. Since the two abundances tend to come from different observations, there is very little overlapping data to adequately test the hypothesis.

Observations in new wavelength regions, such as the extreme ultraviolet, and *in situ* observations that will be provided by the CRAF and Rosetta missions will be needed to identify additional species. Relative abundances must be determined for a reasonable ensemble of comets and for more than the few most easily measured species. Some key questions need to be addressed: (1) How does the ratio of CO/CO_2 vary from comet to comet? (2) Is the apparent variation in CO/H_2O in the coma reflecting a difference in the nuclear ices or a difference in the amount of organic grains? (3) What fraction of other species is released from refractory organic particles rather than from the nucleus? (4) Are NH_3 and CH_4 present at levels higher than 1% in any comets? (5) Is NH_3 the parent of NH and NH_2? (6) Is N substantially depleted? (7) What is the parent of CN? (8) What is the atomic inventory of C, N, and O relative to H? (9) What are the CH_4/CO and NH_3/N_2 ratios? (10) Are rare gases trapped in the ice of the nucleus and if so, what is the ratio of Ar/N_2? (11) Do clathrates exist in comets and if so, how were they formed? (12) Is S_2 present in most comets? (13) Are other sulfur compounds, such as SO, SO_2, and H_2S, present in comets? (14) Most fundamentally, what are the parent species in the ices of the nucleus? (15) What is the D/H ratio and how does it vary in comets? (16) What is the $^{12}C/^{13}C$ ratio? (17) What is the dust-to-gas mass ratio? (18) What is the ratio of organics (CHON) to silicates in dust and in dust-free comets? (19) Are CHON and silicate particle size distributions different from one another and how do they evolve with cometocentric distance? (20) What are the composition and structure of CHON and mixed CHON and silicate particles?

In the future we can hope for a number of facilities that will address some of these questions. The CRAF mission should yield a wealth of new information about the gases in the coma, answering many of the questions above for one single comet. It will also provide us with the composition of one place on the surface layer of a nucleus. The Rosetta/CNSR mission will provide unique information from sampled nucleus material of one other comet. Surveys of many comets will be needed to develop a comprehensive understanding. These surveys will rely on facility instruments, such as the Hubble Space Telescope (HST) and ground-based telescopes, and on proposed facilities, such as the Shuttle Infrared Telescope Facility (SIRTF), the Far-Infrared Space Telescope (FIRST), the Infrared Space Observatory (ISO), Planententeleskop, and the Far-Ultraviolet Spectroscopic Explorer (FUSE/Lyman).

The inhomogeneous nature of the nucleus surface with respect to activity poses an interesting challenge to comet science. Is the inactive surface caused by chemical heterogeneities in the nucleus itself, does a layer of fine dust resettle on the surface creating a dust mantle, or is some other mechanism responsible? The global nature of the chemical composition of the nucleus has been inferred from the coma observations, but the details remain open to debate. Further analysis of the *in situ* measurements of the coma will suggest how intimately the gaseous species are related to the dust and what implications this has for the nature of the nucleus. Basic laboratory work on the gas-phase chemistry and the volatility of large organic molecules and oligomer products, both ionic and neutral, is needed to understand this connection between the gas and solid phase constituents of the coma.

The distribution of the neutral species in the coma is described well by global models out to cometocentric distances of a few hundred thousand kilometers, where the effects of radiation pressure need to be taken into consideration in multi-dimensional models. The effects of energetic ions and neutrals on the chemistry in the inner coma also needs to be incorporated. Energetic ions were detected in the inner coma of P/Halley with energies up to 10 keV. This unexpected result is explained by the work of Eviatar et al. (1989). Neutral molecules with speed of about 1 km/s move radially outward in the coma. Some of them are then ionized by solar photons or by charge exchange with the solar wind. The neutralized solar wind ions (mostly hydrogen atoms) continue toward the inner coma, while at the same time the new ions are picked up by the solar wind. The energetic neutrals can now penetrate the contact surface without any major perturbation by the ions or the piled up magnetic field. Once they are inside the contact surface, the gas density of the coma gas increases, enhancing the probability of yet another charge exchange with a coma ion that was produced by photoionization. Thus the energetic neutrals are converted to the energetic ions that have been detected inside of the contact surface. The importance of this is twofold: There are not only energetic ions as measured by the JPA instrument, there must also be energetic neutrals. These energetic particles can undergo hot chemical reactions. Basic developments of "hot" chemical reactions relevant to comet coma chemistry have been carried out by Roessler and Nebeling (1988).

At large heliocentric distances models of the free molecular flow of coma species directly from the surface will become necessary to investigate the development of the coma as the comet approaches the Sun. The same models will be important for radiation pressure interaction, for the formation of the contact surface, bow shock, and other boundaries, and for analyzing asymmetries in the post-perihelion passage (relative to pre-perihelion) of the comet at large heliocentric distances. When longer time baselines of *in situ* measurements and a more complete spatial mapping of the cometary environment become available from the CRAF and Rosetta missions, it will be important to generalize the current global models to a time-dependent formulation and to three dimensions to further our knowledge of cometary science.

The isotope information obtained by PUMA 1 for carbon is difficult to interpret because of possible interferences. In particular, CH may interfere with the detection of ^{13}C in mass spectrometers. For C the few high ^{12}C/^{13}C ratios, up to 5000, are not easily explained as artifacts. Earlier analyses of ^{13}C^{12}C spectra from several comets indicated an isotope ratio for ^{12}C/^{13}C of about 100 (Vanýsek, 1987) with relatively large error estimates because of the difficulty of isolating the spectra of overlapping lines from other cometary species. The error estimates encompassed the terrestrial ratio of 89. There are two notable exceptions: Lambert and Danks (1983) found a ratio of about 50 from high-resolution spectra of Comet West (1976 VI) and Wyckoff et al. (1989) resolved the ^{13}CN and ^{12}CN rotation lines in P/Halley. From this they found an isotope ratio of 65 ± 9 for ^{12}C/^{13}C. It is dangerous to interpret these results because of possible chemical fractionation effects. Independent confirmations are needed before far reaching conclusions are drawn about cometary origins.

8.5.3 Dust Tail

The structure of the dust coma of Comet Halley proved even more complex than it had been assumed previously. The dust emission from the nucleus is very heterogeneous and time variable. Additional complications arose from the intricate rotational state of the nucleus. The variety of different materials that are contained in comet dust add even more complexity to its dynamics and hence to the structure of the dust coma. Although the foundation for the determination of the dust-to-gas mass ratio has been much improved, its present value of $\chi \simeq 1$ for Comet Halley still includes major uncertainties of about a factor of 2.

The dynamics of dust tails for particles of a constant size are reasonably well understood, but the dynamics of fragmenting particles and the formation of striae, a type of tail inhomogeneity, are still unsolved problems. Sekanina and Farrell (1980) suggested that striae are formed when large particles, simultaneously ejected from the nucleus and subjected to the same repulsive force, fragment at the same time. If this suggestion is correct, striae can be treated as synchrones of the fragments, starting at the place of fragmentation rather than at the comet nucleus. The difficulty arises in explaining the simultaneous destruction of dust particles. Lamy and Koutchmy (1979) have suggested that striae are made up of charged dust particles driven by the interplanetary magnetic field carried along with the solar wind. So far, it is not possible to determine whether the striae formation can be solved in a purely mechanical way, whether electromagnetic forces play an important role, or whether a combination of both effects gives rise to the observed phenomenon.

8.5.4 Coma and Tail Plasma

Cravens (1989) has made a one-dimensional calculation on a fine grid which he used to discuss the three-dimensional fluid behavior of a supersonic ionospheric flow in spherically symmetric expansion in the vicinity of the magnetic cavity boundary. According to his calculations, the abrupt stop of the supersonic flow at the edge of the magnetized region leads to the formation of a shock and a thin, high-density shell for large Mach number. He approximated the thickness of this compressed fluid shell to be between 28 and 280 km, depending on the recombination rate of the dominant ion in the ionosphere. The smaller thickness leads to the formation of a very narrow shell of compressed plasma at a point where, according to Cravens's calculations, there may also be a very steep magnetic field gradient. Additional physics and multi-dimensional calculations on a fine grid are needed to make further progress.

The contact surface divides plasma from two different sources: The mass-loaded solar wind on the outside and the cometary ions on the inside. How stable is such a surface? Ershkovich and Mendis (1983) argued that the ion–neutral frictional force can be extremely destabilizing for the contact surface; hence a fully magnetized ionosphere was predicted. The Giotto observations of a magnetic field-free cavity have raised new questions and provided many new results in this regard. The simplest model for the cometary magnetic cavity assumes that the cavity boundary is a tangential discontinuity separating plasma flowing freely from the direction of the nucleus from the approximately stationary plasma held by the magnetic field permeating the coma. The configuration is subject to various magnetohydrodynamic instabilities, notably the Kelvin-Helmholtz, Rayleigh-Taylor, and fluting instabilities. If the boundary is treated as a simple discontinuity between two uniform regions, ion–neutral friction is destabilizing. In the case of Comet Halley the magnetic cavity does not obey this description since the magnetic field transition is rather broad except for a small "discontinuity" on the inner side. Furthermore, the region within the boundary is completely free of magnetic field which would not be the case if the boundary itself were unstable, since this would permit some leakage of the field into the cavity. However, the description of the cavity formation on the basis of neutral gas friction, which seems to be a good approximation in the case of P/Halley, suggests that there could be a quite different type of instability from those mentioned above; namely, an overturning instability, similar to that which occurs in an unstable atmosphere, resulting from the delicate balance between frictional and magnetic field stresses, including curvature, and allowing for the effects of photoionization and recombination.

The current density in the narrow ramp region can be shown to be far below the critical value required to generate an ion–acoustic instability and therefore we should not expect to find plasma instabilities such as the modified two-stream instability at this location. However, since the exact physical nature of the magnetic ramp is still not clear, numerical simulation experiments addressing the microscopic effects and kinetic processes might reveal other interesting phenomena.

Niedner and Schwingenschuh (1987) have systematically studied the ion tail activity at the time of the Vega encounters of P/Halley using the data from the magnetometer experiment on Vega and observational material from several ground observatories. Their conclusion was that the onset of a disconnection event (DE) observed on March 8 to 10 was related to a reversal of the interplanetary magnetic field (IMF) as detected by the Vega 1 and 2 spacecraft. On the other hand, Saito et al. (1986) have compared the plasma data obtained by the Sakigake spacecraft with ground-based observations and found that no distinct DE can be detected in P/Halley's ion tail during 11 to 14 March, even though Sakigake crossed the neutral sheet of the IMF at least four times. The global dynamics of plasma tails thus remains a field with a number of important questions not yet answered. No satisfactory computer models exist to simulate disconnection events.

Considerable progress has been made on the six objectives listed at the beginning of this chapter. They are primarily based on the successful spacecraft encounters of Comets Giacobini-Zinner and Halley. We must be extremely cautious when generalizing these results to comets universally. The success of the missions, the dangers inherent in generalizations to other comets, and the many open questions are the reasons for the need of further spacecraft investigations as already mentioned in Chap. 1. An unquestionably successful start has been made in expanding our knowledge about the physics and chemistry of comets and all the related implications, but the question about the similarities and differences between short-period and long-period comets will be with us for some time to come.

References

1. Abell, P. I., Fallick, A. E., McNaughton, N. J., Pillinger, C. T. (1981): 'The stable isotope approach.' In *Comets and the Origin of Life,* Ed. C. Ponnamperuma, D. Reidel Publishing Co.; Dordrecht, Holland, p. 129–139.
2. Abergel, A., Bertaux, J. L. (1987): 'Evolution of comet P/Halley in early March 1986 as observed from Vega pictures.' *Astron. Astrophys.* **187**, 829–834.
3. A'Hearn, M. F., Cowan, J. J. (1980): 'Vaporization in comets: The icy grain halo of Comet West.' *Moon and Planets* **22**, 41–52.
4. A'Hearn, M. F., Feldman, P. D., Schleicher, D. G. (1983): 'The discovery of S_2 in Comet IRAS-Araki-Alcock 1983d.' *Astrophys. J.* **274**, L99–L103.
5. A'Hearn, M. F., Schleicher, D. G., Feldman, P. D., Millis, R. L., Thompson, D. T. (1984): 'Comet Bowell 1980b.' *Astron. J.* **89**, 579–591.
6. A'Hearn, M. F., Birch, P. V., Feldman, P. D., Millis, R. L. (1985): 'Comet Encke: Gas production and lightcurve.' *Icarus* **64**, 1–10.
7. A'Hearn, M. F., Hoban, S., Birch, P. V., Bowers, C., Martin, R., Klinglesmith, D. A. III (1986a): 'Cyanogen jets in comet Halley.' *Nature* **324**, 649–651.
8. A'Hearn, M. F., Hoban, S., Birch, P. V., Bowers, C., Martin, R., Klinglesmith, III, D. A. (1986b): 'Gaseous jets in Comet P/Halley.' In *20th ESLAB Symposium on the Exploration of Halley's Comet,* Eds. B. Battrick, E. J. Rolfe, R. Reinhard, ESA SP-250 I, p. 483–486.
9. Aikin, A. C. (1974): 'Cometary coma ions.' *Astrophys. J.* **193**, 263–264.
10. Alfvén, H. (1954): *On the Origin of the Solar System.* At the Clarendon Press; Oxford.
11. Alfvén, H. (1957): 'On the theory of comet tails.' *Tellus* **9**, 92–96:
12. Allamandola, L. J., Sandford, S. A., Wopenka, B. (1987): 'Interstellar polycyclic aromatic hydrocarbons and carbon in interplanetary dust particles and meteorites.' *Science* **237**, 56–59.
13. Allen, M., Delitsky, M., Huntress, W., Yung, Y., Ip, W.-H., Schwenn, R., Rosenbauer, H., Shelley, E., Balsiger, H., Geiss, J. (1987): 'Evidence for methane and ammonia in the coma of comet P/Halley.' *Astron. Astrophys.* **187**, 502–512.
14. Amata, E., Formisano, V. (1985): 'Energization of positive ions in the cometary foreshock region.' *Planet. Space Sci.* **33**, 1243–1250.
15. Anders, E. (1989): 'Prebiotic organic matter from comets and asteroids.' *Nature* **342**, 255–257.
16. Anders, E., Ebihara, M. (1982): 'Solar System abundances of the elements.' *Geochim. Cosmochim. Acta* **46**, 2363–2380.
17. Audouze, J. (1977): 'The Importance of CNO Isotopes in Astrophysics.' In *CNO Isotopes in Astrophysics,* Ed. J. Audouze, D. Reidel Publishing Co.; Dordrecht, Holland, p. 3–11.
18. Axford, W. I. (1964): 'The interaction of solar wind with comets.' *Planet. Space Sci.* **12**, 719–720.
19. Axford, W. I. (1972): 'The interaction of the solar with the interstellar medium.' *Solar Wind II,* NASA SP-308, Eds. C. P. Sonet, P. J. Coleman Jr., J. M. Wilcox, p. 609–650.
20. Axford, W. I., Leer, E., Skadron, G. (1977): 'The acceleration of cosmic rays by shock waves.' *Proc. 15th Int. Conf. Cosmic Rays.,* **11**, 132–135.
21. Bahcall, J. N. (1984): 'Self-consistent determinations of the total amount of matter near the Sun.' *Astrophys. J.* **276**, 169–181.
22. Bailey, M. E. (1983): 'The structure and evolution of the Solar System comet cloud.' *Mon. Not. Roy. Astron. Soc.* **204**, 603–633.
23. Bailey, M. E. (1984): 'The steady-state 1/a-distribution and the problem of cometary fading.' *Mon. Not. Roy. Astron. Soc.* **211**, 347–368.

24. Bailey, M. E. (1986a): 'The near-parabolic flux and the origin of short-period comets.' *Nature* **324**, 350–352.

25. Bailey, M. E. (1986b): 'The mean energy transfer rate to comets in the Oort cloud and implications for cometary origins.' *Mon. Not. Roy. Astron. Soc.* **218**, 1–30.

26. Bailey, M. E., Stagg, C. R. (1988): 'Cratering constraints on the inner Oort cloud: Steady-state models.' *Mon. Not. Roy. Astron. Soc.* **235**, 1–32.

27. Bailey, M. E., Clube, S. V. M., Napier, W. M. (1986): 'The origin of comets.' *Vistas Astron.* **29**, 53–112.

28. Baker, D. N., Feldman, W. C., Gary, S. P., McComas, D. J., Middleditch, J. (1986): 'Plasma fluctuations and large-scale mixing near Comet Giacobini-Zinner.' *Geophys. Res. Lett.* **13**, 271–274.

29. Balsiger, H. (1990): 'Measurements of ion species within the coma of Comet Halley from Giotto.' In *Comet Halley 1986: World-Wide Investigations, Results, and Interpretations*, Eds. J. Mason, P. Moore, Ellis Horwood Ltd.; Chichester, in press.

30. Balsiger, H., Altwegg, K., Bühler, F., Geiss, J., Ghielmetti, A. G., Goldstein, B. E., Goldstein, R., Huntress, W. T., Ip, W.-H., Lazarus, A. J., Meier, A., Neugebauer, M., Rettenmund, U., Rosenbauer, H., Schwenn, R., Sharp, R. D., Shelley, E. G., Ungstrup, E., Young, D. T. (1986): 'Ion composition and dynamics at Comet Halley.' *Nature* **321**, 330–334.

31. Balsiger, H., Altwegg, K., Bühler, F., Fuselier, S. A., Geiss, J., Goldstein, B. E., Goldstein, R., Huntress, W. T., Ip, W.-H., Lazarus, A. J., Meier, A., Neugebauer, M., Rettenmund, U., Rosenbauer, H., Schwenn, R., Shelley, E. G., Ungstrup, E., Young, D. T. (1987): 'The composition and dynamics of cometary ions in the outer coma of Comet P/Halley.' *Astron. Astrophys.* **187**, 163–168.

32. Bame, S. J., Anderson, R. C., Asbridge, J. R., Baker, D. N., Feldman, W. C., Fuselier, S. A., Gosling, J. T., McComas, D. J., Thomsen, M. F., Young, D. T., Zwickl, R. D. (1986): 'Comet Giacobini-Zinner: Plasma description.' *Science* **232**, 356–361.

33. Banks, P. M., Kockarts, G. (1973): *Aeronomy, Part A*. Academic Press, New York, London.

34. Barbosa, D. D. (1989): 'Stochastic acceleration of cometary pickup ions: The classic leaky box model.' *Astrophys. J.* **341**, 493–496.

35. Bar-Nun, A., Herman, G., Laufer, D., Rappaport, M. L. (1985): 'Trapping and release of gases by water ice and implications for icy bodies.' *Icarus* **63**, 317–332.

36. Bar-Nun, A., Lazcano-Araujo, A., Oró, J. (1981): 'Could life have evolved in cometary nuclei?' *Origin Life* **11**, 387–394.

37. Bar-Nun, A., Kleinfeld, I., Kochavi, E. (1988): 'Trapping of gas mixtures by amorphous water ice.' *Phys. Rev.* B **38**, 7749 –7754.

38. Baumgärtel, K., Sauer, K. (1987): 'Fluid simulation of Comet P/Halley's ionosphere.' *Astron. Astrophys.* **187**, 307–310.

39. Beisser, K., Boehnhardt, H. (1987): 'Evidence for the nucleus rotation in streamer patterns of Comet Halley's dust tail.' *Astrophys. Space Sci.* **139**, 5–12.

40. Bell, A. R. (1978): 'The acceleration of cosmic rays in shock fronts - I.' *Mon. Not. Roy. Astron. Soc.* **182**, 147–156.

41. Benz, A. O., Thejappa, G. (1988): 'Radio emission of coronal shock waves.' *Astron. Astrophys.* **202**, 267–274.

42. Biermann, L. (1951): 'Kometenschweife und solare Korpuskularstrahlung.' *Z. Astrophys.* **29**, 274–286.

43. Biermann, L., Michel, K. W. (1978): 'The origin of cometary nuclei in the presolar nebula.' *Moon Planets* **18**, 447–464.

44. Biermann, L., Trefftz, E. (1964): 'Über die Mechanismen der Ionisation und der Anregung in Kometenatmosphären.' *Z. Astrophys.* **59**, 1–28.

45. Biermann, L., Brosowski, B., Schmidt, H. U. (1967): 'The interaction of the solar wind with a comet.' *Solar Phys.* **1**, 254–283.

46. Biermann, L., Giguere, P. T., Huebner, W. F. (1982): 'A model of a comet coma with interstellar molecules in the nucleus.' *Astron. Asrophys.* **108**, 221–226.

47. Biermann, L., Huebner, W. F., Lüst, R. (1983): 'Aphelion clustering of "new" comets: Star tracks through Oort's cloud!' *Proc. Natl. Acad. Sci. USA* **80**, 5151–5155.

48. Biraud, F., Bourgois, G., Crovisier, J., Fillit, R., Gerard, E., Kazes, I. (1974): 'OH observations of comet Kohoutek (1973f) at 18 cm wavelength.' *Astron. Astrophys.* **34**, 163–166.
49. Bobrovnikoff, N. T. (1930): *Halley's comet in its apparition of 1909 – 1911*. University of California Press.
50. Boehnhardt, H., Fechtig, H. (1987): 'Electrostatic charging and fragmentation of dust near P/Giaco-bini-Zinner and P/Halley.' *Astron. Astrophys.* **187**, 824–828.
51. Boice, D. C., Huebner, W. F., Sablik, M. J., Konno, I. (1990): 'On the CH_4/CO ratio in Comet Halley.' *Geophys. Res. Lett.*, in press.
52. Bowell, E., Lume, K. (1979): 'Colorimetry and Magnitudes of Asteroids.' *Asteroids,* Ed. T. Gehrels, University of Arizona Press; Tucson, AZ, p. 132–169.
53. Bradley, J. P. (1988): 'Analysis of chondritic interplanetary dust thin-sections.' *Geochim. Cos-mochim. Acta* **52**, 889–900.
54. Brandt, J. C., Niedner, M. B. (1987): 'Plasma structures in Comets P/Halley and Giacobini-Zinner.' *Astron. Astrophys.* **187**, 281–286.
55. Brandt, J. C., Niedner, M. B., Rahe, J. (1982): 'A worldwide photographic network for wide-field observations of Halley's comet in 1985-1986.' In *Proc. Intern. Conference on Cometary Explo-ration*, **3**, Ed. T. Gombosi, pp.47-52, Central Research Institute for Physics, Hungarian Academy of Sciences.
56. Bredichin, T. (1903): *Mechanische Untersuchungen über Cometenformen in systematischer Darstel-lung*. Ed. R. Jaegermann, St. Petersburg, Voss' Sortiment (G. Hassel); Leipzig, p. 83–97.
57. Bregman, J. D., Campins, H., Witteborn, F. C., Wooden, D. H., Rank, D. M., Allamandola, L. J., Cohen, M., Thielens, A. G. GM. (1987): 'Airborne and groundbased spectrophotometry of Comet P/Halley from 5–13 micrometers.' *Astron. Astrophys.* **187**, 616–620.
58. Breneman, H. H., Stone, E. C. (1985): 'Solar coronal and photospheric abundances from solar energetic particle measurements.' *Astrophys. J.* **299**, L57–L61.
59. Brin, G. D., Mendis, D. A. (1979): 'Dust release and mantle development in comets.' *Astrophys. J.* **229**, 402–408.
60. Brinca, A. L., Tsurutani, B. T. (1987): 'Unusual characteristics of electromagnetic waves excited by cometary newborn ions with large perpendicular energies.' *Astron. Astrophys.* **187**, 311–319.
61. Brownlee, D. E. (1978): 'Microparticle studies by sampling techniques.' In *Cosmic Dust*, Ed. J. A. M. McDonnell, Wiley-Interscience; New York, p. 295–336.
62. Brownlee, D. E. (1979): 'Interplanetary dust.' *Rev. Geophys. Space Phys.* **17**, 1735–1743.
63. Brownlee, D. E., Wheelock, M. M., Temple, S., Bradley, J. P., Kissel, J. (1987): 'A quantitative comparison of Comet Halley and carbonaceous chondrites at the submicron level.' *Lunar Planet. Sci.* **XVIII**, 133–134.
64. Bryant, D. A., Bingham, R., Hall, D. S. (1985): 'Acceleration of auroral electrons by lower-hybrid waves.' *Proc. 7th ESA Symp. on European Rocket and Balloon Programmes and Related Research*, Eds. E. Rolfe and B. Battrick, ESA SP-229, p. 99–102.
65. Burns, J. A., Tedesco, E. F. (1979): 'Asteroidal light curves: Results for rotation and shapes.' In *Asteroids,* Ed. T. Gehrels, University Arizona Press; Tucson, AZ, p. 494–527.
66. Burns, J. A., Lamy, P. L., Soter, S. (1979): 'Radiation forces on small particles in the solar system.' *Icarus* **40**, 1–48.
67. Buti, B., Lakhina, G. S. (1987): 'Stochastic acceleration of cometary ions by lower hybrid waves.' *Geophys. Res. Lett.* **14**, 107–110.
68. Cabot, W., Canuto, V. M., Hubickyj, O., Pollack, J. B. (1987a): 'The role of turbulent convection in the primitive solar nebula. I. Theory.' *Icarus* **69**, 387–422.
69. Cabot, W., Canuto, V. M., Hubickyj, O., Pollack, J. B. (1987b): 'The role of turbulent convection in the primitive solar nebula. II. Results.' *Icarus* **69**, 423–457.
70. Cameron, A. G. W. (1962): 'The formation of the Sun and the planets.' *Icarus* **1**, 13–69.
71. Cameron, A. G. W. (1973): 'Accumulation processes in the primitive solar nebula.' *Icarus* **18**, 407–449.
72. Cameron, A. G. W. (1978): 'Physics of the primitive solar accretion disk.' *Moon Planets* **18**, 5–40.
73. Campins, H., A'Hearn, M. F., McFadden, L.-A. (1987): 'The bare nucleus of Comet Neujman 1.' *Astrophys. J.* **316**, 847–857.

74. Carusi, A., Kresák, L., Perozzi, E., Valsecchi, G. B. (1985a): *Long-term evolution of short-period comets*. Adam Hilger, Bristol.

75. Carusi, A., Kresák, L., Perozzi, E., Valsecchi, G. B. (1985b): 'First results of the integration of motion of short-period comets over 800 years.' In *Dynamics of Comets: Their Origin and Evolution* 115, Eds. A. Carusi, G. B. Valsecchi, D. Reidel Publishing Co.; Dordrecht, Holland, p. 319–340.

76. Carusi, A., Kresák, L., Perozzi, E., Valsecchi, G. B. (1987): 'Long-term resonances and orbital evolutions of Halley-type comets.' In *Interplanetary Matter*, Eds. Z. Ceplecha, P. Pecina, Publ. Astron. Inst. Czechosl. Acad. Sci., No. 67, p. 29–32.

77. Celnik, W. E., Schmidt-Kaler, T. (1987): 'Structure and dynamics of plasma-tail condensations of Comet P/Halley 1986 and inferences on the structure and activity of the cometary nucleus.' *Astron. Astrophys.* 187, 233–248.

78. Ceplecha, Z. (1976): 'Fireballs as an atmospheric source of meteoritic dust.' In *Lecture Notes in Physics* 48, Springer-Verlag, p. 385-388.

79. Ceplecha, Z. (1977): 'Meteoroid populations and orbits.' In *Comets, Asteroids, Meteorites*, Ed. A. H. Delsemme, University of Toledo; Toledo, OH, p. 143–152.

80. Ceplecha, Z., McCrosky, R. E. (1976): 'Fireball end height: A diagnostic for the structure of meteoritic material.' *J. Geophys. Res.* 81, 6257–6275.

81. Chang, S., Des Marais, D., Mack, R., Miller, S. L., Strathearn, G. E. (1983): 'Prebiotic organic synthesis and the origin of life.' In *Earth's Earliest Biosphere: It's Origin and Evolution*. Ed. J. W. Schopf, Princeton University Press; Princeton, NJ, p. 53–92.

82. Chyba, C. F. (1987): 'The cometary contribution to the oceans of primitive Earth.' *Nature* 330, 632–635.

83. Clark, B. C. (1988): 'Primeval procreative comet pond.' *Origin Life* 18, 209–238.

84. Clark, B., Mason, L. W., Kissel, J. (1986): 'Systematics of the "CHON" and other light-element particle populations in Comet Halley.' In *20th ESLAB Symposium on the Exploration of Halley's Comet*, Eds. B. Battrick, E. J. Rolfe, R. Reinhard, ESA SP-250 III, p. 353–358.

85. Clark, B. C., Mason, L. W., Kissel, J. (1987): 'Systematics of the "CHON" and other light-element particle populations in Comet P/Halley.' *Astron. Astrophys.* 187, 779–784.

86. Clube, S. V. M., Napier, W. M. (1982): 'Spiral arms, comets and terrestrial catastrophism.' *Quart. J. Roy. Astron. Soc.* 23, 45–66.

87. Coates, A. J., Johnstone, A. D., Thomsen, M. F., Formisano, V., Amata, E., Wilken, B., Jockers, K., Winningham, J. D., Borg, H., Bryant, D. A. (1987): 'Solar wind flows through the Comet P/Halley bow shock.' *Astron. Astrophys.* 187, 55–60.

88. Cochran, A. L., Barker, E. S., Cochran, W. D. (1980): 'Spectrophotometric observations of P/Schwassmann-Wachmann 1 during outburst.' *Astron. J.* 85, 474–477.

89. Cochran, A. L., Cochran, W. D., Barker, E. S. (1982): 'Spectrophotometry of Comet Schwassmann-Wachmann 1. II. Its color and CO^+ emission.' *Astrophys. J.* 254, 816–822.

90. Combes, M., Moroz, V. I., Crifo, J. F., Bibring, J. P., Coron, N., Crovisier, J., Encrenaz, T., Sanko, N., Grigoryev, A., Bockelée-Morvan, D., Gispert, R., Emerich, C., Lamarre, J. M., Rocard, F., Krasnoplosky, V., Owen, T. (1986): 'Detection of parent molecules in comet Halley from the IKS-Vega experiment.' In *20th ESLAB Symposium on the Exploration of Halley's Comet*, Eds. B. Battrick, E. J. Rolfe, R. Reinhard, ESA SP-250 I, p. 353–358.

91. Combes, M., Moroz, V. I., Crovisier, J., Encrenaz, T., Bibring, J.-P., Grigoriev, A. V., Sanko, N. F., Coron, N., Crifo, J. F., Gispert, R., Bockelée-Morvan, D., Nikolsky, Y. V., Krasnopolsky, V. A., Owen, T., Emerich, C., Lamarre, J. M., Rocard, F. (1988): 'The 2.5 to 12 μm spectrum of Comet Halley from the IKS-Vega experiment.' *Icarus* 76, 404–436.

92. Combi, M. R. (1989): 'The outflow speed of the coma of Halley's comet.' *Icarus*, 81, 41–50.

93. Combi, M. R., Delsemme, A. H. (1980): 'Neutral cometary atmospheres. I. An average random walk model for photodissociation in comets.' *Astrophys. J.* 237, 633–640.

94. Combi, M. R., Smyth, W. H. (1988a): 'Monte Carlo particle-trajectory models for neutral cometary gases. I. Models and equations.' *Astrophys. J.* 327, 1026–1043.

95. Combi, M. R., Smyth, W. H. (1988b): 'Monte Carlo particle-trajectory models for neutral cometary gases. II. The spatial morphology of the Lyman-alpha coma.' *Astrophys. J.* 327, 1044–1059.

96. Combi, M. R., Stewart, A. I. F. Smyth, W. H. (1986): 'Pioneer Venus Lyman-alpha observations of comet P/Giacobin-Zinner and the life expectancy of cometary hydrogen.' *Geophys. Res. Lett.* **13**, 385–388.

97. Cooper, D. M., Nicholls, R. W. (1975): 'Measurements of the electronic transition moments of C_2-band systems.' *J. Quant. Spectrosc. Rad. Transfer* **15**, 139–150.

98. Coplan, M. A., Ogilvie, K. W., A'Hearn, M. F., Bochsler, P., Geiss, J. (1987): 'Ion composition and upstream solar wind observations at comet Giacobini-Zinner.' *J. Geophys. Res.* **92**, 39–46.

99. Coroniti, F. V., Kennel, C. F., Scarf, F. L., Smith, E. J., Tsurutani, B. T., Bame, S. J., Thomsen, M. F., Hynds, R., Wenzel, K. P. (1986): 'Plasma wave turbulence in the strong coupling region at Comet Giacobini-Zinner.' *Geophys. Res. Lett.* **13**, 869–872.

100. Cosmovici, C. B., Ortolani, S. (1984): 'Detection of new molecules in the visible spectrum of comet IRAS-Araki-Alcock (1983d).' *Nature* **310**, 122–124.

101. Cosmovici, C. B., Green, S. F., Hughes, D. W., Keller, H. U., Mack, P., Moreno-Insertis, F., Schmidt, H. U. (1986): 'Groundbased CCD-observations of Comet Halley with the Giotto-HMC-Filters.' In *20th ESLAB Symposium on the Exploration of Halley's Comet*, Eds. B. Battrick, E. J. Rolfe, R. Reinhard, ESA SP-250 II, p. 375–379.

102. Cowan, J. J., A'Hearn, M. F. (1982): 'Vaporization in comets; Outbursts from Comet Schwassmann-Wachmann 1.' *Icarus* **50**, 53–62.

103. Cravens, T. E. (1986): 'The physics of the cometary contact surface.' In *20th ESLAB Symposium on the Exploration of Halley's Comet*, Eds. B. Battrick, E. J. Rolfe, R. Reinhard, ESA SP-250 I, p. 241–246.

104. Cravens, T. E. (1987): 'Ion energetics in the inner coma of Comet Halley.' *Geophys. Res. Lett.* **14**, 983–986.

105. Cravens, T. E. (1990): 'A magnetohydrodynamical model of the inner coma of Comet Halley.' *J. Geophys. Res.*, in press.

106. Cravens, T. E., Körösmezey, A. (1986): 'Vibrational and rotational cooling of electrons by water vapor.' *Planet. Space Sci.* **34**, 961–970.

107. Crifo, J. F. (1987): 'Optical and hydrodynamic implications of Comet Halley dust size distribution.' In *20th ESLAB Symposium on the Exploration of Halley's Comet*, Eds. B. Battrick, E. J. Rolfe, R. Reinhard, ESA SP-250 III, p. 399–408.

108. Crovisier, J. (1989): 'On the photodissociation of water in cometary atmospheres.' *Astron. Astrophys.* **213**, 459–464.

109. Cruikshank, D. P., Hartmann, W. K., Tholen, D. J. (1985): 'Colour, albedo and nucleus size of Halley's comet.' *Nature* **315**, 122–124.

110. Daly, P. W., Sanderson, T. R., Wenzel, K.-P., Cowley, S. W. H., Hynds, R. J., Smith, E. J. (1986): 'Gyroradius effects on the energetic ions in the tail lobes of Comet P/Giacobini-Zinner.' *Geophys. Res. Lett.* **13**, 419–422.

111. Daniels, P. A., Hughes, D. W. (1981): 'The accretion of cosmic dust – a computer symulation.' *Mon. Not. Roy. Astron. Soc.* **195**, 1001–1009.

112. Danks, A. C., Lambert, D. L., Arpigny, C. (1975): 'The $^{12}C/^{13}C$ ratio in comet Kohoutek.' In *Comet Kohoutek*. Ed. G. A. Gary, NASA SP-355, p. 137–143.

113. Davies, J. K., Green, S. F., Stewart, B. C., Meadows, A. J., Aumann, H. H. (1984): 'The IRAS fast-moving object search.' *Nature* **309**, 315–319.

114. Delamere, W. A., Reitsema, H. J., Huebner, W. F., Schmidt, H. U., Keller, H. U., Schmidt, W. K. H., Wilhelm, K., Whipple, F. L. (1986): 'Radiometric observations of the nucleus of Comet Halley.' In *20th ESLAB Symposium on the Exploration of Halley's Comet*, Eds. B. Battrick, E. J. Rolfe, R. Reinhard, ESA SP-250 II, p. 355–357.

115. Delsemme, A. H. (1973): 'Gas and dust in comets.' *Space Sci. Rev.* **15**, 89–101.

116. Delsemme, A. H. (1975): 'Physical interpretation of the brightness variation of Comet Kohoutek.' In *Comet Kohoutek*, Ed. G. A. Gary, NASA SP-355, p. 195–203.

117. Delsemme, A. H. (1981): 'Nature and origin of organic molecules in comets.' In *Comets and the Origin of Life*, Ed. C. Ponnamperuma, D. Reidel Publishing Co.; Dordrecht, Holland, p. 33–42.

118. Delsemme, A. H. (1982): 'Chemical composition of cometary nuclei.' In *Comets*, Ed. L. L. Wilkening, University of Arizona Press; Tucson, AZ, p. 85–130.

119. Delsemme, A. H. (1988): 'The chemistry of comets.' *Phil. Trans. Roy. Soc. London* A 325, 509–523.

120. Delsemme, A. H. (1990): 'Nature and history of the organic compounds in comets: An astrophysical view.' In *Comets in the Post-Halley Era*, Eds. R. L. Newburn, J. Rahe, M. Neugebauer, in press.

121. Delsemme, A. H., Swings, P. (1952): 'Hydrates de gaz dans les noyaux cométaires et les grains interstellaires.' *Ann. d'Astrophys.* 15, 1–6.

122. Delsemme, A. H., Miller, D. C. (1970): 'Physico-chemical phenomena in comets - II. Gas adsorption in the snows of the nucleus.' *Planet. Space Sci.* 18, 717–730.

123. Delsemme, A. H., Miller, D. C. (1971): 'Physico-chemical phenomena in comets - III. The continuum of comet Burnham (1960 II).' *Planet. Space Sci.* 19, 1229–1257.

124. Delsemme, A. H., Rud, D. A. (1973): 'Albedos and cross-sections for the nuclei of Comets 1969 IX, 1970 II and 1971 I.' *Astron. Astrophys.* 28, 1–6.

125. De Pater, I., Palmer, P., Snyder, L. E., Ip, W.-H. (1986): 'VLA observations of comet Halley: The brightness distribution of OH around the comet.' In *20th ESLAB Symposium on the Exploration of Halley's Comet*, Eds. B. Battrick, E. J. Rolfe, R. Reinhard, ESA SP-250 I, p. 409–412.

126. Divine, N., Fechtig, H., Gombosi, T. I., Hanner, M. S., Keller, H. U., Larson, S. M., Mendis, D. A., Newburn Jr., R. L., Reinhard, R., Sekanina, Z., Yeomans, D. K. (1986): 'The Comet Halley dust and gas environment.' *Space Sci. Rev.* 43, 1–104.

127. Dobrovolsky, O. V., Ibadinov, K. I., Aliev, S., Gerasimenko, S. I. (1986): 'Thermal regime and surface structure of periodic comet nuclei.' In *20th ESLAB Symposium on the Exploration of Halley's Comet*, Eds. B. Battrick, E. J. Rolfe, R. Reinhard, ESA SP-250 II, p. 389–394.

128. Dolginov, A. Z., Gnedin, Y. N. (1966): 'A theory of the atmospher of a comet.' *Icarus* 5, 64–74.

129. Dollfus, A., Crussaire, D., Killinger, R. (1986): 'Comet Halley: Dust characterization by photopolarimetry.' In *20th ESLAB Symposium on the Exploration of Halley's Comet*, Eds. B. Battrick, E. J. Rolfe, R. Reinhard, ESA SP-250 II, p. 41–45.

130. Donn, B. (1963): 'The origin and structure of icy cometary nuclei.' *Icarus* 2, 396–402.

131. Donn, B. (1977): 'A comparison of the composition of new and evolved comets.' In *Comets, Asteroids, Meteorites*, Ed. A. H. Delsemme, University of Toledo; Toledo, OH, p. 15–23.

132. Donn, B. (1981): 'Comet nucleus: Some characteristics and a hypothesis on origin and structure.' In *Comets and the Origin of Life*, Ed. C. Ponnamperuma, D. Reidel Publishing Co.; Dordrecht, Holland, p. 21–29.

133. Donn, B., Hughes, D. W. (1986): 'A fractal model of a cometary nucleus formed by random accretion.' In *20th ESLAB Symposium on the Exploration of Halley's Comet*, Eds. B. Battrick, E. J. Rolfe, R. Reinhard, ESA SP-250 III, p. 523–524.

134. Donn, B., Rahe, J. (1982): 'Structure and origin of cometary nuclei.' In *Comets*, Ed. L. L. Wilkening, University of Arizona Press; Tucson, AZ, p. 203–226.

135. Donn, B., Rahe, J., Brandt, J. C. (1986): *Atlas of Comet Halley 1910 II*. NASA SP-488, U. S. Gov. Printing Office; Washington, D. C.

136. Donnison, J. R. (1986): 'The distribution of cometary magnitudes.' *Astron. Astrophys.* 167, 359–363.

137. Draine, B. T. (1985): 'Tabulated optical properties of graphite and silicate grains.' *Astrophys. J. Suppl.* 57, 587–594.

138. Drobyshevskij, E. M. (1981): 'The history of Titan, of Saturn's rings and magnetic field, and the nature of short-period comets.' *Moon Planets* 24, 13–45.

139. Dum, C. T., Chodura, R., Biskamp, D. (1974): 'Turbulent heating and quenching of the ion sound instability.' *Phys. Rev. Lett.* 32, 1231–1234.

140. Duncan, M., Quinn, T., Tremaine, S. (1987): 'The formation and extent of the Solar System comet cloud.' *Astron. J.* 94, 1330–1349.

141. Duncan, M., Quinn, T., Tremaine, S. (1988): 'The origin of short-period comets.' *Astrophys. J.* 328, L69–L73.

142. Duncan, M., Quinn, T., Tremaine, S. (1989): 'The long-term evolution of orbits in the solar system: A mapping approach.' *Icarus* 82, 402–418.

143. Dyson, F. J. (1985): *Origins of Life*. Cambridge University Press; Cambridge.

144. Eaton, N., Davies, J. K., Green, S. F. (1984): 'The anomalous dust tail of Comet P/Tempel 2.' *Mon. Not. Roy. Astron. Soc.* 211, 15–19.

145. Eberhardt, P., Dolder, U., Schulte, W., Krankowsky, D., Lämmerzahl, P., Hoffman, J. H., Hodges, R. R., Berthelier, J. J., Illiano, J. M. (1987a): 'The D/H ratio in water from comet P/Halley.' *Astron. Astrophys.* **187**, 435–437.

146. Eberhardt, P., Krankowsky, D., Schulte, W., Dolder, U., Lämmerzahl, P., Berthelier, J. J., Woweries, J., Stubbeman, U., Hodges, R. R., Hoffman, J. H., Illiano, J. M. (1987b): 'The CO and N_2 abundance in Comet P/Halley.' *Astron. Astrophys.* **187**, 481–484.

147. Eddington, A. S. (1910): 'The envelopes of Comet Morehouse (1908c).' *Mon. Not. Roy. Astron. Soc.* **70**, 442–458.

148. Edenhofer, P., Buschert, H., Bird, M. K., Volland, H., Porsche, H. Brenkle, J. P., Kursinsky, E. R., Mottinger, N. A., Stelzried, C. T. (1986): 'Dust distribution of Comet Halley from the Giotto radio science experiment.' In *20th ESLAB Symposium on the Exploration of Halley's Comet*, Eds. B. Battrick, E. J. Rolfe, R. Reinhard, ESA SP-250 II, p. 215–218.

149. Edenhofer, P., Bird, M. K., Brenkle, J. P., Buschert, H., Kursinski, E. R., Mottinger, N. A., Porsche, H., Stelzried, C. T., Volland, H. (1987): 'Dust distribution of Comet P/Halley's inner coma determined from the Giotto radio-science experiment.' *Astron. Astrophys.* **187**, 712–718.

150. Emerich, C., Lamarre, J. M., Moroz, V. I., Combes, M., Sanko, N. F., Nikolsky, Y. V., Rocard, F., Gispert, R., Coron, N., Bibring, J. P., Encrenaz, T., Crovisier, J. (1987): 'Temperature and size of the nucleus of Comet P/Halley deduced from IKS infrared Vega-1 measurements.' *Astron. Astrophys.* **187**, 839–842.

151. Encrenaz, T. (1986): 'Search for organic molecules in the outer solar system.' *Adv. Space Res.* **6**, 237–246.

152. Ershkovich, A. I., Flammer, K. R. (1988): 'Nonlinear stability of the dayside cometary ionopause.' *Astrophys. J.* **328**, 967–973.

153. Ershkovich, A. I., Mendis, D. A. (1983): 'On the penetration of the solar wind into the cometary ionosphere.' *Astrophys. J.* **269**, 743–750.

154. Ershkovich, A. I., Flammer, K. R., Mendis, D. A. (1986): 'Stability of the sunlit cometary ionopause.' *Astrophys. J.* **311**, 1031–1042.

155. Ershkovich, A. I., McKenzie, J. F., Axford, W. I. (1989): 'Stability of a cometary ionosphere/ionopause determined by ion – neutral friction.' *Astrophys. J.* **344**, 932–939.

156. Everhart, E. (1972): 'The origin of short-period comets.' *Astrophys. Lett.* **10**, 131–135.

157. Everhart, E. (1977): 'The evolution of comet orbits as perturbed by Uranus and Neptune.' In *Comets, Asteroids, Meteorites*, Ed. A. H. Delsemme, University of Toledo; Toledo, OH, p. 99-104.

158. Everhart, E., Marsden, B. G. (1983): 'New original and future cometary orbits.' *Astron. J.* **88**, 135–137.

159. Everhart, E., Marsden, B. G. (1987): 'Original and future cometary orbits. III.' *Astron. J.* **93**, 753–754.

160. Everhart, E., Raghavan, N. (1970): 'Changes in total energy for 392 long-period comets, 1800-1970.' *Astron. J.* **75**, 258–272.

161. Eviatar, A., Goldstein, R., Young, D. T., Balsiger, H., Rosenbauer, H., Fuselier, S. A. (1989): 'Energetic ion fluxes in the inner coma of Comet P/Halley.' *Astrophys. J.* **339**, 545–557.

162. Falchi, A., Gagliardi, L., Palagi, F., Tofani, G., Comoretto, G. (1987): '10.7 GHz continuum observations of Comet P/Halley.' *Astron. Astrophys.* **187**, 462–464.

163. Fanale, F. P., Salvail, J. R. (1984): 'An idealized short-period comet model: Surface insolation, H_2O flux, dust flux, and mantle evolution.' *Icarus* **60**, 476–511.

164. Fanale, F. P., Salvail, J. R. (1986): 'A model of cometary gas and dust production and nongravitational force with application to P/Halley.' *Icarus* **66**, 154–164.

165. Fay Jr., T. D., Wisniewski, W. (1978): 'The light curve of the nucleus of Comet d'Arrest.' *Icarus* **34**, 1–9.

166. Feldman, P. D. (1978): 'A model of carbon production in a cometary coma.' *Astron. Astrophys.* **70**, 547–553.

167. Feldman, P. D. (1982): 'Ultraviolet spectroscopy of comae.' In *Comets*, Ed. L. L. Wilkening, University of Arizona Press; Tucson, AZ, p. 461–479.

168. Feldman, P. D. (1990): 'Ultraviolet spectroscopy of cometary comae.' In *Comets in the Post-Halley Era*, Eds. R. L. Newburn, J. Rahe, M. Neugebauer, in press.

169. Feldman, P. D., Brune, W. H. (1976): 'Carbon production in Comet West 1975n.' *Astrophys. J.* **209**, L45–L48.

170. Feldman, P. D., Weaver, H. A., Festou, M. C., A'Hearn, M. F., Jackson, W. M., Donn, B., Rahe, J., Smith, A. M., Benvenuti, P. (1980): 'IUE observations of the UV spectrum of Comet Bradfield.' *Nature* **286**, 132–135.

171. Feldman, P. D., A'Hearn, M. F., Schleicher, D. G., Festou, M. C., Wallis, M. K., Burton, W. M., Hughes, D. W., Keller, H. U., Benvenuti, P. (1984): 'Evolution of the ultraviolet coma of Comet Austin (1982g).' *Astron. Astrophys.* **131**, 394–398.

172. Feldman, P. D., Festou, M. C., A'Hearn, M. F., Arpigny, C., Butterworth, P. S., Cosmovici, C. B., Danks, A. C., Gilmozzi, R., Jackson, W. M., McFadden, L. A., Patriarchi, P., Schleicher, D. G., Tozzi, G. P., Wallis, M. K., Weaver, H. A., Woods, T. N. (1986): 'IUE Observations of Comet Halley: Evolution of the UV Spectrum between September 1985 and July 1986.' In *20th ESLAB Symposium on the Exploration of Halley's Comet,* Eds. B. Battrick, E. J. Rolfe, R. Reinhard, ESA SP-250 I, p. 325–328.

173. Feldman, P. D., Festou, M. C., A'Hearn, M. F., Arpigny, C., Butterworth, P. S., Cosmovici, C. B., Danks, A. C., Gilmozzi, R., Jackson, W. M., McFadden, L. A., Patriarchi, P., Schleicher, D. G., Tozzi, G. P., Wallis, M. K., Weaver, H. A., Woods, T. N. (1987): 'IUE observations of Comet P/Halley: Evolution of the ultraviolet spectrum between September 1985 and July 1986.' *Astron. Astrophys.* **187**, 325–328.

174. Fernández, J. A. (1980): 'On the existence of a comet belt beyond Neptune.' *Mon. Not. Roy. Astron. Soc.* **192**, 481–491.

175. Fernández, J. A., Ip, W.-H. (1983a): 'Dynamical origin of the short-period comets.' In *Asteroids, Comets, Meteors,* Eds. C.-I. Lagerkvist, H. Rickman, University of Uppsala; Uppsala, p. 387–390.

176. Fernández, J. A., Ip, W.-H. (1983b): 'On the time evolution of the cometary influx in the region of the terrestrial planets.' *Icarus* **54**, 377–387.

177. Fernández, J. A. (1985): 'Dynamical capture and physical decay of short-period comets.' *Icarus* **64**, 308–319.

178. Fernández, J. A. (1990): 'Collisions of comets with meteoroids.' In *Asteroids, Comets, Meteors III,* Eds. C.-I. Lagerkvist, H. Rickman, B. A. Lindblad, M. Lindgren, Uppsala University; Uppsala, p. 309–312.

179. Fernández, J. A., Ip, W.-H. (1987): 'Time-dependent injection of Oort cloud comets into Earth-crossing orbits.' *Icarus* **71**, 46–56.

180. Fertig, J., Schwehm, G. (1984): 'Dust environment models for Comet P/Halley: Support for targeting of the Giotto s/c.' *Adv. Space Res.* **4**, 213–216.

181. Festou, M. C. (1981a): 'The density distribution of neutral compounds in cometary atmospheres. I. Models and equations.' *Astron. Astrophys.* **95**, 69–79.

182. Festou, M. C. (1981b): 'The density distribution of neutral compounds in cometary atmospheres. II. Production rate and lifetime of OH radicals in Comet Kobayashi-Berger-Milon (1975 IX).' *Astron. Astrophys.* **96**, 52–57.

183. Festou, M. C. (1984): 'Aeronomical processes in cometary atmospheres: The carbon compounds' puzzle.' *Adv. Space Res.* **4**, 165–175.

184. Festou, M. C., Feldman, P. D., A'Hearn, M. F., Arpigny, C., Cosmovici, C. B., Danks, A. C., McFadden, L. A., Gilmozzi, R., Patriarchi, P., Tozzi, G. P., Wallis, M. K., Weaver, H. A. (1986): 'IUE observations of Comet Halley during the Vega and Giotto encounters.' *Nature* **321**, 361–365.

185. Festou, M. C., Drossart, P., Lecacheux, J., Encrenaz, T., Puel, F., Kohl-Moreira, J. L. (1987): 'Periodicities in the light curve of P/Halley and the rotation of its nucleus.' *Astron. Astrophys.*, **187**, 575–580.

186. Finson, M. L., Probstein, R. F. (1968a): 'A theory of dust comets. I. Model and equations.' *Astrophys. J.* **154**, 327–352.

187. Finson, M. L., Probstein, R. F. (1968b): 'A theory of dust comets. II. Results for Comet Arend-Roland.' *Astrophys. J.* **154**, 353–380.

188. Fisk, L. A. (1976): 'On the acceleration of energetic particles in the interplanetary medium.' *J. Geophys. Res.* **81**, 4641–4645.

189. Formisano, V., Galeev, A. A., Sagdeev, R. Z. (1982): 'The role of the critical ionization velocity phenomena in the production of inner coma cometary plasma.' *Planet. Space Sci.* **30**, 491–497.

190. Formisano, V., Amata, E., Cattanes, M. B., Torrente, P., Johnstone, A. D., Coates, A., Wilken, B., Jockers, K., Thomsen, M. F., Winningham, D., Borg, H. (1990): 'Plasma flow inside Comet P/Halley.' *Astron. Astrophys.*, in press.

191. Fraundorf, P., Brownlee, D. E., Walker, R. M. (1982): 'Laboratory studies of interplanetary dust.' In *Comets*, Ed. L. L. Wilkening, University of Arizona Press; Tucson, AZ, p. 383–409.

192. Fulle, M. (1989): 'Evaluation of cometary dust parameters from numerical simulations: Comparison with analytical approach and role of anisotropic emissions.' *Astron. Astrophys.* **217**, 283–297.

193. Fuselier, S. A., Shelley, E. G., Balsiger, H., Geiss, J., Goldstein, B. E., Goldstein, R., Ip, W.-H. (1988): 'Cometary H_2^+ and solar wind He^{++} dynamics across the Halley cometopause.' *Geophys. Res. Lett.* **15**, 549–552.

194. Fuselier, S. A., Feldman, W. C., Bame, S. J., Smith, E. J., Scarf, F. L. (1986): 'Heat flux observations and the location of the transition region boundary of Giacobini-Zinner.' *Geophys. Res. Lett.* **13**, 247–250.

195. Galeev, A. A. (1987): 'Encounters with comets: Discoveries and puzzles in cometary plasma physics.' *Astron. Astrophys.* **187**, 12–20.

196. Galeev, A. A., Lipatov, A. S. (1984): 'Plasma processes in cometary atmospheres.' *Adv. Space Res.* **4**, 229–237.

197. Galeev, A. A., Cravens, T. E., Gombosi, T. I. (1985): 'Solar wind stagnation near comets.' *Astrophys. J.* **289**, 807–819.

198. Galeev, A. A., Griov, B. E., Gombosi, T., Gringauz, K. I., Klimov, S. I., Oberz, P., Remizov, A. P., Riedler, W., Sagdeev, R. Z., Savin, S. P., Sokolov, A. Y., Shapiro, V. D., Shevchenko, V. I., Szego, K., Verigin, M. I., Yeroshenko, Y. G. (1986): 'The position and structure of the Comet Halley bow shock: Vega-1 and Vega-2 measurements.' *Geophys. Res. Lett.* **13**, 841–844.

199. Galeev, A. A., Gringauz, K. I., Klimov, S. I., Remizov, A. P., Sagdeev, R. Z., Savin, S. P., Sokolov, A. Y., Verigin, M. I., Szegö, K., Tátrallyay, M., Grard, R., Yeroshenko, Y. G., Mogilevsky, M., Riedler, W., Schwingenschuh, K. (1988): 'Physical processes in the vicinity of the cometopause interpreted on the basis of plasma, magnetic field, and plasma wave data measured on board the Vega 2 spacecraft.' *J. Geophys. Res.* **93**, 7527–7531.

200. Gary, S. P., Smith, C. W., Lee, M. A., Goldstein, M. L., Fooslund, D. W. (1984): 'Electromagnetic ion beam instabilities.' *Phys. Fluids* **27**, 1852–1862.

201. Gary, S. P., Madland, C. D., Omidi, N., Winske, D. (1988): 'Computer simulations of two-pickup-ion instabilities in a cometary environment.' *J. Geophys. Res.* **93**, 9584–9586.

202. Geiss, J. (1988): 'Composition in Halley's Comet: Clues to origin and history of cometary matter' *Rev. Mod. Astron.* **1**, 1–27.

203. Gibson, D. M., Hobbs, R. W. (1981): 'On the microwave emission from comets.' *Astrophys. J.* **248**, 863–866.

204. Giclas, H. L. (1974): 'Photographs of Comet Arend-Roland.' *Sky and Telescope* **47**, 374–374.

205. Glassmeier, K. H., Neubauer, F. M., Acuña, M. H., Mariani, F. (1987): 'Low-frequency magnetic field fluctuations in Comet P/Halley's magnetosheath: Giotto observations.' *Astron. Astrophys.* **187**, 65–68.

206. Glassmeier, K. H., Coates, A. J., Acuña, M. H., Goldstein, M. L., Johnstone, A. D., Neubauer, F. M., Rème, H. (1988): 'Spectral characteristics of low-frequency plasma turbulence upstream of Comet P/Halley.' *J. Geophys. Res.* **94**, 37–48.

207. Gleeson, L. J., Axford, W. I. (1967): 'Cosmic rays in the interplanetary medium.' *Astrophys. J.* **149**, L115–L118.

208. Gol'dyanskii, V. I. (1977): 'Formaldehyde polymer formation mechanism in interstellar space.' *Sov. Phys. Dokl.* **22**, 417–419.

209. Goldreich, P., Ward, W. R. (1973): 'The formation of planetesimals.' *Astrophys. J.* **183**, 1051–1061.

210. Goldstein, B. E., Altwegg, K., Balsiger, H., Fuselier, S. A., Ip, W.-H., Meier, A., Neugebauer, M., Rosenbauer, H., Schwenn, R. (1989): 'Observations of a shock and a recombination layer at the contact surface of Comet Halley.' *J. Geophys. Res.* **94**, 17251–17257.

211. Goldstein, M. L., Wong, H. K. (1987): 'A theory for low-frequency waves observed at Comet Giacobini-Zinner.' *J. Geophys. Res.* **92**, 4695–4700.

212. Goldstein, R. M., Jurgens, R. F., Sekanina, Z. (1984): 'A radar study of Comet Iras-Araki-Alcock 1983d.' *Astron. J.* **89**, 1745–1754.

213. Goldstein, R., Young, D. T., Balsiger, H., Buehler, F., Goldstein, B. E., Neugebauer, M., Rosenbauer, H., Schwenn, R., Shelley, E. G. (1987): 'Hot ions observed by the Giotto ion mass spectrometer at the Comet P/Halley contact surface.' *Astron. Astrophys.* **187**, 220–224.

214. Gombosi, T. I. (1986): 'A heuristic model of the Comet Halley dust size distribution.' In *20th ESLAB Symposium on the Exploration of Halley's Comet,* Eds. B. Battrick, E. J. Rolfe, R. Reinhard, ESA SP-250 II, p. 167–171.

215. Gombosi, T. I. (1987a): 'Charge exchange avalanche at the cometopause.' *Geophys. Res. Lett.* **14**, 1174–1177.

216. Gombosi, T. I. (1987b): 'Preshock region acceleration of implanted cometary H^+ and O^+.' *J. Geophys. Res.* **93**, 35–47.

217. Gombosi, T. I., Houpis, H. L. F. (1986): 'An icy-glue model of cometary nuclei.' *Nature* **324**, 43–46.

218. Gombosi, T. I., Szegö, K., Gribov, B. E., Sagdeev, R. Z., Shapiro, V. D., Shevchenko, V. I., Cravens, T. E. (1983): 'Gas dynamic calculations of dust terminal velocities with realistic dust size distributions.' In *Cometary Exploration,* Ed. T. I. Gombosi. Central Research Institute for Physics, Hungarian Academy of Sciences; Budapest, II, p. 99–111.

219. Gombosi, T. I., Cravens, T. E., Nagy, A. F. (1985): 'Time-dependent dusty gasdynamical flow near cometary nuclei.' *Astrophys. J.* **293**, 328–341.

220. Gombosi, T. I., Nagy, A. F., Cravens, T. E. (1986): 'Dust and neutral gas modeling of the inner atmospheres of comets.' *Rev. Geophys.* **24**, 667–700.

221. Gosling, J. T., Asbridge, J. R., Bame, S. J., Thomsen, M. F., Zwickl, R. D. (1986): 'Large amplitude, low frequency plasma fluctuations at Comet Giacobini-Zinner.' *Geophys. Res. Lett.* **13**, 267–270.

222. Grard, R., Pedersen, A., Trotignon, J.-G., Beghin, C., Mogilevsky, M., Mikhailov, Y., Molchanov, O., Formisano, V. (1986): 'Observations of waves and plasma in the environment of Comet Halley.' *Nature* **321**, 290–292.

223. Gredel, R., van Dishoeck, E. F., Black, J. H. (1989): 'Fluorescent vibration-rotation excitation of cometary C_2.' *Astrophys. J.* **338**, 1047–1070.

224. Greenberg, J. M. (1977): 'From dust to comets.' In *Comets, Asteroids, Meteorites,* Ed. A. H. Delsemme, University of Toledo; Toledo, OH, p. 491–497.

225. Greenberg, J. M. (1982): 'What are comets made of? A model based on interstellar dust.' In *Comets,* Ed. L. L. Wilkening, University of Arizona Press; Tucson, AZ, p. 131–163.

226. Greenberg, J. M. (1983): 'Interstellar dust, comets, comet dust, and carbonaceous meteorites.' In *Asteroids, Comets, Meteors,* Eds. C.-I. Lagerkvist, H. Rickman. University of Uppsala; Uppsala, p. 259–268.

227. Greenberg, J. M. (1986a): 'The chemical composition of comets and possible contributions to planet composition and evolution.' In *The Galaxy and the Solar System.* Ed. R. Smoluchowski, J. N. Bahcall, and M. S. Matthews, University of Arizona Press; Tucson, AZ, p. 103–115.

228. Greenberg, J. M. (1986b): 'Fluffy comets.' In *Asteroids, Comets, Meteors,* Eds. C.-I. Lagerkvist, H. Rickman, University of Uppsala; Uppsala, p. 221–223.

229. Greenberg, J. M. (1990): 'The evidence that comets are made of interstellar dust.' In *Comet Halley 1986: World-Wide Investigations, Results, and Interpretations,* Eds. J. Mason, P. Moore, Ellis Horwood Ltd.; Chichester, in press.

230. Greenberg, J. M., Hage, J. I. (1990): 'From interstellar dust to comets: A unification of observational constraints.' *Astrophys. J.,* in press.

231. Greenstein, J. L. (1958): 'High-resolution spectra of comet Mrkos (1957d).' *Astrophys. J.* **128**, 106–113.

232. Gribov, B. E., Keczkeméty, K., Sagdeev, R. Z., Shapiro, V. D., Shevchenko, V. I., Somogyi, A. J., Szegö, K., Erdös, G., Eroshenko, E. G., Gringauz, K. I., Keppler, E., Marsden, R., Remizov, A. P., Richter, A. K., Riedler, W., Schwingenschuh, K., Wenzel, K. P. (1986): 'Stochastic Fermi

acceleration of ions in the pre-shock region of Comet Halley.' In *20th ESLAB Symposium on the Exploration of Halley's Comet,* Eds. B. Battrick, E. J. Rolfe, R. Reinhard, ESA SP-250 I, p. 271–276.

233. Gringauz, K. I., Gombosi, T. I., Remizov, A. P., Apathy, I., Szemerey, I., Verigin, M. I., Denchikova, L. I., Dyachkov, A. V., Keppler, E., Klimenko, I. N., Richter, A. K., Somogyi, A. J., Szegö, K., Szendrö, S., Tátrallyay, M., Varga, A., Vladimirova, G. A. (1986a): 'First *in situ* plasma and neutral gas measurements at Comet Halley.' *Nature* **321,** 282–285.

234. Gringauz, K. I., Gombosi, T. I., Tátrallyay, M. Verigin, M. I., Remizov, A. P., Richter, A. K., Apáthy, I., Szemerey, I., Dyachkov, A. V., Balakina, O. V., Nagy, A. F. (1986b): 'Detection of a new "chemical" boundary at Comet Halley.' *Geophys. Res. Lett.* **13,** 613–616.

235. Gringauz, K. I., Verigin, M. I., Richter, A. K., Gombosi, T. I., Szegö, K., Tátrallyay, M., Remizov, A. P., Apáthy, I. (1987): 'Quasi-periodic features and the radial distribution of cometary ions in the cometary plasma region of Comet P/Halley.' *Astron. Astrophys.* **187,** 191–194.

236. Grün, E., Pailer, N., Fechtig, H., Kissel, J. (1980): 'Orbital and physical characteristics of micrometeoroids in the inner Solar System as observed by Helios 1.' *Planet. Space. Sci.* **28,** 333–349.

237. Grün, E., Massonne, L., Schwehm, G. (1987a): 'New properties of cometary dust.' In *Symposium on the Diversity and Similarity of Comets,* Eds. E. J. Rolfe, B. Battrick. ESA SP-278, p. 305–314.

238. Grün, E., Kochan, H., Roessler, K., Stöffler, D. (1987b): 'Simulation of cometary nuclei.' In *Symposium on the Diversity and Similarity of Comets,* Eds. E. J. Rolfe, B. Battrick. ESA SP-278, p. 501–508.

239. Grün, E., Benkhoff, J., Fechtig, H., Hesselbarth, P., Klinger, J., Kochan, H., Kohl, H., Krankowsky, D., Lämmerzahl, P., Seboldt, W., Spohn, T., Thiel, K. (1989): 'Mechanisms of dust emission from the surface of a cometary nucleus.' *Adv. Space Res.* **9,** 133–137.

240. Haerendel, G. (1986): 'Plasma flow and critical velocity ionization in cometary comae.' *Geophys. Res. Lett.* **13,** 255–258.

241. Haerendel, G. (1987): 'Plasma transport near the magnetic cavity surrounding Comet Halley.' *Geophys. Res. Lett.* **14,** 673–676.

242. Hahn, G., Rickman, H. (1985): 'Asteroids in cometary orbits.' *Icarus* **61,** 417–442.

243. Hajduk, A. (1987): 'Meteoroids from Comet P/Halley. The comet's mass production and age.' *Astron. Astrophys.* **187,** 925–927.

244. Haldane, J. B. S. (1928): 'The origin of life.' *Rationalist Ann.* **148,** 3–10.

245. Halliday, I. (1987): 'The spectra of meteors from Halley's comet.' *Astron. Astrophys.* **187,** 921–924.

246. Hanner, M. S. (1980): 'Physical characteristics of cometary dust from optical studies.' In *Solid particles in the Solar System,* Eds. I. Halliday, B. A. McIntosh. D. Reidel Publishing Co.; Dordrecht, Boston, London, p. 223–236.

247. Hanner, M. S. (1981): 'On the detectability of icy grains in the comae of comets.' *Icarus* **47,** 342–350.

248. Hanner, M. S. (1983): 'The nature of cometary dust from remote sensing.' In *Cometary Exploration,* Ed. T. I. Gombosi. Central Research Institute for Physics, Hungarian Academy of Sciences; Budapest. II, p. 1–22.

249. Hanner, M. S., Giese, R. H., Weiss, K., Zerull, R. (1981): 'On the definition of albedo and application to irregular particles.' *Astron. Astrophys.* **104,** 42–46.

250. Hanner, M. S., Aitken, D. K., Knacke, R., McCorkle, S., Roche, P. F., Tokunaga, A. T. (1985): 'Infrared spectrophotometry of Comet IRAS-Araki-Alcock (1983d): A bare nucleus revealed.' *Icarus* **62,** 97–109.

251. Hanner, M. S., Newburn, R. L., Spinrad, H., Veeder, G. J. (1987): 'Comet Sugano-Saigusa-Fujikawa (1938V) - a small, puzzling comet.' *Astron. J.* **94,** 1081–1087.

252. Harmon, J. K., Campbell, D. B., Hine, A. A., Shapiro, I. I., Marsden, B. G. (1989): 'Radar observations of Comet IRAS-Araki-Alcock 1983d.' *Astrophys. J.* **338,** 1071–1093.

253. Hartle, R. E., Wu, C. S. (1973): 'Effects of electrostatic instabilities on planetary and interstellar ions in the solar wind.' *J. Geophys. Res.* **78,** 5802–5807.

254. Hartmann, W. K., Cruikshank, D. P. (1984): 'Comet color changes with solar distance.' *Icarus* **57,** 55–62.

255. Hartmann, W. K., Cruikshank, D. P., Degewij, J. (1982): 'Remote comets and related bodies: VJHK colorimetry and surface materials.' *Icarus* **52,** 377–408.

256. Hartmann, W. K., Tholen, D. J., Cruikshank, D. P. (1987): 'The relationship of active comets, "extinct" comets, and dark asteroids.' *Icarus* **69**, 33–50.

257. Haser, L. (1957): 'Distribution d'intensité dans la tête d'une comète. *Bull. Acad. Roy. Belgique*, Classes des Sciences, Ser. 5, **43**, 740–750.

258. Haser, L. (1966): 'Calcul de distribution d'intensité relative dans une tête cométaire.' *Mèm. Soc. Roy. Sci. Liège*, Ser. 5, **12**, 233–241.

259. Havnes, O., Goertz, C. K., Morfill, G. E., Grün, E., Ip, W.-H. (1987): 'Dust charges, cloud potential, and instabilities in a dust cloud embedded in a plasma.' *J. Geophys. Res.* **92**, 2281–2287.

260. Hayashi, C. (1981): 'Structure of the solar nebula, growth and decay of magnetic fields and effects of magnetic and turbulent viscosities on the nebula.' *Prog. Theor. Phys. Suppl.* **70**, 35–53.

261. Heisler, J., Tremaine, S. (1986): 'The influence of the galactic tidal field on the Oort comet cloud.' *Icarus* **65**, 13–26.

262. Hellmich, R. (1981): 'The influence of the radiation transfer in cometary dust halos on the production rates of gas and dust.' *Astron. Astrophys.* **93**, 341–346.

263. Hellmich, R., Schwehm, G. H. (1983): 'Predictions of dust particle number flux and fluence rates for the ESA-Giotto and USSR Vega missions to Comet Halley: A comparison.' In *Cometary Exploration*, Ed. T. I. Gombosi. Central Research Institute for Physics, Hungarian Academy of Sciences; Budapest, III, p. 175–183.

264. Herman, G., Podolak, M. (1985): 'Numerical simulation of comet nuclei. I: Water ice comets.' *Icarus* **61**, 252–266.

265. Herman, G., Weissman, P. R. (1987): 'Numerical simulation of cometary nuclei. III. Internal temperatures of cometary nuclei.' *Icarus* **69**, 314–338.

266. Herter, T., Campins, H., Gull, G. E. (1987): 'Airborne spectrophotometry of P/Halley from 16 to 30 microns.' *Astron. Astrophys.* **187**, 629–631.

267. Hills, J. G. (1981): 'Comet showers and the steady-state infall of comets from the Oort cloud.' *Astron. J.* **86**, 1730–1740.

268. Hills, J. G. (1982): 'The formation of comets by radiation pressure in the outer protosun.' *Astron. J.* **87**, 906–910.

269. Hills, J. G., Sandford, M. T. (1983a): 'The formation of comets by radiation pressure in the outer protosun. II. Dependence on the radiation – coupling.' *Astron. J.* **88**, 1519–1521.

270. Hills, J. G., Sandford, M. T. (1983b): 'The formation of comets by radiation pressure in the outer protosun. III. Dependence on the anisotropy of the radiation field.' *Astron. J.* **88**, 1522–1530.

271. Hoban, S., Samarasinha, N. H., A'Hearn, M. F., Klinglesmith, D. A. (1988): 'An investigation into periodicities in the morphology of CN jets in comet P/Halley.' *Astron. Astrophys.* **195**, 331–337.

272. Hoover, R. B., Hoyle, F., Wickramasinghe, N. C., Hoover, M. J., Al-Mufti, S. (1986): 'Diatoms on earth, comets, Europa and in interstellar space.' *Earth, Moon, Planets* **35**, 19–45.

273. Horányi, M., Gombosi, T. I., Cravens, T. E., Körösmezey, A., Kecskemety, K., Nagy, A. F., Szeö, K. (1984): 'The friable sponge model of a cometary nucleus.' *Astrophys. J.* **278**, 449–455.

274. Houpis, H. L. F., Mendis, D. A. (1980): 'Physicochemical and dynamical processes in cometary ionospheres. I. The basic flow profile.' *Astrophys. J.* **239**, 1107–1118.

275. Houpis, H. L. F., Mendis, D. A. (1981): 'The nature of the solar wind interaction with CO_2/CO-dominated comets.' *Moon Planets* **25**, 95–104.

276. Houpis, H. L. F., Ip, W.-H., Mendis, D. A. (1985): 'The chemical differentiation of the cometary nucleus: The process and its consequences.' *Astrophys. J.* **295**, 654–667.

277. Hoyle, F., Wickramasinghe, C. (1981): 'Comets - a vehicle for panspermia.' In *Comets and the Origin of Life*, Ed. C. Ponnamperuma, D. Reidel Publishing Co.; Dordrecht, Holland, p. 227–239.

278. Hsieh, K. C., Curtis, C. C., Fan, C. Y., Hunten, D. M., Ip, W.-H., Keppler, E., Richter, A. K., Umlauft, G., Afonin, V. V., Erö Jr., J., Somogyi, A. J. (1987): 'Anisotropy of the neutral gas distribution of Comet P/Halley deduced from NGE/Vega 1 measurements.' *Astron. Astrophys.* **187**, 375–379.

279. Huebner, W. F. (1965): 'Über die Gasproduktion der Kometen.' *Z. Astrophys.* **63**, 22–34.

280. Huebner, W. F. (1967): 'Diminution of cometary heads due to perihelion passage.' *Z. Astrophys.* **65**, 185–193.

281. Huebner, W. F. (1970): 'Dust from cometary nuclei.' *Astron. Astrophys.* **5**, 286–297.

282. Huebner, W. F. (1987): 'First polymer in space identified in Comet Halley.' *Science* **237**, 628–630.
283. Huebner, W. F., Boice, D. C. (1989): 'Polymers in comet comae.' In *Solar System Plasma Physics*, Eds. J. H. Waite, Jr., J. L. Burch, R. L. Moore, AGU Monograph **54** p. 453–456.
284. Huebner, W. F., Giguere, P. T. (1980): 'A model of comet comae. II. Effects of solar photodissociative ionization.' *Astrophys. J.* **238**, 753–762.
285. Huebner, W. F., Keady, J. J. (1984): 'First-flight escape from spheres with R^{-2} density distribution.' *Astron. Astrophys.* **135**, 177–180.
286. Huebner, W. F., Weigert., A. (1966): 'Eiskörner in der Koma von Kometen.' *Z. Astrophys.* **64**, 185–201.
287. Huebner, W. F., Snyder, L. E., Buhl, D. (1974): 'HCN radio emission from comet Kohoutek (1973f).' *Icarus* **23**, 580–584.
288. Huebner, W. F., Delamere, W. A., Reitsema, H., Keller, H. U., Wilhelm, K., Whipple, F. L., Schmidt, H. U. (1986): 'Dust – gas interaction deduced from Halley Multicolour Camera observations.' In *20th ESLAB Symposium on the Exploration of Halley's Comet*, Eds. B. Battrick, E. J. Rolfe, R. Reinhard, ESA SP-250 II, p. 363-364.
289. Huebner, W. F., Boice, D. C., Reitsema, H. J., Delamere, W. A., Whipple, F. L. (1988): 'A model for intensity profiles of dust jets neat the nucleus of Comet Halley.' *Icarus* **76**, 78–88.
290. Hughes, D. W. (1985): 'The size, mass, mass loss, and age of Halley's comet.' *Mon. Not. Roy. Astron. Soc.* **213**, 103–109.
291. Hughes, D. W. (1987): 'P/Halley dust characteristics: A comparison between Orionid and Eta Aquarid meteor observations and those from the flyby spacecraft.' *Astron. Astrophys.* **187**, 879–888.
292. Hughes, D. W. (1988): 'Cometary magnitude distribution and the ratio between the numbers of long- and short-period comets.' *Icarus* **73**, 149–162.
293. Hut, P., Alvarez, W., Elder, W. P., Hansen, T., Kauffman, E. G., Keller, G., Shoemaker, E. M., Weissman, P. R. (1987): 'Comet showers as a cause of mass extinctions.' *Nature* **329**, 118–126.
294. Hynds, R. J., Cowley, S. W. H., Sanderson, T. R., Wenzel, K.-P., Van Rooijen, J. J. (1986): 'Observations of energetic ions from Comet Giacobini-Zinner.' *Science* **232**, 361–365.
295. Ibadinov, K. I., Aliev, S. A. (1987): 'Sublimation characteristics of H_2O comet nucleus with CO_2 impurities.' In *Symposium on the Diversity and Similarity of Comets*, Eds. E. J. Rolfe, B. Battrick. ESA SP-278, p. 717–719.
296. Ibadinov, K. I., Aliev, S. A., Rahmonov, A. A. (1987): 'Physical-mechanical properties of matrixes on the comet nuclei surface models.' In *Symposium on the Diversity and Similarity of Comets*, Eds. E. J. Rolfe, B. Battrick. ESA SP-278, p. 713–716.
297. Ioffe, Z. M. (1968): 'Some magnetohydrodynamic effects in comets.' *Sov. Phys. Astron.* **11**, 1044–1047.
298. Ip, W.-H. (1985): 'Solar wind interaction with neutral atmospheres.' In *Proc. ESA Workshop on Future Missions in Solar, Heliospheric, and Space Plasma Physics*, Eds. E. J. Rolfe, B. Battrick, ESA SP-235, p. 65–82.
299. Ip, W.-H. (1986): 'An overview of gas phenomena in Comet Halley.' *Adv. Space Res.* **5**, 233–245.
300. Ip, W.-H. (1988): 'Cometary ion acceleration processes.' *Comp. Phys. Comm.* **49**, 1–7.
301. Ip, W.-H. (1989): 'On charge exchange effect in the vicinity of the cometopause of Comet Halley.' *Astrophys. J.* **343**, 946–952.
302. Ip, W.-H., Axford, W. I. (1982): 'Theories of physical processes in the cometary comae and ion tails.' In *Comets*, Ed. L. L. Wilkening, University of Arizona Press; Tucson, AZ, p. 588–634.
303. Ip, W.-H., Axford, W. I. (1986): 'The acceleration of particles in the vicinity of comets.' *Planet. Space Sci.* **34**, 1061–1065.
304. Ip, W.-H., Axford, W. I. (1987a): 'The formation of a magnetic field free cavity at Comet Halley.' *Nature* **325**, 418–419.
305. Ip, W.-H., Axford, W. I. (1987b): 'A numerical simulation of charged particle acceleration and pitch angle scattering in the turbulent plasma environment of Comet Halley.' *Proc. 20th Intern. Cosmic Ray Conf. (Moscow)*, Vol. 3, SH 4.2-14, p. 233–236.
306. Ip, W.-H., Fernandez, J. A. (1988): 'Exchange of condensed matter among the outer and terrestrial protoplanets and the effect on surface impact and atmospheric accretion.' *Icarus* **74**, 47–61.

307. Ip, W.-H., Mendis, D. A. (1978): 'The flute instability as the trigger mechanism for disruption of cometary plasma tails.' *Astrophys. J.* **223**, 671–675.

308. Ip, W.-H., Spinrad, H., McCarthy, P. (1988): 'A CCD observation of the water ion distribution in the coma of Comet P/Halley near the Giotto encounter.' *Astron. Astrophys.* **206**, 129–132.

309. Ip, W.-H., Schwenn, R., Rosenbauer, H., Balsiger, H., Neugebauer, M., Shelley, E. G. (1987): 'An interpretation of the ion pile-up region outside the ionospheric contact surface.' *Astron. Astrophys.* **187**, 132–136.

310. Ipavich, F. M., Galvin, A. B., Gloeckler, G., Hovestadt, D., Klecker, B., Scholer, M. (1986): 'Comet Giacobini-Zinner: *In situ* observations of energetic heavy ions.' *Science* **232**, 366–369.

311. Irvine, W. M., Knacke, R. F. (1989): 'The chemistry of interstellar gas and grains.' In *Origin and Evolution of Planetary and Satellite Atmospheres*, Ed. S. K. Atreya, J. B. Pollack, M. S. Matthews, University Arizona Press; Tucson, AZ, p. 3–34.

312. Irvine, W. M., Leschine, S. B., Schloerb, F. B. (1980): 'Thermal history, chemical composition and relationship of comets to the origin of life.' *Nature* **283**, 748–749.

313. Irvine, W. M., Leschine, S. B., Schloerb, F. P. (1981): 'Comets and the origin of life.' *6th Int. Conference on the Origin of Life*. Ed. Y. Wolman, D. Reidel Publishing Company; Dordrecht, Holland, p. 748–749.

314. Irvine, W. M., Schloerb, F. P., Hjalmarson, Å., Herbst, E. (1985): 'The chemical state of dense interstellar clouds: An overview.' In *Protostars and Planets II*, Eds. Black, D. C., Matthews, M. S. The University of Arizona Press; Tucson, AZ, p. 579–620.

315. Isenberg, P. A. (1987): 'Energy diffusion of pickup ions upstream of comets.' *J. Geophys. Res.* **92**, 8795–8799.

316. Jessberger, E. K., Kissel, J. (1987): 'Bits and pieces from Halley's Comet.' *Lunar Planet. Sci.* **XVIII**, 466–467.

317. Jessberger, E. K., Kissel, J., Fechtig, H., Krueger, F. R. (1986): 'On the average chemical composition of cometary dust.' In *The Comet Nucleus Sample Return*. Canterbury UK, 15-17 July 1986, ESA SP-249, 27–30.

318. Jessberger, E. K., Kissel, J., Fechtig, H., Krueger, F. R. (1987): 'On the average chemical composition of cometary dust.' In *Physical Processes in Comets, Stars, and Active Galaxies*, Ed. W. Hillebrandt, E. Meyer-Hofmeister, H.-C. Thomas, Springer-Verlag; Berlin, Heidelberg, New York, London, Paris, Tokyo, p. 26–33.

319. Jessberger, E. K., Christoforidis, A., Kissel, J. (1988a): 'Aspects of the major element composition of Halley's dust.' *Nature* **332**, 691–695.

320. Jessberger, E. K., Kissel, J., Rahe, J. (1988b): 'The composition of comets.' In *Origin and Evolution of Planetary and Satellite Atmospheres*, Eds. S. K. Atreya, J. B. Pollack, M. S. Matthews, The University of Arizona Press; Tucson, AZ, p. 167–191.

321. Jewitt, D., Danielson, G. E. (1984): 'Charge-coupled device photometry of Comet P/Halley.' *Icarus* **60**, 435–444.

322. Jewitt, D. C., Meech, K. J. (1986): 'Scattering properties of cometary grains – Comet P/Halley.' In *20th ESLAB Symposium on the Exploration of Halley's Comet*, Eds. B. Battrick, E. J. Rolfe, R. Reinhard, ESA SP-250 II, p. 47–51.

323. Jewitt, D. C., Meech, K. J. (1988): 'Optical properties of cometary nuclei and a preliminary comparison with asteroids.' *Astrophys. J.* **328**, 974–986.

324. Jockers, K. (1981): 'Plasma dynamics in the tail of Comet Kohoutek 1973 XII.' *Icarus* **47**, 397–411.

325. Jockers, K. (1985): 'The ion tail of Comet Kohoutek 1973 XII during 17 days of solar wind gusts.' *Astron. Astrophys. Suppl.* **62**, 791–838.

326. Johnson, R. E., Cooper, J. F., Lanzerotti, L. J., Strazzulla, G. (1987): 'Radiation formation of a non-volatile crust.' *Astron. Astrophys.* **187**, 889–892.

327. Johnstone, A., Coates, A., Kellock, S., Wilken, B., Jockers, K., Rosenbauer, H., Studemann, W., Weiss, W., Formisano, V., Amata, E., Cerulli-Irelli, R., Dobrowolny, M., Terenzi, R., Egidi, A., Borg, H., Hultquist, B., Winningham, J., Gurgiolo, C., Bryant, D., Edwards, T., Feldman, W., Thomsen, M., Wallis, M. K., Biermann, L., Schmidt, H. U., Lüst, R., Haerendel, G., Paschmann, G. (1986): 'Ion flow at comet Halley.' *Nature*, **321**, 344–347.

328. Jokipii, J. R. (1971): 'Propagation of cosmic rays in the solar wind.' *Rev. Geophys. Space Phys.* **9**, 27–87.

329. Joss, P. C. (1973): 'On the origin of short-period comets.' *Astron. Astrophys.* **25**, 271–273.

330. Julian, W. M. (1987): 'Free precession of the Comet Halley nucleus.' *Nature* **326**, 57–58.

331. Kamoun, P. D., Campbell, D. B., Ostro, S. J., Pettengill, G. H., Shapiro, I. I. (1982a): 'Comet Encke: Radar detection of nucleus.' *Science* **216**, 293–296.

332. Kamoun, P. G., Pettengill, G. H., Shapiro, I. I. (1982): 'Radar detectibility of comets.' In *Comets*, Ed. L. L. Wilkening, University of Arizona Press; Tucson, AZ, p. 288–296.

333. Kaneda, E., Takagi, M., Hirao, K., Shimizu, M., Ashihara, O. (1986a): 'Ultraviolet features of Comet Halley observed by Suisei.' In *20th ESLAB Symposium on the Exploration of Halley's Comet*, Eds. B. Battrick, E. J. Rolfe, R. Reinhard, ESA SP-250 I, p. 397–402.

334. Kaneda, E., Ashihara, O., Shimizu, M., Takagi, M., Hirao, K. (1986b): 'Observation of Comet Halley by the ultraviolet imager of Suisei.' *Nature* **321**, 297–299.

335. Kant, I. (1755): *Allgemeine Naturgeschichte und Theorie des Himmels. Universal natural history and theory of the heavens; or an essay on the constitution and mechanical origin of the whole universe treated according to Newton's principles.*

336. Kazimirchak-Polonskaya, E. I. (1972): 'The major planets as powerful transformers of cometary orbits.' In *The Motion, Evolution of Orbits, and Origin of Comets*, Eds. G. A. Chebotarev, E. I. Kazimirchak-Polonskaya, B. G. Marsden, D. Reidel Publishing Co.; Dordrecht, Holland, p. 373–397.

337. Keller, H. U. (1973): 'Hydrogen production rates of Comet Bennett (1969i) in the first half of April 1970.' *Astron. Astrophys.* **27**, 51–57.

338. Keller, H. U. (1976): 'The interpretations of ultraviolet observations of comets.' *Space Sci. Rev.* **18**, 641–684.

339. Keller, H. U. (1983): 'Gas and dust models of the coma.' In *Cometary Exploration*, Ed. T. I. Gombosi. Central Research Institute for Physics, Hungarian Academy of Sciences; Budapest, I, p. 119–137.

340. Keller, H. U., Thomas, G. E. (1975): 'A cometary hydrogen model: Comparison with OGO-5 measurements of Comet Bennett (1970 II).' *Astron. Astrophys.* **39**, 7–19.

341. Keller, H. U., Thomas, N. (1988): 'On the rotation axis of Comet Halley.' *Nature* **333**, 146–148.

342. Keller, H. U., Arpigny, C. B., Bonnet, C. M., Cazes, S., Coradini, M., Cosmovici, C. B., Delamere, W. A., Huebner, W. F., Hughes, D. W., Jamar, C., Malaise, D., Reitsema, H., Schmidt, H. U., Schmidt, W. K. H., Seige, P., Whipple, F. L. Wilhelm, K. (1986a): 'First Halley Multicolour Camera imaging results from Giotto.' *Nature* **321**, 320–326.

343. Keller, H. U., Delamere, W. A., Huebner, W. F., Reitsema, H. J., Schmidt, H. U., Schmidt, W. K. H., Whipple, F. L., Wilhelm, K. (1986b): 'Dust activity of Comet Halley's nucleus.' In *20th ESLAB Symposium on the Exploration of Halley's Comet*, Eds. B. Battrick, E. J. Rolfe, R. Reinhard, ESA SP-250 II, p. 359–362.

344. Keller, H. U., Delamere, W. A., Huebner, W. F., Reitsema, H. J., Schmidt, H. U., Whipple, F. L., Wilhelm, K., Curdt, W., Kramm, J., Rainer, T. N., Arpigny, C., Barbieri, C., Bonnet, R. M., Cazes, S., Coradini, M., Cosmovici, C. B., Hughes, D. W., Jamar, C., Malaise, D., Schmidt, K., Schmidt, W. K. H., Seige, P. (1987a): 'Comet P/Halley's nucleus and its activity.' *Astron. Astrophys.* **187**, 807–823.

345. Keller, H. U., Schmidt, W. K. H., Wilhelm, K., Becker, C., Curdt, W., Engelhardt, W., Hartwig, H., Kramm, J., Rainer, T. N., Meyer, H. J., Schmidt, R., Gliem, F., Krahn, E., Schmidt, H. P., Schwarz, G., Turner, J. J., Boyries, P., Cazes, S., Angrilli, F., Bianchini, G., Fanti, G., Brunello, P., Delamere, W. A. Reitsema, H., Jamar, C., Cucciaro, C. (1987b): 'The Halley multicolour camera.' *J. Phys. E.: Sci. Instr.* **20**, 807–820.

346. Keller, H. U., Kramm, R., Thomas, N. (1988): 'Surface features on the nucleus of Comet Halley.' *Nature* **331**, 227–231.

347. Kennel, C. F., Coroniti, F. V., Scarf, F. L., Tsurutani, B. T., Smith, E. J., Bame, S. J., Gosling, J. T. (1986): 'Plasma waves in the shock interaction regions at Comet Giacobini-Zinner.' *Geophys. Res. Lett.* **13**, 921–924.

348. Khare, B. N., Sagan, C., Arakawa, E. T., Suits, F., Callcot, T. A., Williams, M. W. (1984): 'Optical constants of organic tholins produced in a simulated Titanian atmosphere: From soft x-ray to microwave frequencies.' *Icarus* **60**, 127–137.

349. Kimura, H., Liu, C. P. (1977): 'On the structure of cometary dust tails.' *Chin. Astron.* **1**, 235–264.

350. Kissel, J. (1986): 'The Giotto particulate analyser.' In *Giotto mission – Its scientific investigations,* Eds. R. Reinhard, B. Battrick, ESA SP-1077, p. 67–68.

351. Kissel, J., Krueger, F. R. (1987a): 'The organic component in dust from Comet Halley as measured by the PUMA mass spectrometer on board Vega 1.' *Nature* **326**, 755–760.

352. Kissel, J., Krueger, F. R. (1987b): 'Ion formation by impact of fast dust particles and comparison with related techniques.' *Appl. Phys.*, A **42**, 69–85.

353. Kissel, J., Sagdeev, R. Z., Bertaux, J. L., Angarov, V. N., Audouze, J., Blamont, J. E., Büchler, K., Evlanov, E. N., Fechtig, H., Fomenkova, M. N., von Hoerner, H., Inogamov, N. A., Khromov, V. N., Knabe, W., Krueger, F. R., Langevin, Y., Leonas, V. B., Levasseur-Regourd, A. C., Managadze, G. G., Podkolzin, S. N., Shapiro, V. D., Tabaldyev, S. R., Zubkov, B. V. (1986a): 'Composition of Comet Halley dust particles from Vega observations.' *Nature* **321**, 280–282.

354. Kissel, J., Brownlee, D. E., Büchler, K., Clark, B. C., Fechtig, H., Grün, E., Hornung, K., Igenbergs, E. B., Jessberger, E. K., Krueger, F. R., Kuczera, H., McDonnell, J. A. M., Morfill, G. M., Rahe, J., Schwehm, G. H., Sekanina, Z., Utterback, N. G., Völk, H. J., Zook, H. A. (1986b): 'Composition of Comet Halley dust particles from Giotto observations.' *Nature* **321**, 336–337.

355. Kitamura, Y. (1986): 'Axisymmetric dusty gas jet in the inner coma of a comet.' *Icarus* **66**, 241–257.

356. Kitamura, Y. (1987): 'Axisymmetric dusty gas jet in the inner coma of a comet. II. The case of isolated jets.' *Icarus* **72**, 555–567.

357. Klimov, S., Savin, S., Aleksevich, Y., Avanesova, G., Balebanov, V., Baiikhin, M., Galeev, A. A., Gribov, B., Mozdrachev, M., Smirnov, V., Sokolov, A., Vaisberg, O., Oberc, P., Krawczyk, Z., Grzedzielski, S., Juchniewicz, J., Nowak, K., Orlowski, D., Parfianovich, B., Wozniak, D., Zbyszynski, Z., Voita, Y., Triska, P. (1986): 'Extremely-low-frequency plasma waves in the environment of Comet Halley.' *Nature* **321**, 292–294.

358. Klinger, J. (1981): 'Some consequences of a phase transition of water ice on the heat balance of comet nuclei.' *Icarus* **47**, 320–324.

359. Klinger, J. (1983): 'Classification of cometary orbits based on the concept of orbital mean temperature.' *Icarus* **55**, 169–176.

360. Klinger, J. (1985): 'Composition and structure of the comet nucleus and its evolution on a periodic orbit.' In *Ices in the Solar System,* Eds. J. Klinger, D. Benest, A. Dollfus, R. Smoluchowski, D. Reidel Pulishing Co.; Dordrecht, Boston, Lancaster, p. 407–417.

361. Knacke, R. F., Brooke, T. Y., Joyce, R. R. (1986): 'Observations of 3.2-3.6 micron emission features in Comet Halley.' *Astrophys. J.* **310**, L49–L54.

362. Kömle, N. I., Ip, W.-H. (1987): 'Anisotropic non-stationary gas flow dynamics in the coma of Comet P/Halley.' *Astron. Astrophys.* **187**, 405–410.

363. Konno, I., Wyckoff, S. (1989): 'Atomic and molecular abundances in Comet Giacobini-Zinner.' *Adv. Space Res.* **9**, 163–168.

364. Konno, I., Wyckoff, S., Wehinger, P. A. (1990): 'Observations of Comet P/Giacobini-Zinner before, during, and after the ICE spacecraft encounter.' *Astrophys. J.*, in press.

365. Körösmezey, A., Cravens, T. E., Gombosi, T. I., Nagy, A. F., Mendis, D. A., Szegö, K., Gribov, B. E., Sagdeev, R. Z., Shapiro, V. D., Shevchenko, V. I. (1986): 'A model of inner cometary ionospheres.' In *20th ESLAB Symposium on the Exploration of Halley's Comet,* Eds. B. Battrick, E. J. Rolfe, R. Reinhard, ESA SP-250 I, p. 235–239.

366. Korth, A., Richter, A. K., Loidl, A., Anderson, K. A., Carlson, C. W., Curtis, D. W., Lin, R. P., Rème, H., Sauvaud, J. A., d'Uston, C., Cotin, F., Cros, A., Mendis, D. A. (1986): 'Mass spectra of heavy ions near Comet Halley.' *Nature* **321**, 335–336.

367. Kozai, Y. (1979): 'Secular perturbations of asteroids and comets.' In *Dynamics of the Solar System,* Ed. R. L. Duncombe, D. Reidel Publishing Co.; Dordrecht, Holland, p. 231–237.

368. Krankowsky, D. (1990): 'The composition of comets.' In *Comets in the Post-Halley Era,* Eds. R. L. Newburn, J. Rahe, M. Neugebauer, in press.

369. Krankowsky, D., Eberhardt, P. (1990): 'Evidence for the composition of ices in the nucleus of Comet Halley.' In *Comet Halley 1986: World-Wide Investigations, Results, and Interpretations*, Eds. J. Mason, P. Moore, Ellis Horwood Ltd.; Chichester, in press.

370. Krankowsky, D., Lämmerzahl, P., Herrwerth, I., Woweries, J., Eberhardt, P., Dolder, U., Herrmann, U., Schulte, W., Berthelier, J. J., Illiano, J. M., Hodges, R. R., Hoffman, J. H. (1986): '*In situ* gas and ion measurements at Comet Halley.' *Nature* **321**, 326–329.

371. Krasnopolsky, V. A., Moroz, V. I., Krysko, A. A., Tkachuk, A. Y., Moreels, G., Clairmidi, J., Parisot, J. P., Gogoshev, M., Gogosheva, T. (1987a): 'Properties of dust in Comet P/Halley meaured by the Vega-2 three-channel spectrometer.' *Astron. Astrophys.* **187**, 707–711.

372. Krasnopolsky, V. A., Tkachuk, A. Y., Moreels, G., Gogoshev, M. (1987b): 'Water vapor and hydroxyl distributions in the inner coma of Comet Halley measured by the Vega 2 three channel spectrometer TKS.' In *Symposium on the Diversity and Similarity of Comets*, Eds. E. J. Rolfe, B. Battrick. ESA SP-278, p. 185–190.

373. Krasnopolsky, V. A., Tkachuk, A. Y., Moreels, G., Gogoshev, M. (1988): 'Water vapor and hydroxyl distributions in the inner coma of comet P/Halley measured by the Vega 2 three-channel spectrometer TKS.' *Astron. Astrophys.* **203**, 175–182.

374. Kresák, L. (1974): 'The aging and the brightness decrease of comets.' *Bull. Astron. Inst. Czechosl.* **25**, 87–112.

375. Kresák, L. (1977): 'On the differences between the new and old comets.' *Bull. Astron. Inst. Czechosl.* **28**, 346–355.

376. Kresák, L. (1981): 'Evolutionary aspects of the splits of cometary nuclei.' *Bull. Astron. Inst. Czechosl.* **32**, 19–40.

377. Kresák, L. (1982): 'On the reality and genetic association of comet groups and pairs.' *Bull. Astron. Inst. Czechosl.* **33**, 150–160.

378. Kresák, L. (1985): 'The aging and lifetimes of comets.' In *Dynamics of Comets: Their Origin and Evolution* 115, Eds. A. Carusi, G. B. Valsecchi, D. Reidel Publishing Co.; Dordrecht, Holland, p. 279-302.

379. Kresák, L. (1987a): 'The 1808 apparition and the long-term physical evolution of periodic comet Grigg-Skjellerup.' *Bull. Astron. Inst. Czechosl.* **38**, 65–75.

380. Kresák, L. (1987b): 'Dormant phases in the aging of periodic comets.' *Astron. Astrophys.* **187**, 906–908.

381. Kresák, L. (1990): 'The ageing of comet Halley and other periodic comets.' In *Comet Halley 1986: World-Wide Investigations, Results, and Interpretations*, Eds. J. Mason, P. Moore, Ellis Horwood Ltd.; Chichester, in press.

382. Kresák, L., Kresáková, M. (1987): 'The absolute total magnitudes of periodic comets and their variations.' In *Symposium on the Diversity and Similarity of Comets*, Eds. E. J. Rolfe, B. Battrick. ESA SP-278, p. 37–42.

383. Kresák, L., Kresáková, M. (1989): 'The absolute magnitudes of periodic comets. I. Catalogue.' *Bull. Astron. Inst. Czechosl.* **40**, 269–284.

384. Krimsky, G. F. (1977): 'A regular mechanism for the acceleration of charged particles on the front of a shock wave.' *Dokl. Akad. Nauk SSSR* **234**, 1306–1308.

385. Krueger, F. R., Kissel, J. (1987): 'The chemical composition of the dust of Comet P/Halley as measured by "PUMA" on board Vega-1.' *Naturwiss.* **74**, 312–316.

386. Kührt, E. (1984): 'Temperature profiles and thermal stresses in cometary nuclei.' *Icarus* **60**, 512–521.

387. Kuiper, G. P. (1951): 'Origin of the Solar System.' In *Astrophysics*, Ed. J. A. Hynek, McGraw Hill Book Co.; New York, p. 357–424.

388. Kushner, D. (1981): 'Extreme environments: Are there any limits to life?' In *Comets and the Origin of Life*, Ed. C. Ponnamperuma, D. Reidel Publishing Co.; Dordrecht, Holland, p. 241–248.

389. Lagrange, J. L. (1814): 'Sur l'origine des comètes.' *Additions à la Connaissance des Temps*, 211–223.

390. Lakhina, G. S., Buti, B. (1988): 'Coherent radiation mechanism for cometary kilometric radiation.' *Astrophys. J.* **327**, 1020–1025.

391. Lambert, D. L., Danks, A. C. (1983): 'High resolution spectra of C_2 Swan bands from Comet West 1976 VI.' *Astrophys. J.* **268**, 428–446.

392. Lämmerzahl, P., Krankowsky, D., Hodges, R. R., Stubbemann, U., Woweries, J., Herrwerth, I., Berthelier, J. J., Illiano, J. M., Eberhardt, P., Dolder, U., Schulte, W., Hoffman, J. H. (1987): 'Expansion velocity and temperatures of gas and ions measured in the coma of Comet P/Halley.' *Astron. Astrophys.* **187**, 169–173.

393. Lamy, P. (1985): 'Ground-based observations of the dust emission from Comet Halley.' *Adv. Space Res.* **5**, 317–323.

394. Lamy, P. (1986): 'Cometary dust: Observational evidences and properties.' In *Asteroids, Comets, Meteors II*, Eds. C.-I. Lagerkvist, B. A. Lindblad, H. Lundstedt, H. Rickman, Uppsala University; Uppsala, p. 373–388.

395. Lamy, P., Koutchmy, S. (1979): 'Comet West 1975n part II: Study of the striated tail.' *Astron. Astrophys.* **72**, 50–54.

396. Lamy, P. L., Grün, E., Perrin, J. M. (1987): 'Comet P/Halley: Implications of the mass distribution function for the photopolarimetric properties of the dust coma.' *Astron. Astrophys.* **187**, 767–773.

397. Landgraf, W. (1986): 'The uncertainty of position predictions for Comet Halley in March 1986.' *Astron. Astrophys.* **157**, 245–251.

398. Langevin, Y., Kissel, J., Bertaux, J.-L., Chassefiere E. (1987a): 'Impact ionization mass spectrometry of the cometary grains on board Giotto, Vega 1, and Vega 2 spacecraft; preliminary statistical analysis of spectra in compressed modes.' *Lunar Planet. Sci.* **XVIII**, 533–534.

399. Langevin, Y., Kissel, J., Bertaux, J.-L., Chassefiere, E. (1987b): 'First statistical analysis of 5000 mass spectra of cometary grains obtained by PUMA 1 (Vega-1), and PIA (Giotto). Impact ionization mass spectrometers in the compressed modes.' *Astron. Astrophys.* **187**, 779–784.

400. Laplace, P. S. (1805): *Théorie des cometes; méchanique céleste.* **4**, chez Couvcier; Paris, New York.

401. Larson, H. P., Davis, D. S., Mumma, M. J., Weaver, H. A. (1986): 'Velocity-resolved observations of water in Comet Halley.' In *20th ESLAB Symposium on the Exploration of Halley's Comet*, Eds. B. Battrick, E. J. Rolfe, R. Reinhard, ESA SP-250 I, p. 335–340.

402. Larson, H. P., Mumma, M. J., Weaver, H. A. (1987): 'Kinematic properties of the neutral gas outflow from comet P/Halley.' *Astron. Astrophys.* **187**, 391–397.

403. Larson, S. M. (1980): 'CO⁺ in Comet Schwassmann-Wachmann 1 near minimum brightness.' *Astrophys. J.* **238**, L47–L48.

404. Larson, S. M., Sekanina, Z. (1984): 'Coma morphology and dust-emission pattern of periodic Comet Halley. I. High-resolution images taken at Mount Wilson in 1910.' *Astron. J.* **89**, 571–578.

405. Larson, S. M., Sekanina, Z. (1985): 'Coma morphology and dust-emission pattern of periodic comet Halley. III. Additional high-resolution images taken in 1910.' *Astron. J.* **90**, 823–826, and 917–923.

406. Larson, S., Sekanina, Z., Levy, D., Tapia, S., Senay, M. (1987): 'Comet P/Halley near-nucleus phenomena in 1986.' *Astron. Astrophys.* **187**, 639–644.

407. Laufer, D., Kochavi, E., Bar-Nun, A. (1987): 'Structure and dynamics of amorphous water ice.' *Phys. Rev.* **B 36**, 9219–9227.

408. Lazcano-Araujo, A., Oró, J. (1981): 'Cometary material and the origin of life on Earth.' In *Comets and the Origin of Life*, Ed. C. Ponnamperuma, D. Reidel Publishing Co.; Dordrecht, Holland, p. 191–225.

409. Lebofsky, L. A., Sykes, M. V., Tedesco, E. F., Veeder, G. J., Matson, D. L., Brown, R. H., Gradie, J. C., Feierabend, M. A., Rudy, R. J. (1986): 'A refined "standard" thermal model for asteroids based on observations of 1 Ceres and 2 Pallas.' *Icarus* **68**, 239–251.

410. Le Borgne, J. F., Leroy, J. L., Arnaud, J. (1987): 'Polarimetry of Comet P/Halley: Continuum versus molecular bands.' *Astron. Astrophys.* **187**, 526–530.

411. Lee, M. A. (1982): 'Coupled hydormagnetic wave excitation and ion acceleration upstream of the Earth's bow shock.' *J. Geophys. Res.* **87**, 5063–5080.

412. Lee, M. A. (1989): 'Ultra low frequency waves at comets.' *Plasma Waves and Instabilities at Comets and in Magnetospheres*, Eds. B. T. Tsurutani and H. Oya, AGU Monograph **53**, 13–29.

413. Lee, M. A., Ip, W.-H. (1987): 'Hydromagnetic wave excitation by ionized interstellar hydrogen and helium in the solar wind.' *J. Geophys. Res.* **92**, 11041–11052.

414. Lee, T., Papanastassiou, D. A., Wasserburg, G. J. (1976): 'Demonstration of ²⁶Mg excess in Allende and evidence for ²⁶Al.' *Geophys. Res. Lett.* **3**, 109–112.

415. Lees, L. (1964): 'Interaction between the solar plasma wind and the geomagnetic cavity.' *Amer. Inst. Aero. Astronaut. J.* **2**, 1576–1582.

416. Léger, A. (1987): 'Desorption mechanism of gases from interstellar grains and PAH molecules.' In *Astrochemistry*. Eds. M. S. Vardya, S. P. Tarafdar, D. Reidel Publishing Company; Dordrecht, Holland, p. 539–543.

417. Léger, A., Jura, M., Omont, A. (1985): 'Desorption from interstellar grains.' *Astron. Astrophys.* **144**, 147–160.

418. Lehninger, A. L. (1975): *Biochemistry: The Molecular Basis of Cell Structure and Function.* Second edition. Worth Publishers; New York.

419. Leibowitz, E. M., Brosch, N. (1986): 'Periodic light variations in the near nucleus zone of Comets P/Halley and P/Giacobini-Zinner.' In *20th ESLAB Symposium on the Exploration of Halley's Comet,* Eds. B. Battrick, E. J. Rolfe, R. Reinhard, ESA SP-250 I, p. 605–607.

420. Levasseur-Regourd, A. L., Weinberg, J. L., Bertaux, J. L., Dument, R., Festou, M., Giese, R. H., Giovane, F., Lamy, P., Le Blanc, J. M., Llebaria, A. (1986): '*In-itu* photopolarimetric measurements of dust and gas in the coma of Halley's comet.' *Adv. Space Res.* **5**, 197–199.

421. Levin, B. J. (1943a): 'Gas evolution from the nucleus of a comet as related to the variatons in its absolute brightness.' *Compt. Rend. Acad. Sci. URSS.* **38**, 72–74.

422. Levin, B. Y. (1943b): 'The emission of gases by the nucleus of a comet and the variation of its absolute brightness.' *Astron. Zh.* **20**, 37–48.

423. Levine, J. S. (1985a): 'The photochemistry of the early atmosphere.' In *The Photochemistry of Atmospheres: Earth, the other Planets, and Comets.* Academic Press, Inc.; Orlando, San Diego, New York, London, Toronto, Montreal, Sydney, Tokyo, p. 3–38.

424. Levine, J. S. (1985b): *The Photochemistry of Atmospheres: Earth, the other Planets, and Comets.* Academic Press, Inc.; Orlando, San Diego, New York, London, Toronto, Montreal, Sydney, Tokyo, Appendix I.

425. Lewis, J. S., Prinn, R. G. (1980): 'Kinetic inhibition of CO and N_2 reduction in the solar nebula.' *Astrophys. J.* **238**, 357–364.

426. Lindblad, B. A. (1985): 'Do comet groups exist?' In *Dynamics of Comets: Their Origin and Evolution* **115**, Eds. A. Carusi, G. B. Valsecchi, D. Reidel Publishing Co.; Dordrecht, Holland, p. 353-363.

427. Luhmann, J. G. (1986): 'The solar wind interaction with Venus.' *Space Sci. Rev.* **44**, 241–306.

428. Lüst, R. (1984): 'The distribution of the aphelion directions of long-period comets.' *Astron. Astrophys.* **141**, 94–100.

429. Lüst, R. (1985): 'Some remarks about the aphelion distribution of long period comets on the sky.' In *Dynamics of Comets: Their Origin and Evolution* **115**, Eds. A. Carusi, G. B. Valsecchi, D. Reidel Publishing Co.; Dordrecht, Holland, p. 105–111.

430. Luu, J. X., Jewitt, D. (1988): 'A two-part search for slow-moving objects.' *Astron. J.* **95**, 1256–1262.

431. Lyttleton, R. A. (1948): 'On the origin of comets.' *Mon. Not. Roy. Astron. Soc.* **108**, 465–475.

432. Lyttleton, R. A. (1953): *The comets and their origin.* Cambridge University Press; Cambridge.

433. Maas, D., Goeller, J. R., Grün, E., Lange, G., McDonnell, J. A. M., Nappo, S., Perry, C., Zarnecki, J. C. (1989): 'Cometary dust particles detected by the DIDSY-IPM-P sensor on board Giotto.' *Adv. Space Res.* **9**, 247–252.

434. Magee-Sauer, K., Scherb, F., Roesler, F. L., Harlander, J., Lutz, B. L. (1989): 'Fabry-Perot observations of NH_2 emission from Comet Halley.' *Icarus* **82**, 50–60.

435. Malaise, D. J. (1970): 'Collisional effects in cometary atmospheres. I. Model atmospheres and synthetic spectra.' *Astron. Astrophys.* **5**, 209–227.

436. Marconi, M. L., Mendis, D. A. (1982a): 'The photochemical heating of the cometary atmosphere.' *Astrophys. J.* **260**, 386–394.

437. Marconi, M. L., Mendis, D. A. (1982b): 'The photochemistry and dynamics of a dusty cometary atmosphere.' *Moon Planets* **27**, 27–47.

438. Marconi, M. L., Mendis, D. A. (1983): 'The atmosphere of a dirty-clathrate cometary nucleus: A two-phase, multifluid model.' *Astrophys. J.* **273**, 381–396.

439. Marconi, M. L., Mendis, D. A. (1984): 'The effects of the diffuse radiation fields due to multiple scattering and thermal raradiation by dust on the dynamics and thermodynamics of a dusty cometary atmosphere.' *Astrophys. J.* **287**, 445–454.

440. Marconi, M. L., Mendis, D. A. (1988): 'On the ammonia abundance in the coma of Halley's comet.' *Astrophys. J.* **330**, 513–517.

441. Marsden, B. G., Sekanina, Z. 1973): 'On the distribution of "original" orbits of comets of large perihelion distance.' *Astron. J.* **78**, 1118–1124.

442. Marsden, B. G., Sekanina, Z., Yeomans, D. K. (1973): 'Comets and nongravitational forces. V.' *Astron. J.* **78**, 211–225.

443. Marsden, B. G., Sekanina, Z., Everhart, E. (1978): 'New osculating orbits for 110 comets and analysis of original orbits for 200 comets.' *Astron. J.* **83**, 64–71.

444. Massonne, L. (1985): 'Coma morphology and dust emission pattern of Comet Halley.' *Adv. Space Res.* **5**, 187–196.

445. Massonne, L., Grün, E. (1986): 'Structures in Halley's dust coma.' In *20th ESLAB Symposium on the Exploration of Halley's Comet,* Eds. B. Battrick, E. J. Rolfe, R. Reinhard, ESA SP-250 III, p. 319–322.

446. Massonne, L., Fertig, J., Grün, E., Schwehm, G. (1985): 'A cometary dust environmental model for the generation of synthetic images.' In *Asteroids, Comets, Meteors II,* Eds. C.-I. Lagerkvist, B. A. Lindblad, H. Lundstedt, H. Rickman, Uppsala University; Uppsala, p. 407–410.

447. Mazets, E. P., Aptekar, R. L., Golenetskii, S. V., Guryan, Y. A., Dyachkov, A. V., Ilyinskii, V. N., Panov, V. N., Petrov, G. G., Savvin, A. V., Sagdeev, R. Z., Sokolov, I. A., Khavenson, N. G., Shapiro, V. D., Shevchenko, V. I. (1986a): 'Comet Halley dust environment from Sp-2 detector measurements.' *Nature* **321**, 276–278.

448. Mazets, E. P., Sagdeev, R. Z., Aptekar, R. L., Golenetskii, S. V., Guryan, Y. A., Dyachkov, A. V., Ilyinskii, V. N., Panov, V. N., Petrov, G. G., Savvin, A. V., Sokolov, I. A., Frederiks, D. D., Khavenson, N. G., Shapiro, V. D., Shevchenko, V. I. (1986b): 'Dust in Comet Halley from Vega observations.' In *20th ESLAB Symposium on the Exploration of Halley's Comet,* Eds. B. Battrick, E. J. Rolfe, R. Reinhard, ESA SP-250 II, p. 3–10.

449. Mazets, E. P., Sagdeev, R. Z., Aptekar, R. L., Golenetskii, S. V., Guryan, Y. A., Dyachkov, A. V., Ilyinskii, V. N., Panov, V. N., Petrov, G. G., Savvin, A. V., Sokolov, I. A., Frederiks, D. D., Khavenson, N. G., Shapiro, V. D., Shevchenko, V. I. (1987): 'Dust in Comet P/Halley from Vega observations.' *Astron. Astrophys.* **187**, 699–706.

450. McComas, D. J., Gosling, J. T., Bame, S. J., Slavin, J. A., Smith, E. J., Steinberg, J. T. (1987): 'The Giacobini-Zinner magnetotail configuration and current sheet.' *J. Geophys. Res.* **92**, 1139–1152.

451. McDonnell, J. A. M., Alexander, W. M., Burton, W. M., Bussoletti, E., Clark, D. H., Grard, R. J. L., Grün, E., Hanner, M. S., Hughes, D. W., Igenbergs, E., Kuczera, H., Lindblad, B. A., Mandeville, J.-C., Minafra, A., Schwehm, G. H., Sekanina, Z., Wallis, M. K., Zarnecki, J. C., Chakaveh, S. C., Evans, G. C., Evans, S. T., Firth, J. G., Littler, A. N., Massonne, L., Olearczyk, R. E., Pankiewicz, G. S., Stevenson, T. J., Turner, R. F. (1986a): 'Dust density and mass distribution near Comet Halley from Giotto observations.' *Nature* **321**, 338–341.

452. McDonnell, J. A. M., Kissel, J., Grün, E., Grard, R. J. L., Langevin, Y., Olearczyk, R. E., Perry, C. H., Zarnecki, J. C. (1986b): 'Giotto's dust impact detection system DIDSY and particulate impact analyser PIA: Interim assessment of ths dust distribution and properties within the coma.' In *20th ESLAB Symposium on the Exploration of Halley's Comet,* Eds. B. Battrick, E. J. Rolfe, R. Reinhard, ESA SP-250 II, p. 25–38.

453. McDonnell, J. A. M., Alexander, W. M., Burton, W. M., Bussoletti, E., Evans, G. C., Evans, S. T., Firth, J. G., Grard, R. J. L., Green, S. F., Grün, E., Hanner, M. S., Hughes, D. W., Igenbergs, E., Kissel, J., Kuczera, H., Lindblad, B. A., Langevin, Y., Mandeville, J.-C., Nappo, S., Pankiewicz, G. S. A., Perry, C. H., Schwehm, G. H., Sekanina, Z., Stevenson, T. J., Turner, R. F., Weishaupt, U., Wallis, M. K., Zarnecki, J. C. (1987): 'The dust distribution within the inner coma of Comet P/Halley 1982i: Encounter by Giotto's impact detectors.' *Astron. Astrophys.* **187**, 719–741.

454. McIntosh, B. A., Hajduk, A. (1983): 'Comet Halley meteor stream: A new model.' *Mon. Not. Roy. Astron. Soc.* **205**, 931–943.

455. McKay, C. P. (1986): 'Exobiology and future Mars missions: The search for Mars' earliest bio-sphere.' *Adv. Space Res.* **6**, 269–285.

456. McKay, C. P., Squyres, S. W., Reynolds, R. T. (1986): 'Methods for computing comet core tem-peratures.' *Icarus* **66**, 625–629.

457. McKay, C. P., Scattergood, T. W., Pollack, J. B., Borucki, W. J., Van Ghyseghem, H. T. (1988): 'High-temperature shock formation of N and organics on primordial Titan.' *Nature* **332**, 520–522.

458. McKeegan, K. D., Walter, R. M., Zinner, A. (1985): 'Ion microprobe isotopic measurements of individual interplanetary dust particles.' *Geochim. Cosmochim. Acta* **49**, 1971–1987.

459. McKeegan, K. D., Swan, P., Walker, R. M., Wopenka, B., Zinner, E. (1987): 'Hydrogen isotopic variation in interplanetary dust particles.' *Lunar Planet. Sci.* **XVIII**, 627–628.

460. McKenna-Lawlor, S., Kirsch, E., O'Sullivan, D., Thompson, A., Wenzel, K.-P. (1986): 'Energetic ions in the environment of Comet Halley.' *Nature* **321**, 344–347.

461. McKenna-Lawlor, S., Wilken, B., Daly, P., Ip, W.-H., Kirsch, E., Coates, A., Johnstone, A., Thomp-son, A., O'Sullivan, D., Wenzel, K.-P. (1987): 'Energy spectra of pickup ions recorded during the encounter of Giotto with Comet Halley.' In *Symposium on the Diversity and Similarity of Comets,* Eds. E. J. Rolfe, B. Battrick. ESA SP-278, p. 133–137.

462. Meakin, P., Donn, B. (1988): 'Aerodynamic properties of fractal grains: Implications for the pri-mordial solar nebula.' *Astrophys. J.* **329**, L39–L41.

463. Meech, K. J., Belton, M. J. S. (1989): '(2060) Chiron.' *IAU Circular No. 4770.*

464. Meech, K. J., Jewitt, D. (1987): 'Comet Bowell at record heliocentric distance.' *Nature* **328**, 506–509.

465. Meisel, D. D., Morris, C. S. (1976): 'Comet brightness parameters: Definition, determination, and correlations.' In *The Study of Comets,* Eds. B. Donn, M. Mumma, W. Jackson, M. A'Hearn, R. Harrington, NASA; Washington, DC. NASA SP-393, p. 410-444.

466. Mendis, D. A., Brin, G. D. (1977): 'The monochromatic brightness variations of comets – II. The core-mantle model.' *Moon Planets* **17**, 359–372.

467. Mendis, D. A., Wickramasinghe, N. C. (1975): 'Composition of cometary dust: The case against silicates.' *Astrophys. Space Sci.* **37**, L13–L16.

468. Mendis, D. A., Hill, J. R., Houpis, H. L. F., Whipple, F. L. (1981): 'On the electrostatic charging of the cometary nucleus.' *Astrophys. J.* **249**, 787–797.

469. Mendis, D. A., Houpis, H. L. F., Marconi, M. L. (1985): 'The physics of comets.' *Fund. Cosmic Phys.* **10**, 1–380.

470. Meyer-Vernet, N. (1982): 'Flip-flop of electric potential of dust grains in space.' *Astron. Astrophys.* **105**, 98–106.

471. Meyer-Vernet, N., Couturier, P., Hoang, S., Perche, C., Steinberg, J. L., Fainberg, J., Meetre, C. (1986): 'Plasma diagnosis from thermal noise and limits on dust flux or mass in Comet Giacobini-Zinner.' *Science* **232**, 370–374.

472. Michels, D. J., Sheeley Jr., N. R., Howard, R. A., Koomen, M. J. (1982): 'Observations of a comet on collision course with the Sun.' *Science* **215**, 1097–1102.

473. Mignard, F., Remy, F. (1985): 'Dynamical evolution of the Oort cloud. II. A theoretical approach.' *Icarus* **63**, 20–30.

474. Miller, S. L. (1953): 'A production of amino acids under possible primitive Earth conditions.' *Science* **117**, 528–529.

475. Millis, R. L., Schleicher, D. G. (1986): 'Rotational period of Comet Halley.' *Nature* **324**, 646–649.

476. Millis, R. L., A'Hearn, M. F., Campins, H. (1988): 'An investigation of the nucleus and coma of Comet P/Arend-Rigaux.' *Astrophys. J.* **324**, 1194–1209.

477. Millman, P. M.(1972): 'Cometary meteoroids.' In *From plasma to planet,* Ed. A. Elvius, John Wiley & Sons; New York, London, Sydney, p. 157–168.

478. Millman, P. M. (1977): 'The chemical composition of cometary meteoroids.' In *Comets, Asteroids, Meteorites,* Ed. A. H. Delsemme, University of Toledo; Toledo, OH, p. 127–132.

479. Mitchell, D. L., Lin, R. P., Anderson, K. A., Carlson, C. W., Curtis, D. W., Korth, A., Richter, A. K., Rème, H., Sauvaud, J. A., d'Uston, C., Mendis, D. A. (1986): 'Derivation of heavy (10–210 amu) ion composition and flow parameters for the Giotto PICCA instrument.' In *20th ESLAB Symposium*

on the Exploration of Halley's Comet, Eds. B. Battrick, E. J. Rolfe, R. Reinhard, ESA SP-250 I, p. 203–205.

480. Mitchell, D. L., Lin, R. P., Anderson, K. A., Carlson, C. W., Curtis, D. W., Korth, A., Rème, H., Sauvaud, J. A., d'Uston, C., Mendis, D. A. (1989): 'Complex organic ions in the atmosphere of Comet Halley.' *Adv. Space Res.* **9**, 35–39.

481. Mitchell, G. F., Prasad, S. A., Huntress, W. T. (1981): 'Chemical model calculations of C_2, C_3, CH, CN, OH, and NH_2 abundances in cometary comae.' *Astrophys. J.* **244**, 1087–1093.

482. Möbius, E., Papadopoulos, K., Piel, A. (1987): 'On the turbulent heating and the threshold condition in the critical ionization velocity interaction.' *Planet. Space Sci.* **35**, 345–352.

483. Mogilevsky, M., Mikhilov, Y., Molchanov, O., Grard, R., Pedersen, A., Trotignon, J. G., Beghin, C., Formisano, V., Shapiro, V., Shevchenko, V. (1987): 'Identification of boundaries in the cometary environment from an electric field measurements.' *Astron. Astrophys.* **187**, 80–82.

484. Möhlmann, D., Börner, H., Danz, M., Elter, G., Mangoldt, T., Rubbert, B., Weidlich, U. (1986): 'Physical properties of P/Halley – derived from Vega-images.' In *20th ESLAB Symposium on the Exploration of Halley's Comet,* Eds. B. Battrick, E. J. Rolfe, R. Reinhard, ESA SP-250 I, p. 339–340.

485. Möhlmann, D., Danz, M., Börner, H. (1987): 'Properties of the nucleus of P/Halley.' In *Symposium on the Diversity and Similarity of Comets,* Eds. E. J. Rolfe, B. Battrick. ESA SP-278, p. 481–492.

486. Morfill, G. E., Tscharnuter, W., Völk, H. J. (1985): 'Dynamical and chemical evolution of teh protoplanetary nebula,' In *Protostars and Planets II,* Eds. D. C. Black, M. S. Matthews, University of Arizona Press; Tucson, AZ, p. 493–533.

487. Moroz, V. I., Combes, M., Bibring, J. P., Coron, N., Crovisier, J., Encrenaz, T., Crifo, J. F., Sanko, N., Grigoryev, A. V., Bockelée-Morvan, D., Gispert, R., Nikolsky, Y. V., Emerich, C., Lamarre, J. M., Rocard, F., Krasnopolsky, V. A., Owen, T. (1987): 'Detection of parent molecules in Comet P/Halley from the IKS-Vega experiment.' *Astron. Astrophys.* **187**, 513–518.

488. Mount, G. H., Rottman, G. J. (1985): 'Solar absolute spectral irradiance 118–300 nm: July 25, 1983.' *J. Geophys. Res.* **90**, 13031–13036.

489. Mukai, T., Mukai, S., Kikuchi, S. (1986a): 'Role of small grains in the visible polarization of Comet Halley.' In *20th ESLAB Symposium on the Exploration of Halley's Comet,* Eds. B. Battrick, E. J. Rolfe, R. Reinhard, ESA SP-250 II, p. 59–63.

490. Mukai, T., Miyake, W., Terasawa, T., Kitayaman, Hirao, K. (1986b): 'Plasma observation by Suisei of solar-wind interaction with Comet Halley.' *Nature* **321**, 299–303.

491. Mukhin, L. M., Evlanov, E. N., Fomenkova, M. N., Khromov, V. N., Kissel, J., Priludski, O. F., Zubkov, B. V., Sagdeev, R. Z. (1987): 'Different types of dust particles in Halley's Comet.' *Lunar Planet. Sci.* **XVIII**, 674–675.

492. Mumma, M. J., Reuter, D. C. (1989): 'On the identification of formaldehyde in Halley's Comet.' *Astrophys. J.* **344**, 940–948.

493. Mumma, M. J., Weaver, H. A., Larson, H. P., Davis, D. S., Williams, M. (1986): 'Detection of water vapor in Halley's comet.' *Science* **232**, 1523–1528.

494. Mumma, M. J., Weaver, H. A., Larson, H. P. (1987): 'The ortho – para ratio of water vapor in Comet P/Halley.' *Astron. Astrophys.* **187**, 419–424.

495. Ness, N. F., Donn, B. (1966): 'Concerning a new theory of type I comet tails.' *Mèm. Soc. Roy. Liège,* Ser. 5, **12**, 141–144.

496. Neubauer, F. M. (1987): 'Giotto magnetic-field results on the boundaries of the pile-up region and the magnetic cavity.' *Astron. Astrophys.* **187**, 73–79.

497. Neubauer, F. M. (1988): 'The ionopause transition and boundary layers at Comet Halley from Giotto magnetic field observations.' *J. Geophys. Res.* **93**, 7272–7281.

498. Neubauer, F. M., Glassmeier, K. H., Pohl, M., Raeder, J., Acuna, M. H., Burlaga, L. F., Ness, N. F., Musmann, G., Mariani, F., Wallis, M. K., Ungstrup, E., Schmidt, H. U. (1986): 'First results from the Giotto magnetometer experiment at Comet Halley.' *Nature* **321**, 352–355.

499. Neugebauer, M., Lazarus, A. J., Altwegg, K., Balsiger, H., Goldstein, B. E., Goldstein, R., Neubauer, F. M., Rosenbauer, H., Schwenn, R., Shelley, E. G., Unstrup, E. (1987): 'The pickup of cometary protons by the solar wind.' *Astron. Astrophys.* **187**, 21–24.

500. Neugebauer, M., Russell, C. T., Smith, E. J. (1974): 'Observations of the internal structure of the magnetopause.' *J. Geophys. Res.* **79**, 499–510.

501. Newburn, R. L. (1981): 'A semi-empirical photometric theory of cometary gas and dust production: Application to P/Halley's gas production rates.' In *The Comet Halley Dust and Gas Environment*, Eds. B. Battrick, E. Swallow, ESA SP-174, p. 3–18.

502. Newburn Jr., R. L., Spinrad H. (1989): 'Spectrophotometry of 25 comets: Post-Halley updates for 17 comets plus new observations for eight additional comets.' *Astron. J.* **97**, 552–569.

503. Ney, E. P. (1974): 'Multiband photometry of Comets Kohoutek, Bennett, Bradfield and Encke.' *Icarus* **23**, 551–560.

504. Niedner Jr., M. B., Brandt, J. C. (1978): 'Interplanetary gas. XXIV. Plasma tail disconnection events in comets: Evidence for magnetic field line reconnection at interplanetary sector boundaries.' *Astrophys. J.* **234**, 723–732.

505. Niedner Jr., M. B., Schwingenschuh, K. (1987): 'Plasma-tail activity at the time of the Vega encounters.' *Astron. Astrophys.* **187**, 103–108.

506. O'Dell, R. (1971): 'Nature of particulate matter in comets as determined from infrared observations.' *Astrophys. J.* **166**, 675–681.

507. O'Dell, C. R. (1973): 'A new model for cometary nuclei.' *Icarus* **19**, 137–146.

508. O'Dell, C. R., Robinson, R. R., Krishna Swamy, K. S., McCarthy, P. J., Spinrad, H. (1988): 'C_2 in Comet Halley: Evidence for its being third generation and resolution of the vibrational population discrepancy.' *Astrophys. J.* **334**, 476–488.

509. Ogino, T., Walker, R. J., Ashour-Abdalla, M. (1986): 'An MHD simulation of the interaction of the solar wind with the outflowing plasma from a comet.' *Geophys. Res. Lett.* **13**, 929–932.

510. Omidi, N., Winske, D. (1987): 'A kinetic study of solar wind mass loading and cometary bow shocks.' *J. Geophys. Res.* **92**, 13409–13426.

511. Omidi, N., Winske, D., Wu, C. S. (1986): 'The effects of heavy ions on the formation and structure of cometary bow shocks.' *Icarus* **66**, 165–180.

512. Oparin, A. I. (1938): *The origin of life*. Tranlated and annotated by S. Morgulis, MacMillan; New York.

513. Oparin, A. I. (1953): *The origin of life*. Tranlated and annotated by S. Morgulis, Dover (second edition); New York.

514. Oort, J. H. (1950): 'The structure of the cloud of comets surrounding the Solar System, and a hypothesis concerning its origin.' *Bull. Astron. Inst. Neth.* **11**, 91–110.

515. Oort, J. H., Schmidt, M. (1951): 'Differences between new and old comets.' *Bull. Astron. Inst. Neth.* **11**, 259–269.

516. Öpik, E. (1932): 'Note on stellar perturbations of nearly parabolic orbits.' *Proc. Am. Acad. Arts Sci.* **67**, 169–182.

517. Öpik, E. (1963): 'Survival of cometary nuclei and the asteroids.' *Adv. Astron. Astrophys.* **2**, 219–262.

518. Öpik, E. (1966a): 'The stray bodies in the solar system II: The cometary origin of meteorites.' *Adv. Astron. Astrophys.* **4**, 302–336.

519. Öpik, E. J. (1966b): 'Sun-grazing comets and tidal disruption.' *Irish Astron. J.* **7**, 141–161.

520. Öpik, E. J. (1971): 'Comet families and transneptunian planets.' *Irish Astron. J.* **10**, 35–92.

521. Öpik, E. (1973): 'Comets and the formation of planets.' *Astrophys. Space Sci.* **21**, 307–398.

522. Oró, J. (1961): 'Comets and the formation of biochemical compounds on the primitive earth.' *Nature* **190**, 389–390.

523. Ostro, S. J. (1985): 'Radar observations of asteroids and comets.' *Publ. Astron. Soc. Pac.* **97**, 877–884.

524. Overbeck, V. R., McKay, C. P., Scattergood, T. W., Carle, G. C., Valentin, J. R. (1989): 'The role of cometary particle coalescence in chemical evolution.' *Origins Life Evol. Biosphere* **19**, 39–55.

525. Oya, H., Morioka, A., Miyake, W., Smith, E. J., Tsurutani, B. T. (1986): 'Discovery of cometary kilometric radiation and plasma waves at Comet Halley.' *Nature* **321**, 307–310.

526. Peale, S. J. (1989): 'On the density of Halley's comet.' *Icarus* **82**, 36–49.

527. Pirronello, V., Brown, W. L., Lanzerotti, L. J., Marcantonio, K. J., Simmons, E. H. (1982): 'Formaldehyde formation in a H_2O/CO_2 ice mixture under irradiation by fast ions.' *Astrophys. J.* **262**, 636–640.

528. Podolak, M., Herman, G. (1985): 'Numerical simulations of comet nuclei. II. The effect of the dust mantle.' *Icarus* **61**, 267–277.

529. Pollack, J. B., Cuzzi, J. N. (1980): 'Scattering by nonspherical particles of size comparable to a wavelength: A new semi-empirical theory and its application to tropospheric aerosols.' *J. Atmos. Sci.* **37**, 868–881.

530. Pollack, J. B., Yung, Y. L. (1980): 'Origin and evolution of planetary atmospheres.' *Ann. Rev. Earth Planet. Sci.* **8**, 425–487.

531. Ponnamperuma, C., Ochiai, E. (1982): 'Comets and origin of life.' In *Comets*, Ed. L. L. Wilkening, University of Arizona Press; Tucson, AZ, p. 696–703.

532. Prialnik, D., Bar-Nun, A. (1987): 'On the evolution and activity of cometary nuclei.' *Astrophys. J.* **313**, 893–905.

533. Prialnik, D., Bar-Nun, A. (1988): 'The formation of a permanent dust mantle and its effect on cometary activity.' *Icarus* **74**, 272–283.

534. Prialnik, D., Bar-Nun, A., Podolak, M. (1987): 'Radiogenic heating of comets by ^{26}Al and implications for their time of formation.' *Astrophys. J.* **319**, 993–1002.

535. Prinn, R. G., Fegley Jr., B. (1989): 'Solar nebula chemistry: Origin of planetary, satellite, and cometary volatiles.' In *Origin and Evolution of Planetary and Satellite Atmospheres*, Ed. S. K. Atreya, J. B. Pollack, M. S. Matthews, University Arizona Press; Tucson, AZ, p. 78–136.

536. Probstein, R. F. (1969): 'The dusty gasdynamics of comet heads.' In *Problems of hydrodynamcis and continuum mechanics*, Ed. M. A. Lavrent'ev, Soc. Industr. Appl. Math.; Philadelphia, p. 568–583.

537. Quinn, T., Tremaine, S., Duncan, M (1990): 'Planetary perturbations and the origin of short-period comets,' *Astrophys. J.*, in press.

538. Rahe, J. (1981): 'Comets.' In *Landolt-Börnstein, Neue Serie, Band 2: Astronomie und Astrophysik*, Eds. K. Schaifers, H. H. Voigt, Springer-Verlag, Berlin, Heidelberg, New York, p. 202–228.

539. Rahe, J., Donn, B., Wurm, K. (1969): *Atlas of Cometary Forms*, NASA SP-198, U. S. Gov. Printing Office; Washington, DC.

540. Reinhard, R. (1986): 'The Giotto encounter with Comet Halley.' *Nature* **321**, 313–318.

541. Reitsema, H. J., Delamere, W. A., Williams, A. R., Boice, D. C., Huebner, W. F., Whipple, F. L. (1989): 'Dust distribution in the inner coma of Comet Halley: Comparison with models.' sl Icarus **81**, 31–40.

542. Rème, H., Sauvaud, J. A., d'Uston, C., Cotin, F., Cros, A., Anderson, K. A., Carlson, C. W., Curtis, D. W., Lin, R. P., Mendis, D. A., Korth, A., Richter, A. K. (1986): 'Comet Halley – solar wind interaction from electron measurements aboard Giotto.' *Nature* **321**, 349–352.

543. Remy, F., Mignard, F. (1985): 'Dynamical evolution of the Oort cloud. I. A Monte Carlo simulation.' *Icarus* **63**, 1–19.

544. Retterer, J. M., Chang, T., Jasperse, J. R. (1983): 'Ion acceleration in the supraauroral region: A Monte Carlo model.' *Geophys. Res. Lett.* **10**, 583–586.

545. Richardson, I. G., Cowley, S. W. H., Hynds, R. J., Sanderson, T. R., Wenzel, K.-P., Daly, P. W. (1986): 'Three-dimensional energetic ion bulk flows at Comet P/Giacobini-Zinner.' *Geophys. Res. Lett.* **13**, 415–418.

546. Richardson, I. G., Cowley, S. W. H., Hynds, R. J., Tranquille, C., Sanderson, T. R., Wenzel, K. P. (1987): 'Observations of energetic water-group ions at Comet Giacobini-Zinner: Implications for ion acceleration processes.' *Planet. Space Sci.* **35**, 1323–1345.

547. Richter, K., Keller, H. U. (1988): 'Anomalous dust tail of Comet Kohoutek (1973 XII) near perihelion.' *Astron. Astrophys.* **206**, 136–142.

548. Rickman, H. (1985): 'Interrelations between comets and asteroids.' In *Dynamics of Comets: Their Origin and Evolution* **115**, Eds. A. Carusi, G. B. Valsecchi, D. Reidel Publishing Co.; Dordrecht, Holland, p. 149-172.

549. Rickman, H. (1986): 'Masses and Densities of Comets Halley and Kopff.' In *ESA Workshop on Comet Nucleus Sample Return Mission* ESA SP-249, 195–205.

550. Rickman, H. (1989): 'The nucleus of comet Halley: Surface structure, mean density, gas and dust production.' *Adv. Space Res.* **9**, 59–71.

551. Rickman, H., Fernández, J. A. (1986): 'Formation and blowoff of a cometary dust mantle.' In *The Comet Nucleus Sample Return Mission*, Ed. O. Melita, ESA SP-249, p. 185–194.

552. Rickman, H., Kamél, L., Festou, M. C., Froeschlé, C. (1987): 'Estimates of masses, volumes and densities of short-period comet nuclei.' In *Symposium on the Diversity and Similarity of Comets*, Eds. E. J. Rolfe, B. Battrick. ESA SP-278, p. 471-481.

553. Riedler, W., Schwingenschuh, K., Yeroshenko, Y. G., Styashkin, V. A., Russell, C. T. (1986): 'Magnetic field observations in Comet Halley's coma.' *Nature* **321**, 288–289.

554. Rietmeijer, F. J. M. (1988): 'A quantitative comparison of fine-grained chondritic interplanetary dust and Comet Halley dust.' *Lunar Planet. Sci.* **XIX**, 980–981.

555. Roemer, E. (1966a): 'Cometary nuclei; introductory report.' In *Nature et origine des comètes, Mèm. Soc. Roy. Sci. Liège*, Ser. 5, **12**, 15–22.

556. Roemer, E. (1966b): 'The dimensions of cometary nuclei.' In *Nature et origine des comètes, Mèm. Soc. Roy. Sci. Liège*, Ser. 5, **12**, 23–32.

557. Roemer, E. (1973): 'Comet Notes.' *Mercury* **2**, 17–19.

558. Roessler, K., Nebeling, B. (1988): 'High energy and radiation chemistry in space.' *Lunar Planet. Sci.* **XIX**, 994–995.

559. Russell, C. T., Saunders, M. A., Phillips, J. L. (1986): 'Near-tail reconnection as the cause of cometary tail disconnections.' *J. Geophys. Res.* **91**, 1417–1423.

560. Ryan Jr., M. P., Draganić, I. G. (1986): 'An estimate of the contribution of high energy cosmic-ray protons to the absorbed dose inventory of a cometary nucleus.' *Astrophys. Space Sci.* **125**, 49–67.

561. Safronov, V. S. (1972): *Evolution of the Protoplanetary Cloud and the Formation of the Earth and the Planets*. U. S. Dept. Commerce, Natl. Tech. Information Service; Springfield, VA 22151.

562. Safronov, V. S. (1977): 'Oort's cometary cloud in the light of modern cosmogony.' In *Comets, Asteroids, Meteorites*, Ed. A. H. Delsemme, University of Toledo; Toledo, OH, p. 483–484.

563. Sagdeev, R. Z., Avanesov, G., Shamis, V. A., Ziman, Y. L., Krasikov, V. A., Tarnopolsky, V. A., Kuzmin, A. A., Szeg, K., Merenyi, E., Smith, B. A. (1986a): 'TV experiment in Vega mission: image processing technique and some results.' In *20th ESLAB Symposium on the Exploration of Halley's Comet*, Eds. B. Battrick, E. J. Rolfe, R. Reinhard, ESA SP-250 II, p. 295–305.

564. Sagdeev, R. Z., Avanesov, G. A., Ziman, Y. L., Moroz, V. I., Tarnopolsky, V. I., Zhukov, B. S., Shamis, V. A., Smith, B., Tóth, I. (1986b): 'TV experiment of the Vega mission: Photometry of the nucleus and the inner coma.' In *20th ESLAB Symposium on the Exploration of Halley's Comet*, Eds. B. Battrick, E. J. Rolfe, R. Reinhard, ESA SP-250 II, p. 317-326.

565. Sagdeev, R. Z., Krasikov, V. A., Shamis, V. A., Tarnopolski, V. I., Szegö, K., Tóth, I., Smith, B., Larson, S., Merényi, E. (1986c): 'Rotation period and spin axis of Comet Halley.' In *20th ESLAB Symposium on the Exploration of Halley's Comet*, Eds. B. Battrick, E. J. Rolfe, R. Reinhard, ESA SP-250 II, p. 335–338.

566. Sagdeev, R. Z., Szabó, F., Avanesov, G. A., Cruvellier, P., Szabó, L., Szegö, K., Abergel, A., Balazs, A., Barinov, I. V., Bertaux, J.-L., Blamont, J., Detaille, M., Demarelis, E., Dul'nev, G. N., Endröczy, G., Gardos, M., Kanyo, M., Kostenko, V. I., Krasikov, V. A., Nguyen-Trong, T., Nyitrai, Z., Reny, I., Rusznyak, P., Shamis, V. A., Smith, B., Sukhanov, K. G., Szabó, F., Szalai, S., Tarnopolsky, V. I., Toth, I., Tsukanova, G., Valníček, B. I., Varhalmi, L., Zaiko, Y. K., Zatsepin, S. I., Ziman, Y. L., Zsenei, M., Zhukov, B. S. (1986d): 'Television observations of Comet Halley from Vega spacecraft.' *Nature* **321**, 262-266.

567. Sagdeev, R. Z., Kissel, J., Evlanov, E. N., Mukhin, L. M., Zubkov, B. V., Prilutskii, O. F., Fomenkova, M. N. (1986e): 'Elemental composition of the dust component of Halley comet: Preliminary analysis.' In *20th ESLAB Symposium on the Exploration of Halley's Comet*, Eds. B. Battrick, E. J. Rolfe, R. Reinhard, ESA SP-250 III, p. 349–352.

568. Sagdeev, R. Z., Shapiro, V. D., Shevchenko, V. I., Szegö, K. (1986f): 'MHD turbulence in the solar wind – comet interaction region.' *Geophys. Res. Lett.* **13**, 85–88.

569. Sagdeev, R. Z., Elyasberg, P. E., Moroz, V. I. (1988): 'Is the nucleus of Comet Halley a low density body?' *Nature* **331**, 240–242.

570. Saito, T., Yumoto, K., Hirao, K., Nakagawa, T., Saito, K. (1986): 'Interaction between Comet Halley and the interplanetary magnetic field observed by Sakigake.' *Nature* **321**, 303–307.

571. Sanderson, T. R., Wenzel, K.-P., Daly, P. W., Cowley, S. W. H., Hynds, R. J., Smith, E. J., Bame, S. J., Zwickl, R. D. (1986): 'The interaction of heavy ions from Comet P/Giacobini-Zinner with the solar wind.' *Geophys. Res. Lett.* **13**, 411–414.

572. Saunders, R. S., Fanale, F. P., Parker, T. J., Stephens, I .B., Sutton, S. (1986): 'Properties of filamentary sublimation residues from dispersions of clay in ice.' *Icarus* **66**, 94–104.

573. Sandford, S. A., Allamandola, L. J., Tielens, A. G. G. M., Valero, G. J. (1988): 'Laboratory studies of the infrared spectral properties of CO in astrophysical ices.' *Astrophys. J.* **329**, 498–510.

574. Sauer, K., Motschmann, U., Roatsch, T. (1990): 'Plasma boundaries at Comet Halley.' *Ann. Geophys.*, in press.

575. Scarf, F. L., Coroniti, F. V., Kennel, C. F., Gurnett, D. A., Ip, W.-H., Smith, E. J. (1986): 'Plasma wave observations at Comet Giacobini-Zinner.' *Science* **232**, 377–381.

576. Scattergood, T. W., McKay, C. P., Borucki, W. J., Giver, L. P., van Ghyseghem, H., Parris, J. E., Miller, S. L. (1988): 'Production of organic compounds in plasmas: A comparison among electric sparks, laser-induced plasmas, and UV light.' *Icarus* **81**, 413–428.

577. Schleicher, D. G., A'Hearn, M. F. (1982): 'OH fluorescence in comets: Fluorescence efficiency of the ultraviolet bands.' *Astrophys. J.* **258**, 864–877.

578. Schleicher, D., A'Hearn, M. F. (1986): 'Comets P/Giacobini-Zinner and P/Halley at high dispersion.' In *New Insights in Astrophysics, 8 Years of UV Astronomy with IUE*, ESA SP-263, p. 31–33.

579. Schleicher, D. G., A'Hearn, M. F. (1988): 'The fluorescence of cometary OH.' *Astrophys. J.* **331**, 1058–1077.

580. Schleicher, D. G., Millis, R. L., Tholen, D., Lark, N., Birch, P. V., Martin, R., A'Hearn, M. F. (1986): 'The variability of Halley's comet during the Vega, Planet-A, and Giotto encounters.' In *20th ESLAB Symposium on the Exploration of Halley's Comet*, Eds. B. Battrick, E. J. Rolfe, R. Reinhard, ESA SP-250 I, p. 565-567.

581. Schleicher, D. G., Millis, R. L. Birch, P. V. (1987): 'Photometric observations of Comet P/Giacobini-Zinner.' *Astron. Astrophys.* **187**, 531–538.

582. Schloerb, F. P. (1988): 'Collisional quenching of cometary emission in the 18-cm OH transitions.' *Astrophys. J.* **332**, 524–532.

583. Schloerb, F. P., Kinzel, W. M., Swade, D. A., Irvine, W. M. (1987): 'Observations of HCN in comet P/Halley.' *Astron. Astrophys.* **187**, 475–480.

584. Schmidt, J. F. J. (1863): 'Donati's comet 1858.' *Pub. Athens Obs.* Ser. 1, **1**, 1–74.

585. Schmidt, M. (1951): 'The variation of the total brightness of comets with heliocentric distance.' *Bull. Astron. Inst. Neth.* **11**, 253–258.

586. Schmidt, H. U., Wegmann, R. (1982): 'Plasma flow and magnetic fields in comets.' In *Comets*, Ed. L. L. Wilkening, University of Arizona Press; Tucson, AZ, p. 538–560.

587. Schmidt, H. U., Wegmann, R., Huebner, W. F., Boice, D. C. (1988): 'Cometary gas and plasma flow with detailed chemistry.' *Comp. Phys. Comm.* **49**, 17–59.

588. Schmidt-Voigt, M. (1987): 'Time dependent MHD models for the cometary magnetosphere.' In *Symposium on the Diversity and Similarity of Comets*, Eds. E. J. Rolfe, B. Battrick. ESA SP-278, p. 127–131.

589. Schmitt, B., Klinger, J. (1987): 'Different trapping mechanisms of gases by water ice and their relevance for comet nuclei.' In *Symposium on the Diversity and Similarity of Comets*, Eds. E. J. Rolfe, B. Battrick. ESA SP-278, p. 613–619.

590. Schmitt, B., Espinasse, S., Grim, R. J. A., Greenberg, J. M., Klinger, J. (1989): 'Laboratory studies of cometary ice analogues.' In *Physics and Mechanics of Cometary Materials*, Eds. J. Hunt, T. D. Guyenne, ESA SP-302, p. 65–69.

591. Schopf, J. W. (1983): *Earth's Earliest Biosphere: It's Origin and Evolution*. Princeton University Press; Princeton, NJ.

592. Schwartz, A. W., Joosten, H., Voet, A. B. (1982): 'Prebiotic adenine synthesis via HCN oligomerization in ice.' *Biosystems* **15**, 191–193.

593. Schwarz, G., Craubner, H., Delamere, W. A., Göbel, M., Gonano, M., Huebner, W. F., Keller, H. U., Kramm, R., Mikusch, E., Reitsema, H. J., Whipple, F. L., Wilhelm, K. (1987): 'Detailed analysis of a surface feature on Comet P/Halley.' *Astron. Astrophys.* **187**, 847–851.

594. Schwenn, R., Ip, W.-H., Rosenbauer, H., Balsiger, H., Bühler, F., Goldstein, R., Meier, A., Shelley, E. G. (1987): 'Ion temperature and flow profiles in comet P/Halley's close environment.' *Astron. Astrophys.* **187**, 160–162.

595. Schwingenschuh, K., Riedler, W., Yeroshenko, Y., Phillips, J. L., Russell, C. T., Luhmann, J. G., Fedder, J. A. (1987): 'Magnetic field draping in the Comet Halley coma: Comparison of Vega observations with computer simulations.' *Geophys. Res. Lett.* **14**, 640–643.

596. Sekanina, Z. (1972): 'A model for the nucleus of Encke's comet.' In *The Motion, Evolution of Orbits, and Origin of Comets*, Eds. G. A. Chebotarev, E. I. Kazimirchak-Polonskaya, B. G. Marsden, D. Reidel Publishing Co.; Dordrecht, Holland, p. 301–307.

597. Sekanina, Z. (1973): 'Existence of icy comet tails at large distances from the Sun.' *Astrophys. Lett.* **14**, 175–180.

598. Sekanina, Z. (1976a): 'A continuing controversy: Has the cometary nucleus been resolved?' In *The Study of Comets*, Eds. B. Donn, M. Mumma, W. Jackson, M. A'Hearn, R. Harrington, NASA; Washington, DC. NASA SP-393, p. 537–585.

599. Sekanina, Z. (1976b): 'Progress in our understanding of cometary dust tails.' In *The Study of Comets*, Eds. B. Donn, M. Mumma, W. Jackson, M. A'Hearn, R. Harrington, NASA SP-393, p. 893–942.

600. Sekanina, Z. (1977): 'Relative motions of fragments of the split comets. I. A new approach.' *Icarus* **30**, 574–594.

601. Sekanina, Z. (1981a): 'Rotation and precession of cometary nuclei.' *Ann. Rev. Earth Planet Sci.* **9**, 113–145.

602. Sekanina Z. (1981b): 'Distribution and activity of discrete emission areas on the nucleus of periodic Comet Swift-Tuttle.' *Astron. J.* **86**, 1741-1773.

603. Sekanina, Z. (1982): 'The problem of split comets in review.' In *Comets*, Ed. L. L. Wilkening, University of Arizona Press; Tucson, AZ, p. 251–287.

604. Sekanina, Z. (1984): 'Disappearance and disintegration of comets.' *Icarus* **58**, 81–100.

605. Sekanina, Z. (1985): 'Light variations of periodic Comet Halley beyond 7 AU.' *Astron. Astrophys.* **148**, 299–308.

606. Sekanina, Z. (1987a): 'Nucleus of Comet Halley as a torque-free rigid rotator.' *Nature* **325**, 326–328.

607. Sekanina, Z. (1987b): 'Anisotropic emission from comets: Fans versus jets. I. Concept and Modeling.' In *Symposium on the Diversity and Similarity of Comets*, Eds. E. J. Rolfe, B. Battrick. ESA SP-278, p. 315–322.

608. Sekanina, Z. (1987c): 'Anisotropic emission from comets: Fans versus jets. II. Periodic Comet Tempel 2.' In *Symposium on the Diversity and Similarity of Comets*, Eds. E. J. Rolfe, B. Battrick. ESA SP-278, p. 323–336.

609. Sekanina, Z. (1988a): 'Nucleus of Comet IRAS-Araki-Alcock (1983 VII).' *Astron. J.* **95**, 1876–1894.

610. Sekanina, Z. (1988b): 'Outgassing asymmetry of periodic Comet Encke. I. Apparitions 1924–1984.' *Astron. J.* **95**, 911–924 and 970–971.

611. Sekanina, Z., Ferrell, J. A. (1978): 'Comet West 1976 VI: Discrete bursts of dust, split nucleus, flare-ups, and particle evaporation.' *Astron. J.* **83**, 1675–1689.

612. Sekanina, Z., Ferrell, J. A. (1980): 'The striated dust tail of Comet West 1976 VI as a particle fragmentation phenomenon.' *Astron. J.* **85**, 1538–1554.

613. Sekanina, Z., Larson, S. M. (1984): 'Coma morphology and dust-emission pattern of periodic Comet Halley. II. Nucleus spin vector and modeling of major dust features in 1910.' *Astron. J.* **89**, 1408–1425.

614. Sekanina, Z., Larson, S. M. (1986): 'Coma morphology and dust-emission pattern of periodic Comet Halley. IV. Spin vector refinement and map of discrete dust sources for 1910.' *Astron. J.* **92**, 462–482.

615. Sekanina, Z., Miller, F. D. (1973): 'Comet Bennett 1970 II.' *Science* **179**, 565–567.

616. Sharma, O. P., Patel, V. L. (1986): 'Low-frequency electromagnetic waves driven by gyrotropic gyrating ion beams.' *J. Geophys. Res.* **91**, 1529–1534.

617. Shelley, E. G., Fuselier, S. A., Balsiger, H., Drake, J. F., Geiss, J., Goldstein, B. E., Goldstein, R., Ip, W.-H., Lazarus, A. J. (1987): 'Charge exchange of solar wind ions in the coma of Comet P/Halley.' *Astron. Astrophys.* **187**, 304–306.

618. Sheeley Jr., N. R., Howard, R. A., Koomen, M. J., Michels, D. J. (1982): 'Coronograph observations of two new Sun-grazing comets.' *Nature* **300**, 239–242.

619. Shimizu, M. (1975): 'Ion chemistry in the cometary atmosphere.' *Astrophys. Space Sci.* **36**, 353–361.

620. Shimizu, M. (1976): 'The structure of cometary atmospheres – I: Temperature distribution.' *Astrophys. Space Sci.* **40**, 149–155.

621. Shoemaker, E. M., Wolfe, R. E. (1982): 'Cratering time scales for the Galilean satellites.' In *Satellites of Jupiter,* Ed. D. Morrison, University of Arizona Press; Tucson, AZ, p. 277–329.

622. Shoemaker, E. M., Wolfe, R. E. (1984): 'Evolution of the Uranus-Neptune planetesimal swarm.' *Lunar Planet. Sci.* **XV**, 780–781.

623. Shtejns, K. A. (1972): 'Diffusion of comets from parabolic into nearly parabolic orbits.' In *The Motion, Evolution of Orbits, and Origin of Comets,* Eds. G. A. Chebotarev, E. I. Kazimirchak-Polonskaya, B. G. Marsden, D. Reidel Publishing Co.; Dordrecht, Holland, p. 347-351.

624. Shul'man L. M. (1972): 'The evolution of cometary nuclei.' In *The Motion Evolution of Orbits and Origin of Comets,* Eds. G. A. Chebotarev, E. I. Kazimirchak-Polonskaya, B. G. Marsden, D. Reidel Publishing Co.; Dordrecht, Holland, p. 271–276.

625. Shul'man, L. M. (1981): 'A two-layer model of the cometary nucleus.' *Astrometria Astrofisica* **45**, 21–34.

626. Shul'man, L. M. (1983): 'Have cometary nuclei any internal sources of energy?' In *Proc. Intern. Conference on Cometary Exploration,* **1**, Ed. T. Gombosi, pp.55–58, Central Research Institute for Physics, Hungarian Academy of Sciences.

627. Simonelli, D. P., Pollack, J. B., McKay, C. P., Reynolds, R. T., Summers, A. L. (1989): 'The carbon budget in the outer solar nebula.' *Icarus* **82**, 1–35.

628. Simpson, J. A., Rabinowitz, D., Tuzzolino, A. J., Ksanfomality, L. V., Sagdeev, R. Z. (1986a): 'Halley's comet coma dust particle mass spectra, flux distributions and jet structures derived from measurements on the Vega-1 and Vega-2 spacecraft.' In *20th ESLAB Symposium on the Exploration of Halley's Comet,* Eds. B. Battrick, E. J. Rolfe, R. Reinhard, ESA SP-250 II, p. 11–16.

629. Simpson, J. A., Sagdeev, R. Z., Tuzzolino, A. J., Perkins, M. A., Ksanfomality, L. V., Rabinowitz, D., Lentz, G. A., Afonin, V. V., Eroe, J., Keppler, E., Kosorokov, J., Petrova, E., Szab, L., Umlauft, G. (1986b): 'Dust counter and mass analyser (DUCMA) measurements of Comet Halley's coma from Vega spacecraft.' *Nature* **321**, 278–280.

630. Simpson, J. A., Rabinowitz, D., Tuzzolino, A. J., Ksanformality, L. V., Sagdeev, R. Z. (1987): 'The dust coma of comet P/Halley: Measurements on the Vega-1 and Vega-2 spacecraft.' *Astron. Astrophys.* **187**, 742–752.

631. Singh, P. D., Dalgarno, A. (1987): 'Photodissociation lifetimes of CH and CD radicals in comets.' In *Symposium on the Diversity and Similarity of Comets,* Eds. E. J. Rolfe, B. Battrick. ESA SP-278, p. 177–180.

632. Siscoe, G. L., Slavin, J. A., Smith, E. J., Tsurutani, B. T., Jones, D. E., Mendis, D. A. (1986): 'Statics and dynamics of Giacobini-Zinner magnetic tail.' *Geophys. Res. Lett.* **13**, 287–290.

633. Sleep, N. H., Zahnle, K. J., Kasting, J. F., Morowitz, H. J. (1989): 'Annihilation of ecosystems by large asteroid impacts on the early Earth.' *Nature* **342**, 139–142.

634. Smith, B. A., Larson, S. M., Szegö , K., Sagdeev, R. Z. (1987): 'Rejection of a proposed 7.4-day rotation period of the Comet Halley nucleus.' *Nature* **326**, 573–574.

635. Smith, B., Szegö, K., Larson, S., Merényi, E., Tóth, I., Sagdeev, R. Z., Avanesov, G. A., Krasikov, V. A., Shamis, V. A., Tarnapolsky, V. I. (1986): 'The spatial distribution of dust jets seen at Vega-2 flyby.' In *20th ESLAB Symposium on the Exploration of Halley's Comet,* Eds. B. Battrick, E. J. Rolfe, R. Reinhard, ESA SP-250 II, p. 327–332.

636. Smith, E. J., Tsurutani, B. T., Slavin, J. A., Jones, D. E., Siscoe, G. L., Mendis, D. A. (1986): 'International cometary explorer encounter with Giacobini-Zinner: Magnetic field observations.' *Science* **232**, 382–385.

637. Smoluchowski, R. (1981a): 'Amorphous ice and the behaviour of cometary nuclei.' *Astrophys. J.* **244**, L31–L34.

638. Smoluchowski, R. (1981b): 'Heat content and evolution of cometary nuclei.' *Icarus* **47**, 312–319.

639. Smoluchowski, R. (1985): 'Amorphous and porous ices in cometary nuclei.' In *Ices in the Solar System,* Eds. J. Klinger, D. Benest, A. Dollfus, R. Smoluchowski, D. Reidel Pulishing Co.; Dordrecht, Boston, Lancaster, p. 397-406.

640. Snyder, L. E., Palmer, P., de Pater, I. (1989): 'Radio detection of formaldehyde emission from comet Halley.' *Astron. J.* **97**, 246–253.

641. Šolc, M., Vanýsek, V., Kissel, J. (1986): 'Carbon stable isotopes in comets after encounters with P/Halley.' In *20th ESLAB Symposium on the Exploration of Halley's Comet*, Eds. B. Battrick, E. J. Rolfe, R. Reinhard, ESA SP-250 III, p. 373–376.

642. Šolc, M., Jessberger, E. K., Hsiung, P., Kissel, J. (1987a): 'Halley dust composition.' In *Proc. 10th regional meeting of the IAU*, Prag, p. 47–51.

643. Šolc, M., Vanýsek, V., Kissel, J. (1987b): 'Carbon-isotope ratio in PUMA 1 spectra of P/Halley dust.' *Astron. Astrophys.* **187**, 385–387.

644. Somogyi, A. J., Gringauz, K. I., Szegö, K., Szabó, L., Kozma, G., Remizov, A. P., Erö Jr., J., Klimenko, I. N., T.-Szücs, I., Verigin, M. I., Windberg, J., Cravens, T. E., Dyachkov, A., Erdös, G., Faragó, M., Gombosi, T. I., Kecskeméty, K., Keppler, E., Kovács Jr., T., Kondor, A., Logachev, Y. I., Lohonyai, L., Marsden, R., Redl, R., Richter, A. K., Stolpovskii, V. G., Szabó, J., Szentpétery, I., Szepesváry, A., Tátrallyay, M., Varga, A., Vladimirova, G. A., Wenzel, K. P., Zarándy, A. (1986): 'First observations of energetic particles near Comet Halley.' *Nature* **321**, 285–288.

645. Somogyi, A. J., Axford, W. I., Erdös, G., Ip, W.-H., Shapiro, V. D., Shevchenko, V. I. (1990): 'Particle acceleration in the plasma fields near Comet Halley.' In *Comet Halley 1986: World-Wide Investigations, Results, and Interpretations*, Eds. J. Mason, P. Moore, Ellis Horwood Ltd.; Chichester, in press.

646. Stagg, C. R., Bailey, M. E. (1989): 'Stochastic capture of short-period comets.' *Mon. Not. Roy. Astron. Soc.* **241**, 507–541.

647. Stawikowski, A., Greenstein, J. L. (1964): 'The isotope ratio C^{12}/C^{13} in a comet.' *Astrophys. J.* **140**, 1280–1291.

648. Stefanik, R. P. (1966): 'On thirteen split comets.' *Mém. Roy. Soc. Sci. Liège* **12**, 29–32.

649. Stern, S. A. (1988): 'Collisions in the Oort cloud.' *Icarus* **73**, 499–507.

650. Stern, S. A., Shull, J. M. (1988): 'The influence of supernovae and passing stars on comets in the Oort cloud.' *Nature* **332**, 407–411.

651. Stewart, A. I. F. (1987): Pioneer Venus measurements of H, O, and C production in Comet P/Halley near perihelion.' *Astron. Astrophys.* **187**, 369–374.

652. Štohl, J. (1986): 'On meteor contribution by short-period comets.' In *20th ESLAB Symposium on the Exploration of Halley's Comet*, Eds. B. Battrick, E. J. Rolfe, R. Reinhard, ESA SP-250 II, p. 225–228.

653. Stothers, R. B. (1988): 'Structure of Oort's comet cloud inferred from terrestrial impact craters.' *Observatory* **108**, 1–9.

654. Strazzulla, G., Pirronello, V., Foti, G. (1983): 'Destruction of ice grains in T Tauri stars.' *Astrophys. J.* **271**, 255–258.

655. Sussman, G.J., Wisdom, J. (1988): 'Numerical evidence that the motion of Pluto is chaotic.' *Science* **241**, 433–437.

656. Svoreň, J. (1982): 'Maximum sizes of cometary nuclei.' In *Cometary Exploration*, Ed. T. I. Gombosi. Central Research Institute for Physics, Hungarian Academy of Sciences; Budapest, I, p. 31–37.

657. Swings, P. (1941): 'Complex structure of cometary bands tentatively ascribed to the contour of the solar spectrum.' *Lick Obs. Bull.* **19**, 131–136.

658. Sykes, M. V., Lebofsky, L. A., Hunten, D. M. Low, F. J. (1986): 'The discovery of dust trails in the orbits of periodic comets.' *Science* **232**, 1115–1117.

659. Tholen, D. J., Cruikshank, D. P., Hammel, H. B., Piscitelli, J. R., Hartmann, W. K., Lark, N. (1986): 'A comparison of the continuum colors of P/Halley, other comets, and asteroids.' In *20th ESLAB Symposium on the Exploration of Halley's Comet*, Eds. B. Battrick, E. J. Rolfe, R. Reinhard, ESA SP-250 III, p. 503–507.

660. Tholen, D. J., Hartmann, W. K., Cruikshank, D. P. (1988): '(2060) Chiron.' *IAU Circular No. 4554*.

661. Thomas, N., Keller, H. U. (1987a): 'Comet P/Halley's near-nucleus jet activity.' In *Symposium on the Diversity and Similarity of Comets*, Eds. E. J. Rolfe, B. Battrick. ESA SP-278, p. 337–342.

662. Thomas, N., Keller, H. U. (1987b): 'Fine dust structures in the emission of Comet P/Halley observed by the Halley Multicolour Camera on board Giotto.' *Astron. Astrophys.* **187**, 843–846.

663. Thomas, N., Keller, H. U. (1988): 'Global distribution of dust in the inner coma of Comet P/Halley observed by the Halley Multicolour Camera.' In *Proceedings of the Symposium Dust in the Universe*, Eds. M. E. Bailey, D. A. Williams, Cambridge University Press; Cambridge, p. 540–541.

664. Thomas, N., Boice, D. C., Huebner, W. F., Keller, H. U. (1988): 'Intensity profiles of dust near extended sources on Comet Halley.' *Nature* **332**, 51–52.

665. Thompson, W. R., Murray, B. G., Khare, B. N., Sagan, C. (1988): 'Coloration and darkening of methane clathrate and other ices by charged particle irradiation: Applications to the outer solar system.' *J. Geophys. Res.* **92**, 14933–14948.

666. Thomsen, M. F., Bame, S. J., Feldman, W. C., Gosling, J. T., McComas, D. J., Young, D. T. (1986): 'The comet/solar wind transition region at Giacobini-Zinner.' *Geophys. Res. Lett.* **13**, 393–396.

667. Thorne, R. M., Tsurutani, B. T. (1987): 'Resonant interactions between cometary ions and low frequency electromagnetic waves.' *Planet. Space Sci.* **35**, 1501–1511.

668. Tokunaga, A. T., Golish, W. F., Griep, D. M., Kaminski, C., Hanner, M. S. (1986): 'The NASA infrared telescope facility Comet Halley monitoring program. I. Preperihelion results.' *Astron. J.* **92**, 1183–1190.

669. Toptyghin, I. N. (1980): 'Acceleration of particles by shocks in a cosmic plasma.' *Space Sci. Rev.* **26**, 157–213.

670. Torbett, M. V. (1986): 'Injection of Oort cloud comets to the inner solar system by galactic tidal fields.' *Mon. Not. Roy. Astron. Soc.* **223**, 885–895.

671. Tóth, I., Szegö, K., Kondor, A. (1987): 'Dust photometry in the near nucleus region of Comet Halley from Vega-2 observations.' In *Symposium on the Diversity and Similarity of Comets*, Eds. E. J. Rolfe, B. Battrick. ESA SP-278, p. 343–347.

672. Tscharnuter, W. M. (1980): '1D, 2D, and 3D collapse of interstellar clouds.' *Space Sci. Rev.* **27**, 235–246.

673. Tsuda, Y. (1961): 'Solid state polymerization of formaldehyde induced by ionizing radiation.' *J. Polymer Sci.* **49**, 369–376.

674. Tsurutani, B. T., Smith, E. J. (1986): 'Strong hydromagnetic turbulence associated with Comet Giacobini-Zinner.' *Geophys. Res. Lett.* **13**, 259–262.

675. Tsurutani, B. T., Thorne, R. M., Smith, E. J., Gosling, J. T., Matsumoto, H. (1987): 'Steepend magnetosonic waves at Comet Giacobini-Zinner.' *J. Geophys. Res.* **92**, 11074–11082.

676. Ulich, B. L., Conklin, E. K. (1974): 'Detection of methyl cyanide in Comet Kohoutek.' *Nature* **248**, 121–122.

677. Vaisberg, O. L., Smirnov, V. N., Gorn, L. S., Iovlev, M. V., Balikchin, M. A., Klimov, S. I., Savin, S. P., Shapiro, S. P., Shevchenko, V. I. (1986a): 'Dust coma structure of Comet Halley from SP-1 detector measurements.' *Nature* **321**, 274–276.

678. Vaisberg, O., Smirnov, V., Omelchenko, A. (1986b): 'Spatial distribution of low-mass dust particles ($m \leq 10^{-10}$ g) in Comet Halley coma. In *20th ESLAB Symposium on the Exploration of Halley's Comet*, Eds. B. Battrick, E. J. Rolfe, R. Reinhard, ESA SP-250 II, p. 17–23.

679. Vaisberg, O. L., Zastenker, G., Smirnov, V., Khazanov, B., Omelchenko, A., Fedorov, A., Zakharov, D. (1987a): 'Spatial distribution of heavy ions in Comet P/Halley's coma.' *Astron. Astrophys.* **187**, 183–190.

680. Vaisberg, O. L., Smirnov, V., Omelchenko, A., Gorn, L., Iovlev, M. (1987b): 'Spatial and mass distribution of low-mass dust particles ($m < 10^{-10}$ g) in Comet P/Halley's coma.' *Astron. Astrophys.* **187**, 753–760.

681. Valtonen, M. J., Innanen, K. A. (1982): 'The capture of interstellar comets.' *Astrophys. J.* **255**, 307–315.

682. Van Dishoeck, E. F., Dalgarno, A. (1984): 'The dissociation of OH and OD in comets by solar radiation.' *Icarus* **59**, 305–313.

683. Van Flandern, T. A. (1978): 'A former asteroidal planet as the origin of comets.' *Icarus* **36**, 51–74.

684. Vanýsek, V. (1987): 'Isotopic abundances in comets.' In *Astrochemistry*. Eds. M. S. Vardya, S. P. Tarafdar, D. Reidel Publishing Company; Dordrecht, Holland, p. 461–467.

685. Vanýsek, V., Rahe, J. (1978): 'The $^{12}C/^{13}C$ isotope ratio in comets, stars and interstellar matter.' *Moon Planets* **18**, 441–447.

686. Vanýsek, V., Wickramasinghe, N. C. (1975): 'Formaldehyde polymers in comets.' *Astrophys. Space Sci.* **33**, L19–L28.

687. Vanýsek, V., Valnicek, B., Sudova, J. (1988): 'Flux ratio of C_2 Swan bands in the innermost atmosphere of comet Halley.' *Nature* **333**, 435–436.

688. Verigin, M. I., Axford, W. I., Gringauz, K. I., Richter, A. K. (1987): 'Acceleration of cometary plasma in the vicinity of Comet Halley associated with an interplanetary magnetic field polarity change.' *Geophys. Res. Lett.* **14**, 987–990.
689. Verniani, F. (1969): 'Structure and fragmentation of meteoroids.' *Space Sci. Rev.* **10**, 230–261.
690. Verniani, F. (1973): 'An analysis of the physical parameters of 5759 faint radio meteors.' *J. Geophys. Res.* **78**, 8429–8462.
691. Vorontsov-Velyaminov, B. (1946): 'Structure and mass of cometary nuclei.' *Astrophys. J.* **104**, 226–233.
692. Vsekhsvyatskij, S. K. (1958): *Fizicheskie Kharakteristiki Komet*; Moscow. Translated as *Physical Characteristics of Comets*, Israel Program for Scientific Translations; Jerusalem, (1964).
693. Vsekhsvyatskij, S. K. (1963): 'Absolute magnitudes of 1954–1960 comets.' *Sov. Astron.* **6**, 849–854.
694. Vsekhsvyatskij, S. K. (1964): 'Physical characteristics of 1961–1963 comets.' *Sov. Astron.* **8**, 429–431.
695. Vsekhsvyatskij, S. K. (1966): 'Comet cosmogony of Lagrange and the problem of the solar system.' *Mèm. Soc. Roy. Sci. Liège*, Ser. 5, **12**, 495-515.
696. Vsekhsvyatskij, S. K. (1967): 'Physical characteristics of comets observed during 1961–1965.' *Sov. Astron.* **10**, 1034–1041.
697. Vsekhsvyatskij, S. K. (1972): 'The origin and evolution of the comets and other small bodies in the Solar System.' In *The Motion, Evolution of Orbits, and Origin of Comets*, Eds. G. A. Chebotarev, E. I. Kazimirchak-Polonskaya, B. G. Marsden, D. Reidel Publishing Co.; Dordrecht, Holland, p. 413-418.
698. Vsekhsvyatskij, S. K. Il'ichishina, N. I. (1971): 'Absolute magnitudes of comets, 1965–1969.' *Sov. Astron.* **15**, 310–313.
699. Walker, J. C. G. (1977): *Evolution of the Atmosphere*. MacMillan Publishing Co.; New York.
700. Wallace, L. V., Miller, F. D. (1958): 'Isophote configurations for model comets.' *Astron. J.* **63**, 213–219.
701. Wallis, M. K. (1971): 'Shock-free deceleration of the solar wind?' *Nature* **233**, 23–25.
702. Wallis, M. K. (1980): 'Radiogenic melting of primordial comet interiors.' *Nature* **284**, 431–433.
703. Wallis, M. K. (1982): 'Dusty gasdynamics in real comets'. In *Comets*, Ed. L. L. Wilkening, University of Arizona Press; Tucson, AZ, p. 357–369.
704. Wallis, M. K., Dryer, M. (1976): 'Sun and comets as sources in an external flow.' *Astrophys. J.* **205**, 895–899.
705. Wallis, M. K., Johnstone, A. D. (1982): 'Implanted ions and the draped cometary field.' In *Cometary Exploration*, Ed. T. I. Gombosi. Central Research Institute for Physics, Hungarian Academy of Sciences; Budapest, I, p. 307–311.
706. Wallis, M. K., Macpherson, A. K. (1981): 'On the outgassing and jet thrust of snowball comets.' *Astron. Astrophys.* **98**, 45–49.
707. Wallis, M. K., Ong, R. S. B. (1975): 'Strongly-cooled ionizing plasma flows with application to Venus.' *Planet. Space Sci.* **23**, 713–721.
708. Wasson, J. T. (1985): *Meteorites, Their Record of Early Solar System History*. Freeman & Co.; New York, NY.
709. Wegmann, R., Schmidt, H. U., Huebner, W. F., Boice, D. C. (1987): 'Cometary MHD and chemistry.' *Asron. Astrophys.* **187**, 339–350.
710. Weidenschilling, S. J. (1988): 'Formation processes and time scales for meteorite parent bodies.' In *Meteorites in the early Solar System*, Eds. J. F. Kerridge, M. S. Matthews, University Arizona Press; Tucson, AZ, p. 348–371.
711. Weidenschilling, S. J., Donn, B., Meakin, P. (1989): 'The physics of planetesimal formation.' *Formation and Evolution of Planetary Systems*, Eds. H. A. Weaver, F. Paresce, L. Danly, Cambrige University Press; Cambridge, p. 117–136.
712. Weigert, A. (1959): 'Halo production in Comet 1925 II.' *Astron. Nachr.* **285**, 117–128.
713. Weissman, P. R. (1979): 'Physical and dynamical evolution of long-period comets.' In *Dynamics of the Solar System*, Ed. R. L. Duncombe, D. Reidel Publishing Co.; Dordrecht, Holland, p. 277–282.
714. Weissman, P. R. (1980): 'Physical loss of long-period comets.' *Astron. Astrophys.* **85**, 191–196.

715. Weissman, P. R. (1985): 'Dynamical evolution of the Oort cloud.' In *Dynamics of Comets: Their Origin and Evolution* 115, Eds. A. Carusi, G. B. Valsecchi, D. Reidel Publishing Co.; Dordrecht, Holland, p. 87–96.

716. Weissmann, P. R. (1986a): 'Are cometary nuclei primordial rubble piles?' *Nature* **320**, 242–244.

717. Weissman, P. R. (1986b): 'Post-perihelion brightening of Halley's comet: A case of nuclear summer.' In *20th ESLAB Symposium on the Exploration of Halley's Comet,* Eds. B. Battrick, E. J. Rolfe, R. Reinhard, ESA SP-250 III, p. 517–522.

718. Weissman, P. R. (1990): 'Physical processing of cometary nuclei since their formation.' In *Comet Halley 1986: World-Wide Investigations, Results, and Interpretations,* Eds. J. Mason, P. Moore, Ellis Horwood Ltd.; Chichester, in press.

719. Weissman, P. R., Kieffer, H. H. (1981): 'Thermal modeling of cometary nuclei.' *Icarus* **47**, 302–311.

720. Weissman, P. R., A'Hearn, M. F., McFadden, L. A., Rickman, H. (1990): 'Evolution of comets into asteroids.' In *Asteroids II,* Eds. R. Binzel, T. Gehrels, M. S. Matthews, University of Arizona Press; Tucson, AZ, in press.

721. West, R. M. (1988): 'Halley's comet (Part I): Ground-based observations.' In *IAU Highlights of Astronomy* **8**, Ed. D. McNally, Kluwer Academic Publishing; Dordrecht, Boston, London, p. 3–16.

722. Westfall, R. S. (1980): *Never at Rest: A Biography of Isaac Newton.* Cambridge University Press; Cambridge.

723. Wetherill, G. W. (1988): 'Where do the Apollo objects come from?' *Icarus* **76**, 1–18.

724. Wetherill, G. W., ReVelle, D. O. (1982): 'Relationships between comets, large meteors, and meteorites'. In *Comets,* Ed. L. L. Wilkening, University of Arizona Press; Tucson, AZ, p. 297–319.

725. Whipple, F. L. (1950): 'A comet model I. The acceleration of Comet Encke.' *Astrophys. J.* **111**, 375–394.

726. Whipple, F. L. (1951): 'A comet model II. Physical relations for comets and meteors.' *Astrophys. J.* **113**, 464–474.

727. Whipple, F. L. (1955): 'A comet model III. The zodiacal light.' *Astrophys. J.* **121**, 750–770.

728. Whipple, F. L. (1962): 'On the distribution of semimajor axes among comet orbits.' *Astron. J.* **67**, 1–9.

729. Whipple, F. L. (1963): 'On the structure of the cometary nucleus.' In *The Solar System IV: The Moon, Meteorites, and Comets,* Eds. B. M. Middlehurst, G. P. Kuiper. University of Chicago Press; Chicago, p. 639–663.

730. Whipple, F. L. (1977): 'The constitution of cometary nuclei.' In *Comets, Asteroids, Meteorites,* Ed. A. H. Delsemme, University of Toledo; Toledo, OH, p. 25–36.

731. Whipple, F. L. (1978a): 'Rotation period of Comet Donati.' *Nature* **273**, 134–135.

732. Whipple, F. L. (1978b): 'Cometary brightness variation and nucleus structure.' *Moon Planets* **18**, 343–359.

733. Whipple, F. L. (1980): 'Rotation and outbursts of Comet P/Schwassmann-Wachmann 1.' *Astron. J.* **85**, 305–313.

734. Whipple, F. L. (1981a): 'On observing comets for nuclear rotation.' In *Modern Observational Techniques for Comets,* NASA/JPL report 81–68; Pasadena p. 191–201.

735. Whipple, F. L. (1981b): 'The nature of comets.' In *Comets and the Origin of Life,* Ed. C. Ponnamperuma, D. Reidel Publishing Co.; Dordrecht, Holland, p. 1–20.

736. Whipple, F. L. (1985): *The Mystery of Comets.* Smithsonian Institute; Cambridge, MA.

737. Whipple, F. L. (1986): 'The cometary nucleus – current concepts'. In *20th ESLAB Symposium on the Exploration of Halley's Comet,* Eds. B. Battrick, E. J. Rolfe, R. Reinhard, ESA SP-250 II, p. 281–288.

738. Whipple, F. L. (1990): 'The forest and the trees.' In *Comets in the Post-Halley Era,* Eds. R. L. Newburn, J. Rahe, M. Neugebauer, in press.

739. Whipple, F. L., Douglas-Hamilton, D. H. (1966): 'Brightness changes in periodic comets.' *Mèm. Soc. Roy. Sci. Liège,* Ser. 5, **12**, 469–480.

740. Whipple, F. L., Sekanina, Z. (1979): 'Comet Encke: Precession of the spin axis, nongravitational motion and sublimation.' *Astron. J.* **84**, 1894–1909.

741. Whipple, F. L., Stefanik, R. P. (1966): 'On the physics and splitting of cometary nuclei.' *Mèm. Soc. Roy. Sci. Liège,* Ser. 5, **12**, 33–52.

742. Wickramasinghe, N. C. (1974): 'Formaldehyde polymers in interstellar space.' *Nature* **252**, 462–463.

743. Wickramasinghe, N. C. (1975): 'Polyoxymethylene polymers as interstellar grains.' *Mon. Not. Roy. Astron. Soc.* **170**, 11P–16P.

744. Wilhelm, K. (1987): 'Rotation and precession of Comet Halley.' *Nature* **327**, 27–30.

745. Wilhelm, K., Cosmovici, C. B., Delamere, A. W., Huebner, W. F., Keller, H. U., Reitsema, H., Schmidt, H. U., Whipple, F. L. (1986): 'A three-dimensional model of the nucleus of Comet Halley.' In *20th ESLAB Symposium on the Exploration of Halley's Comet*, Eds. B. Battrick, E. J. Rolfe, R. Reinhard, ESA SP-250 II, p. 367–369.

746. Wilken, B., Johnstone, A. D., Coates, A., Borg, H., Amata, E., Formisano, V., Jockers, K., Rosenbauer, H., Stüdemann, W., Thomsen, M. F., Winningham, J. D. (1987): 'Pick-up ions at Comet P/Halley's bow shock: Observations with the IIS spectrometer on Giotto.' *Astron. Astrophys.* **187**, 153–159.

747. Wilkening, L. L. (1982): *Comets*. University of Arizona Press; Tucson, AZ.

748. Winske, D., Wu, C. S., Li, Y. Y., Mou, Z. Z., Guo, S. Y. (1985a): 'Coupling of newborn ions to the solar wind by electromagnetic instabilities and their interaction with the bow shock.' *J. Geophys. Res.* **90**, 2713–2726.

749. Winske, D., Tanaka, M., Wu, C. S., Quest, K. B. (1985b): 'Coupling of newborn ions to the solar wind by electromagnetic instabilities and their interaction with the bow shock.' *J. Geophys. Res.* **90**, 123–136.

750. Wisdom, J. (1985): 'Meteorites may follow a chaotic route to Earth.' *Nature* **315**, 731–733.

751. Wood, J. (1979): *The Solar System*. Prentice Hall; Englewood Cliffs, NJ.

752. Wood, J. A., Chang, S. (1985): *The Cosmic History of Biogenic Elements and Compounds*. NASA SP-476.

753. Woods, T. N., Feldman, P. D., Dymond, K. F. (1987): 'The atomic carbon distribution in the coma of Comet P/Halley.' *Astron. Astrophys.* **187**, 380–384.

754. Woods, T. N., Feldman, P. D., Dymond, K. F., Sahnow, D. J. (1986): 'Rocket ultraviolet spectroscopy of Comet Halley and abundance of carbon monoxide and carbon.' *Nature* **324**, 436–438.

755. Wu, C. S., Davidson, R. C. (1972): 'Electromagnetic instabilities produced by neutral particle ionization in interplanetary space.' *J. Geophys. Res.* **77**, 5399–5406.

756. Wu, C. S., Hartle, R. E. (1974): 'Further remarks on plasma instabilities produced by ions born in the solar wind.' *J. Geophys. Res.* **79**, 283–285.

757. Wu, C. S., Gaffey, J. D., Liberman, B. (1981): 'Statistical acceleration of electrons by lower-hybrid turbulence.' *J. Plasma Phys.* **25**, 391–401.

758. Wu, Z.-J. (1987): 'Calculation of the shape of the contact surface at Comet Halley.' In *Symposium on the Diversity and Similarity of Comets*, Eds. E. J. Rolfe, B. Battrick. ESA SP-278, p. 69–73.

759. Wurm, K. (1934): 'Beitrag zur Deutung der Vorgänge in Kometen. I.' *Z. Astrophys.* **8**, 281–291.

760. Wurm, K. (1935): 'Beitrag zur Deutung der Vorgänge in Kometen. II.' *Z. Astrophys.* **9**, 62–78.

761. Wurm, K. (1943): 'Die Natur der Kometen.' *Mitt. Hamburger Sternwarte* **8**, Nr. 51.

762. Wyatt, S. P. (1969): 'The electrostatic charge of interplanetary grains.' *Planet. Space Sci.* **17**, 155–171.

763. Wyckoff, S. (1982): 'Overview of comet observations.' In *Comets*, Ed. L. L. Wilkening, University of Arizona Press; Tucson, AZ, p. 3–55.

764. Wyckoff, S., Lindholm, E. (1989): 'On the carbon and nitrogen isotope abundance ratios in Comet Halley.' *Adv. Space Res.* **9**, 151–155.

765. Wyckoff, S., Wehinger, P. A., Spinrad, H., Belton, M. J. S. (1988): 'Abundances in Comet Halley at the time of the spacecraft encounters.' *Astrophys. J.* **325**, 927–938.

766. Wyckoff, S., Lindholm, E., Wehinger, P. A., Peterson, B. A., Zucconi, J.-M., Festou, M. C. (1989): 'The $^{12}C/^{13}C$ abundance ratio in Comet Halley.' *Astrophys. J.* **339**, 488–500.

767. Yabushita, S. (1983): 'Distribution of cometary binding energies based on the assumption of steady state.' *Mon. Not. Roy. Astron. Soc.* **204**, 1185–1191.

768. Yamamoto, T. (1985): 'Formation history and environment of cometary nuclei.' In *Ices in the Solar System*, Eds. J. Klinger, D. Benest, A. Dollfus, R. Smoluchowski, D. Reidel Pulishing Co.; Dordrecht, Boston, Lancaster, p. 205–219.

769. Yamamoto, T., Ashihara, O. (1985): 'Condensation of ice particles in the vicinity of a cometary nucleus.' *Astron. Astrophys.* **152**, L17–L20.

770. Yang, J., Epstein, S. (1983): 'Interstellar organic matter in meteorites.' *Geochim. Cosmochim. Acta* **47**, 2199–2216.

771. Yavnel, A. A. (1977): 'Chemical composition of meteoric and meteoritic matter.' In *Comets, Asteroids, Meteorites,* Ed. A. H. Delsemme, University of Toledo; Toledo, OH, p. 133–135.

772. Yeomans, D. K. (1985): 'The dynamical history of Comet Halley.' In *Dynamics of Comets: Their Origin and Evolution* **115**, Eds. A. Carusi, G. B. Valsecchi, D. Reidel Publishing Co.; Dordrecht, Holland, p. 389–398.

773. Zank, G. P. (1988): 'Oscillatory cosmic ray-shock structures.' *Astrophys. Space Sci.* **140**, 301–324.

774. Zank, G. P. (1990): 'Structure of the Halley bow shock.' *Plan. Space Sci.*, in press.

775. Zinner, E., Epstein, S. (1987): 'Heavy carbon in individual oxide grains from the Murchison meteorite.' *Earth Planet. Sci. Lett.* **84**, 359–368.

776. Zinner, E., McKeegan, K. D., Walker, R. M. (1983): 'Laboratory measurements of D/H ratios in interplanetary dust.' *Nature* **305**, 119–121.

Subject Index

accommodation coefficient   254, 267
accretion *see also* agglomeration, aggregation
   15, 48, 180, 247–248, 250, 252, 259–260,
   262–265, 268, 281, 324–325
acetonitrile   320
active area *or* region   5, 7, 8, 13, 20–21, 28,
   34–35, 38–41, 44, 47, 50, 52–54, 62, 67–68,
   72–75, 110, 123, 129, 130, 134, 135, 139–140
   172, 250–251, 260, 262, 290, 302, 306–307,
   310, 326–327
activity   1, 2, 8, 13–18, 20–23, 26–27, 29,
   35–36, 39–41, 43–47, 50, 53–54, 56, 59–62,
   65–66, 68, 72–74, 83, 86, 111, 123, 130–131,
   134, 157, 173, 186–187, 212, 214, 246, 249,
   261, 268–269, 283, 286–287, 291–292, 294,
   300, 307, 310–312, 326–328, 331
– curve   61, 296
adenine   164, 320, 322
adsorption   24, 314
advection   46
age   14, 63–65, 108, 143–145, 147, 149, 172,
   241, 248–249, 261, 269, 272, 277, 292–294
agglomeration *see also* accretion, aggregation
   8, 24–25, 48, 55, 68
aggregation *see also* accretion, agglomeration
   2, 12, 15, 68, 252, 257, 263, 266, 302, 316–317
albedo   7, 11–12, 13, 17–18, 20–23, 26–27, 36,
   43, 45, 59, 66, 150, 158, 239, 260, 300–301,
   310
Alfvén wave   7, 215, 227–229, 233
amino acid *see also* complex molecules   318,
   320, 323
angular momentum   34–35, 239, 262, 270–271,
   273–276
antitail   114, 143–145, 148
aphelion   2, 4, 69, 72, 239–240, 247, 268–271,
   274, 281–283, 302
Apollo–Amor asteroid   277, 301
Aquarids   63
argument of perihelion   235, 243, 275
Aristotle   14
Arpigny   114
ascending node   235
asteroid   2, 3, 9, 22–23, 36, 69, 158, 176, 249,
   268–269, 277, 292, 301, 319

atomic ions *see also* molecular ions:
   C^+ 87, 189, 192–193
   $^{13}C^+$ 166
   H^+ 189, 219–220
   He^{2+} 194, 218
   Mg^+ 102
   $^{24}Mg^{++}$ 166
   Na^+ 102–103
   S^+ 192

Bessel   70, 83, 113
Birkle   134, 185
Brahe   14
Bredichin   113
brightness   16–18, 21–22, 31, 34–40, 69, 72,
   86, 90, 93, 95, 100, 114, 122, 127, 132, 146,
   150, 153–154, 158, 242, 244, 249, 295–296,
   298, 310, 326
–, coma   34–35, 90, 114, 127, 132, 153–154,
   326
–, nucleus   16–18, 21–22, 31, 36–40, 69, 310
–, variability   17, 22, 34–38, 72, 122, 127, 150,
   326
brown dwarf   268
Brownlee particle   63–64
Burgers–Korteweg–de Vries equation   232

Callisto   246
capture   3, 10–11, 116, 239, 243, 248–249,
   267, 271, 277, 280–281, 283, 287–289, 297,
   299–300
catalyst   11, 266
"central depression"   36–39, 42, 259
"chain of hills"   38–39
charge exchange *see* chemical reactions
Charon   10, 268
chemical kinetics   252, 256, 264, 266–267
chemical reactions *or* networks   196–197, 285,
   287, 329
– –, anomalous ionization *see also* critical velocity
   ionization   230
– –, charge exchange   71, 86, 89–90, 181, 189,
   192, 194–195, 206–207, 231, 310, 328–329